Spatial Analysis with R

Statistics, Visualization, and Computational Methods

Second Edition

Spatial Analysis with R

Statistics, Visualization, and Computational Methods

Second Edition

Tonny J. Oyana

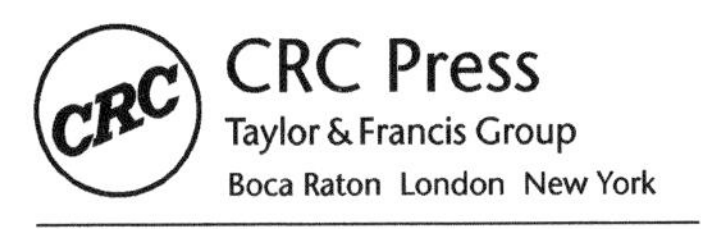

CRC Press is an imprint of the
Taylor & Francis Group, an **informa** business

MATLAB® is a trademark of The MathWorks, Inc. and is used with permission. The MathWorks does not warrant the accuracy of the text or exercises in this book. This book's use or discussion of MATLAB® software or related products does not constitute endorsement or sponsorship by The MathWorks of a particular pedagogical approach or particular use of the MATLAB® software.

First edition published 2015
by CRC Press
6000 Broken Sound Parkway NW, Suite 300, Boca Raton, FL 33487-2742

and by CRC Press
2 Park Square, Milton Park, Abingdon, Oxon, OX14 4RN

CRC Press is an imprint of Taylor & Francis Group, LLC

International Standard Book Number-13: 978-0-367-86085-1 (Hardback)
International Standard Book Number-13: 978-1-003-02164-3 (eBook)

Typeset in Palatino
by Nova Techset Private Limited, Bengaluru & Chennai, India

Visit the eResources: www.crcpress.com/9780367860851

Spatial knowledge is boundless, so start your epic journey now. Learn about geospatial data methods and tools and equip yourself with practical skills in spatial analysis.

Tonny J. Oyana

Contents

Preface

This book fosters a problem-based learning collaborative approach among students and stresses the mastering of spatial analysis knowledge and skills through a combined use of fundamental theories, concepts, and the practical application of geospatial data tools, techniques, and strategies in environmental studies. Since the publication of the first edition, many new developments have taken shape, such as the rigorous implementation of new tools and methods for spatial analysis using R, growth and expansion of artificial intelligence, machine learning and deep learning algorithms with a spatial perspective, and increased interdisciplinary use of spatial analysis. Other developments are in citizen science and the development of new analytical strategies, concepts and algorithms, methods, and cloud-based platforms and tools to serve numerous mobile applications with spatial perspectives.

Currently, we see a deepening of understanding around big data, locational analytics, and its increased use across a variety of contexts. Spatial data now drives the world of goods and services, human and animal movements, unmanned systems (ground and aerial vehicles, drones, aircraft, and spacecraft), health, real estate, business, indoor and outdoor emergency response and planning, and many other applications.

A new wave of increased and renewed interest for simpler ways to use, visualize, describe, and present spatial data, find relationships in spatial data, and discover patterns in spatial data is further inspiring work in spatial analysis. We can accomplish these spatial tasks easily by ensuring that we have the most accurate, robust, and unbiased data to work with. Spatial techniques therefore assure us with the production of actionable spatial data, information, and knowledge to support decisions.

The overarching objectives of this book are (1) to offer readers a theoretical/methodological foundation in spatial analysis using traditional, contemporary, and emerging computational approaches; and (2) to encourage readers to apply the critical knowledge and skills to appropriately analyze and interpret geographic data. To achieve these objectives, the second edition draws from traditional statistical methods, spatial statistics, visualization, and computational methods and algorithms, with the primary goal of supporting the growing field of geographic information science and training the next generation of geospatial analysts and data scientists. Spatial analytical concepts are introduced together with a series of active learning activities to enable readers to better understand, analyze, and synthesize spatial patterns, distributions, and relationships.

By offering problem-based exercises and case studies, the second edition provides state-of-the-art comprehensive coverage of topics in exploratory and spatial descriptive methods, hypothesis testing, spatial regression, hotspot analysis, geostatistics, spatial modeling, and data science. The ability to understand data and the methodological limitations associated with spatial analytical techniques will have a strong bearing on how geographers draw conclusions from statistical tests and the degree of certainty, validity, and translatability of findings drawn from their research. Through the use of the data, methods, and tools presented in this book, learners should be equipped with the knowledge and skills to (1) identify and characterize the types of nonspatial and spatial data; (2) construct testable hypotheses that require inferential statistical analysis; (3) preprocess spatial data, identify the relevant explanatory variables, and choose the appropriate statistical tests based on the data properties; (4) understand and interpret spatial data summaries and relevant statistical measures; (5) demonstrate competence in exploring, visualizing, summarizing, analyzing, and optimizing spatial data and clearly presenting the results using maps, charts, reports, pictures, infographics, analytical metadata, animations and three-dimensional visualization sciences, and on-the-fly dashboards; and (6) be proficient in the use of the primary analytical packages.

In summary, this second edition provides learners and researchers with the following key features:

- A good balance between concepts and practicums of spatial statistics with a comprehensive coverage of the most important approaches to understand spatial data, analyze spatial relationships and spatial patterns, and predict spatial processes
- Illustration of different concepts using data from environmental and social sciences, as well as a conceptual foundation on which to successfully apply statistics to geographic work
- A breadth that encompasses the analysis of point, areal, and geostatistical data with an updated chapter devoted to Big data, data mining, artificial intelligence, and how to develop actionable knowledge
- Implementation of practical exercises and worked out examples using R
- A wide range of hands-on spatial analysis worktables and lab exercises; instruction on how to develop deep learning architectures and algorithms and how to build prototypes of learning solutions to show proofs of concepts, as well as how to scale integrated enterprise services
- Integration of both technical and practical aspects; for instance, included are expanded materials on spatiotemporal methods, visual analytics methods, data science, and computational methods

- New datasets, insights, and excellent illustrations, combining statistical concepts with computer-based GIS exercises, as well as offering classroom-tested materials and lessons

Data for this textbook can be downloaded or accessed using this link: https://www.taylorfrancis.com/books/9781003021643

MATLAB® is a registered trademark of The MathWorks, Inc. For product information, please contact:

The MathWorks, Inc.
3 Apple Hill Drive
Natick, MA, 01760-2098 USA
Tel: 508-647-7000
Fax: 508-647-7001
E-mail: info@mathworks.com
Web: www.mathworks.com

Acknowledgments

This book represents the culmination of years of dedicated research, teaching, and active engagement of students in the classroom, collaborative partnerships, and feedback from colleagues from various institutions. I express my deepest appreciation and gratitude to all who have contributed to this effort in different ways and at various stages of development. I am particularly grateful to the students in the Spatial Analysis classes (2007–2014) at the Southern Illinois University Carbondale (SIUC), who inspired the development of the problem sets, and to the students in the Advanced Statistics and Spatial Analysis class, for providing the setting in which a number of foundational theories and principles presented in this book have been tested and applied. My colleagues at various institutions, including Samuel Adu-Prah, Luke Achenie, and Guangxing Wang, as well as teaching assistants (especially Alassane Barro, Chris Weston, Chen Wang, Jingjing Li, and Clara Mundia) also provided valuable feedback on the initial drafts of this book. The contributions by Damalie Oyana, specifically the significant role she played in developing the computer science algorithms used in Chapter 9 of this book, are also noteworthy.

My respective families have been very supportive throughout the writing of this book. Specifically, I thank my wife, Damalie Oyana, and lovely kids, Owen Oyana and Phoebe Oyana, for their unconditional support and encouragement, and my parents, Jerry and Phoebe Mary Oyana, both now deceased, who cultivated the values of discipline, hard work, focus, endurance, work ethic, and integrity and gave me a chance at finding the best educational opportunities in the world. To other family members, brothers and sisters, friends, and relatives, thank you so much for the support and encouragement you have offered to me during this incredible journey.

My respective institutions have also provided rich intellectual environments for me to thrive in and succeed as a scholar, and for this I am deeply appreciative. I especially thank SIUC, my former academic home for nearly 12 years; University of Tennessee, Memphis for four years; and the University of Buffalo, Buffalo, New York, where I earned my PhD and significantly enhanced my studiousness and work ethic. Special thanks and appreciation to colleagues at Makerere University, specifically Paul Mukwaya, Bob Nakileza, and Yazidhi Bamutaze, the University of Tennessee Health Science Center, and the Spatial Analytics and Informatics Core Research Collaboration Group. I also recognize the contributions of my colleague in the first edition of this book, now deceased, Florence Margai. The progress made in the second edition could not have happened without the initial efforts. In this second edition, I offer my special thanks to my two programming assistants, Janet Namutebi and Jesse Kisembe, who worked with me to write and implement R statistical programming code. Finally, I wish to acknowledge the anonymous reviewers and a number of people for their editorial skills, notably Alysha V. Baratta, who patiently edited the very first draft; Jennifer Winans, who edited the second draft; Irma Britton, the senior editor; Rebecca Pringle (editorial assistant); and Paul Boyd (production editor). Thank you all for working with us on this book project.

Author

Dr. Tonny J. Oyana received his PhD from the University of Buffalo, Buffalo, New York, in 2003. He did his postdoctoral training at the Department of Internal Medicine at the University of Buffalo with Dr. Jameson Lwebuga-Mukasa. He currently serves as the College Principal at the Makerere University College of Computing and Information Sciences, Kampala, Uganda. He is a professor of GIS and Spatial Analysis. He has served for over 20 years in several academic positions at the Southern Illinois University Carbondale and University of Tennessee Health Science Center, Memphis. His research focuses on establishing whether there is a link between environmental health and exposure, and advancing GIS methods, algorithm design, computational intelligence, analytical reasoning, and spatial analytical methods. He has a significant research interest in interweaving topics and fields and is currently expanding his extensive knowledge of data science to solve complex societal problems. He has mentored over ten PhD and 38 master degree students and three resident fellows/physicians, and his research has been funded by multiple agencies. He developed four computational algorithms (FES-k-means, MIL-SOM, Flexible Genetic Algorithm, and Reaction-Diffusion Mechanistic Model) and streamlined Diggle's method in ClusterSeer, a disease detection software. And he has taught GIS, Data Science, and Spatial Analysis courses for over two decades.

Dr. Oyana has authored 100 scientific publications, including more than 50 journal articles, two books, more than 24 refereed conference proceedings, ten book chapters, and ten book reviews; presented over 100 papers at regional, national, and international conferences; and written more than 20 technical reports.

Dr. Oyana earned his Master of Science in GIS at the National University of Ireland, Galway, Ireland, in 1996, and Bachelor of Science in Education at the University of Dar-es-Salaam, Tanzania, in 1993. He is an internationally recognized expert in GIS/GPS, spatial data science, algorithm design, and spatial analytical methods and strategies, with over 25 years of proven research and educational leadership with a strong track record of key accomplishments across a wide array of initiatives in North America, South America, Asia, Europe, and Africa.

1

Understanding the Context and Relevance of Spatial Analysis

Learning Objectives

1. Define and describe spatial analysis.
2. Describe the trends and significant developments in spatial analysis.
3. Define, describe, and illustrate key spatial concepts.
4. Learn about the unique properties of spatial data and inherent challenges.

Introduction

In conventional terms, geographers regard spatial analysis as a broad and comprehensive undertaking that entails the use of well-established analytical/visualization tools and procedures to analyze and synthesize locationally referenced data. These are formal techniques and analytical approaches for studying entities through their topological, geometric, or geographic properties. The approaches are rigorous and are drawn from statistical, mathematical, and geographic principles to conduct a systematic examination of spatial patterns and processes, including the exploration of interactions between space and time. Studying the locational and distributional arrangement of objects, people, events, and processes in space, and the underlying factors that account for these arrangements are some of the analytical goals of a geospatial data scientist. The work requires a place-based mindset with emphasis on uncovering spatial patterns and spatial linkages and examining spatial behaviors and complex interactions within and across locations that result in these distributional patterns.

Engaging in spatial analysis typically requires the use of quantitative data in a digital format, but increasingly data scientists are devising interesting and creative ways to integrate qualitative and contextual data into the analysis. Once a research project is defined with the articulation of a clear set of goals, objectives, and research questions, the data scientist begins by systematically choosing the appropriate units of observation from which to collect the data, the spatial scales at which they will be measured, and the variables and means by which the data values will be assigned to those variables.

The field of spatial analysis is inspired by a strong logical positivist tradition that involves inductive and deductive reasoning, hypothesis testing, and model building. It develops and advances geographic knowledge by investigating empirical events that occur in space, time, or both space and time. It consists of one or more of the following activities: (1) the analysis of numerical spatial data, (2) the development of spatial theory, and (3) the construction and testing of mathematical models of spatial processes (Fotheringham et al. 2000). Through spatial analysis, knowledge about spatial patterns and processes can be obtained, a large-scale dataset can be separated into several smaller components and meaningful information can be extracted, and a set of hypotheses can be derived and tested, thus culminating in empirical evidence. In addition, we can examine the role of randomness in generating observed spatial patterns of data, test hypotheses about such patterns, account for spatial variability, measure spatial autocorrelation and the underlying structure of the data, confirm the presence of outliers (if any), provide information about the explanatory factors or determinants through estimates of the model parameters, and provide a framework in which predictions can be made about the spatial impacts of various actions.

As an example, let us assume that you are working with a local food bank agency, and efforts are underway to develop urban community gardens, a new initiative deemed to be effective in combating food insecurity in urban areas. A lingering concern in the community is soil quality with the strong likelihood of environmental contaminants such as lead in the soil. To explore this, a sampling design strategy is formulated to collect soil samples. Using the Global Positioning System (GPS), a total of 150 samples are taken from various parts of the city. The samples are tested for lead contaminants along with other variables such as organic content, distance from major highways, soil moisture, alkalinity, and so on. The data are integrated into a geographic information system (GIS) with preexisting databases garnered through other devices such as land use and land cover maps from satellite imagery, housing quality data, roadways, and demographic data generated from the U.S. Census. As a geospatial data scientist, how would you go about organizing and integrating the soil quality data into the GIS? What are the key properties of the soil quality data? Are there any unique challenges associated with the spatial data? Are the soil samples adequate and spatially representative of the study area? What methods would be ideal for analyzing the dataset for presentation to the food bank? These questions call for a comprehensive understanding of the underlying spatial data structure, the data distribution, variable properties, and potential limitations that accompany a typical research project. Spatial data structures consist of features such as points, lines, areal polygons, surfaces, or other objects that may be associated with valuable geographic information including potential records pertaining to other dimensions such as time (Samet 1990, 1995). Each feature in the database is specifically associated with locational information and the attribute value characterizing the nature of the observation. As shown in Table 1.1, a number of methods have been developed to handle point, line, areal, and surface data structures (Fotheringham and Rogerson 1994; Bailey and Gatrell 1995; O'Sullivan and Unwin 2010). These data structures have a strong bearing on the methods of analysis. For example, a commonly used approach called point pattern analysis is purposely designed to assess whether the geographic distribution of geographic points is random or not, and to describe the type of pattern so that it can be used to infer about the underlying processes that are responsible for the observed structure (Legendre 1993). Line pattern analysis is based on a topological approach to study a network of connections among points; surface patternanalysis is concerned with spatially continuous phenomena, where one or several variables are attached to observation points, and each point is considered to represent its surrounding portion of space (Legendre 1993).

From Data to Information, to Knowledge, and Wisdom

Spatial analysis enables data scientists to utilize a specialized set of skills, tools, methods, algorithms, and analytical strategies to better understand the distributional patterns, events, and processes that are captured in spatial and temporal data. Through spatial analysis, we can visually explore and manipulate data, create subsets or stratify the data based on a set of meaningful criteria, compare and contrast attributes that are measured across various entities, and use the analytical findings to test hypotheses. Through these endeavors, we derive new knowledge and gain substantial insights that add to our spatial thinking and evidence-based line of reasoning. The ongoing uses of geospatial technologies and methods produce cumulative knowledge about events and processes, and ultimately the collective wisdom to generate, support, or affirm an underlying theory. For example, we know from empirical observations that cumulative exposures from particulate matter and chemical sources place a heavy burden on the environment. If this statement is true, the use of geospatial data technologies to study the life trajectories of particulates and chemical sources may significantly advance our knowledge and understanding of their potential impacts on human health and the environment. Figure 1.1 shows a visual representation of a

TABLE 1.1

A List of Spatial Concepts and Techniques

Method	Scale	Spatial Pattern	Examples of Techniques	Contribution	Limitation	Cited Sources
Cell Count	Dichotomous	Point, Area	Quadrat	Identification of large-scale clusters	Low power, Inaccurate expected cell counts	Cuzick and Edwards (1990); Wallet and Dussent (1997); Wallet and Dussent (1998); Zurita and Bellocq (2010)
Adjacencies of Cells with High Count	Continuous	Area, Point	Geary's *C*, Moran's *I*, spatial autocorrelation	Identification of large-scale clusters	May ignore nonadjacent areas	Moran (1948); Barnes et al. (1987); Ohno et al. (1979); Marshall (1991); Munasinghe and Morris (1996); Miller (2004)
Distance-Based Methods	Case Points, Case-Control	Point	NN, *K* Function, Kernel Density Estimation	Identification of local area clusters	NN ignores larger spatial scale Edge effects Spatial dependency problems	Clark and Evans (1954, 1955); Haggett et al. (1977); Lewis (1980); Ripley (1981); Gatrell et al. (1996); Diggle (1990); Diggle and Rowlingson (1994); Lawson (1989); Waller and Jacquez (1995); Bithell (1995); Bithell (1999); Rogerson (2005)
Geographical Analysis Machine (GAM)	Dichotomous			Adjusts for population distribution	Computer intensive Difficult to assess its statistical properties	Openshaw et al. (1987); Besag and Newell (1991); Openshaw et al. (1988); Sabel et al. (2000); Conley et al. (2005); Garzon et al. (2006)
Bayesian Methods	Continuous	Area		For smoothening regional estimates toward a local mean rate of neighboring observation	Difficult to sustain because it ignores spatial configuration	Efron and Morris (1975, 1976); Longford (1993); Kleinbuam et al. (1998); Besag and Newell (1991); Besag et al. (1991); Besag and Green (1993); Besag et al. (1995); Diggle et al. (1998); Excoffier et al. (2005); Guillot et al. (2005)
Mapping and Estimation Methods	Continuous	Surface	Spline, Kriging, Trend Surface	Prediction and estimation of unknown values Measures spatial dependency	May ignore other vital factors Error-prone Involves complex computations	Legendre and Fortin (1989); Legendre (1993); Rossi et al. (1992); Matheron (1963); Myers (1994, 1997); Glantz (2012)
Predictive Analytics and Spatial Analysis Tools, Methods, and Strategies for Data Science	Discrete Continuous			Explore, identify, discover, and predict spatial patterns	Data may be inconsistent	Marmion et al. (2009); Guo et al. (2006); Chen et al. (2008); Andrienko and Andrienko (2010); Ye et al. (2014); Hall et al. (2014); Kothur et al. (2014)

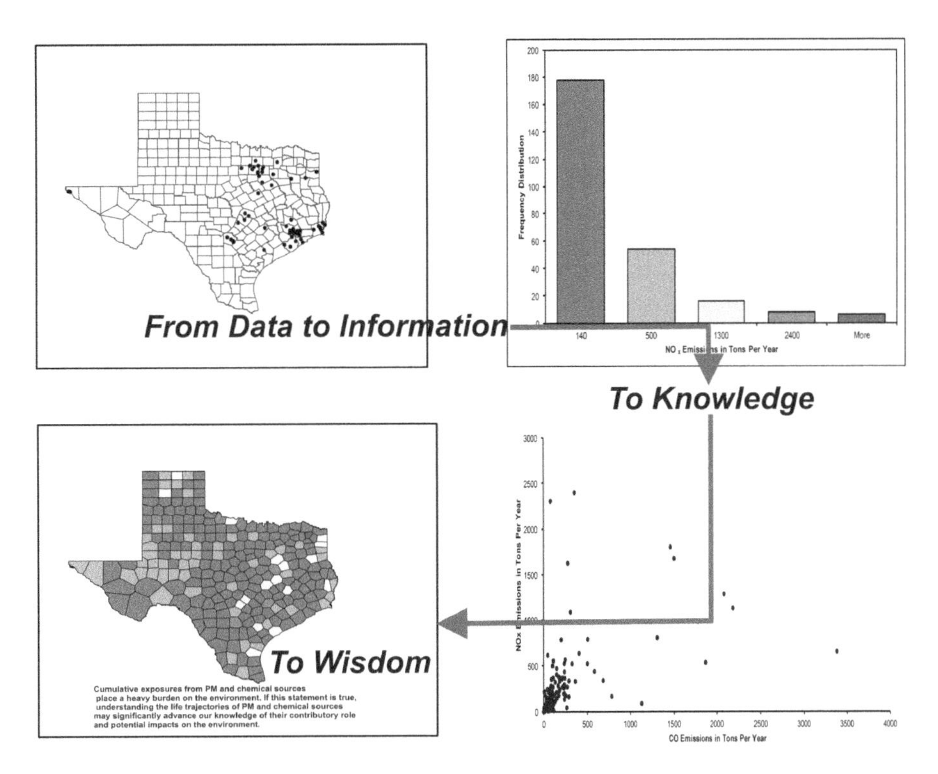

FIGURE 1.1
The transformation of data to information, and to knowledge and wisdom.

data transformation process. Throughout this book, we are going to learn how spatial data can be transformed using different analytical strategies, methods, algorithms, or tools into valuable information, knowledge, and wisdom.

Spatial Analysis Using a GIS Timeline

The evolution of spatial analysis has been fueled by five major events: (1) the 1950s quantitative revolution in the United States, (2) trends in regional science, (3) spatial statistics (including both geostatistics and stochastic modeling), (4) computational techniques (geocomputation), and more recently, (5) the emerging field of data science. Also, noteworthy and of historical significance is John Snow's groundbreaking analysis of spatial patterns of the 1854 cholera outbreak around the Broad Street water pump in London. John Snow's work is a classic embodiment of this field, and the story of spatial analysis cannot be complete without referencing his contributions. When reviewing the trends and significant milestones over the course of several decades, it is apparent that spatial analysis gained prominence around the time GIS was created, mostly during the quantitative revolution in the late 1950s and throughout the 1960s and 1970s. Early innovators considered the potential and impact of the quantitative revolution on GIS, GPS, and spatial analysis. The approaches in spatial analysis came under intense scrutiny in the 1970s by behavioral and Marxist geographers and suffered a major setback. It was not until the 1990s that these approaches reemerged with strong interests among geographers. Nelson (2012) in his review of *Trends in Spatial Statistics* notes the four major developments since the quantitative revolution as (1) new data sources, (2) increased understanding and advancement in spatial autocorrelation, (3) creation of local spatial methods, and (4) expansion of spatial science beyond geography. One way of discerning the emerging trends and powerful influence of spatial analysis in scientific research is through the number of articles published between 1984 and 2012, showing the exponential growth in the uses and applications of spatial analytical tools and technologies. This information was compiled using a variety of sources. The Web of Knowledge that references multiple databases had 15,561 search results based on the two keywords (spatial analysis OR GIS); Anselin's 1995 article on Local Indicators of Spatial Association (LISA) was the most cited with over 3750 citations. As demonstrated by these electronic searches, there is a substantial interest in spatial analysis using GIS. Going by the mean number of published articles, there has been a significant increase annually with 98 times more articles in 2012 than 1984. In addition, using key words GIS, spatial analysis (has maximum hits), or spatial statistics, Google Scholar returned 2,620,000 hits in 2012 on this topic alone. When a similar search was done in 2019, Google Scholar returned only 1,610,000 hits. Although the 2019 hits suggest a drop from what was observed in 2012, this drop can be explained by Google Scholar's efforts in significantly improving their Search algorithms. Even with this drop, it is quite evident from these articles that there has been increased interest in GIS, GPS, and spatial analysis in the last 25 years (Figure 1.2a and b).

Following is a summary of the number of articles related to spatial analysis from 1995 to 2000

1995 (1 article)

Journal of Environmental Planning and Management (1 article)

1996 (18 articles) :
- *Geographical Information Systems* (12)
- *International Journal of Geographical Information Systems* (3)
- *Journal of Multilingual and Multicultural Development* (1)
- *Journal of Property Research* (1)
- *Urban Studies* (1)

1997 (25 articles)
- *International Journal of Geographical Information Science* (16)
- *International Journal of Remote Sensing* (7)
- *Journal of Environmental Planning and Management* (1)
- *Urban Studies* (1)

1998 (25 articles)
- *International Journal of Geographical Information Science* (12)
- *International Journal of Remote Sensing* (9)
- *Journal of Environmental Planning and Management* (1)
- *Behavior and Information Technology* (1)
- *International Journal of Water Resources Development* (1)
- *Human Ecology* (1)

1999 (17 articles)
- *International Journal of Geographical Information Science* (10)
- *International Journal of Remote Sensing* (2)
- *Journal of Sustainable Tourism* (2)
- *Marine Geodesy* (3)
- *Urban Studies* (1)

2000 (16 articles)
- *International Journal of Geographical Information Science* (8)
- *International Journal of Remote Sensing* (4)
- *Urban Studies* (1)
- *Outlook on Agriculture* (1)
- *Society and Natural Resources* (2)
- *Process Safety and Environmental Protection* (1)
- *Australian Geographer* (1)

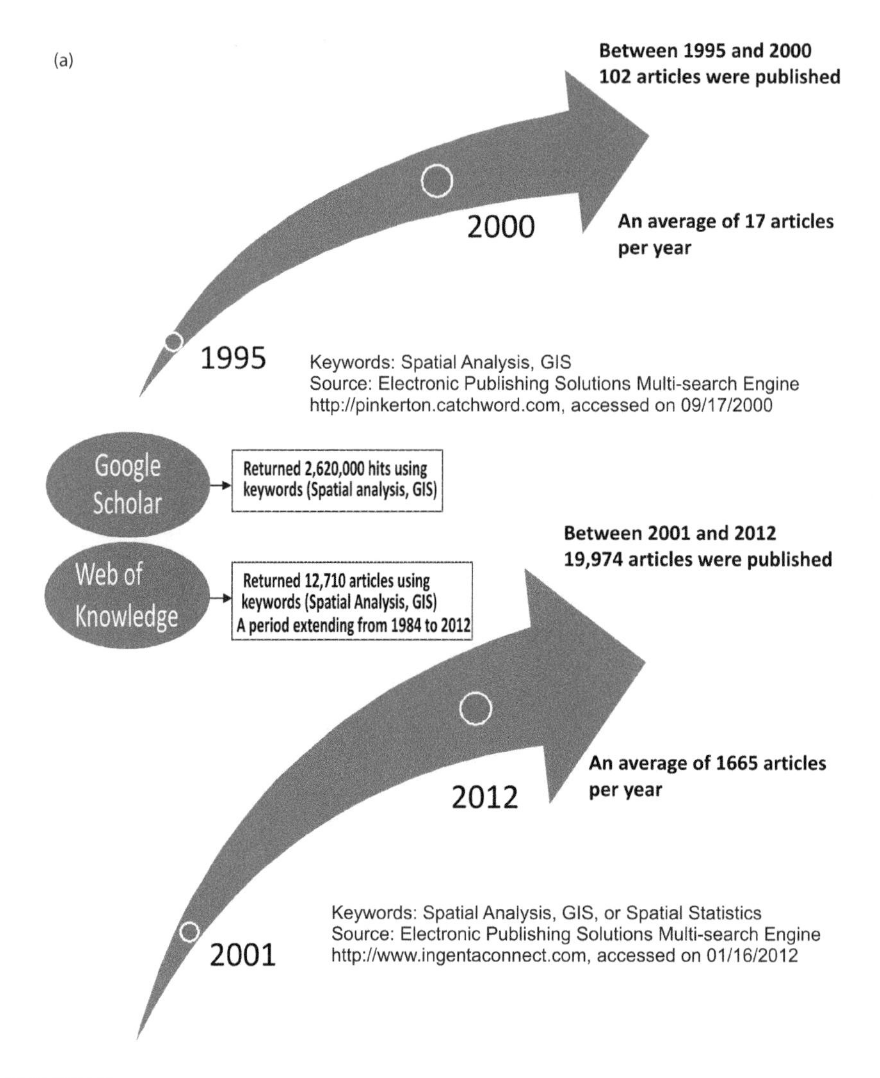

FIGURE 1.2
(a) Trends in the uses and applications of methods in spatial analysis. *(Continued)*

Spatial Analysis in the Post-1990s Period

Significant progress has been made toward the improvement of spatial analytical capabilities of GIS in the last 20 years. Many advanced spatial analytical routines, such as principal components analysis, spatial statistics and geostatistics, and spatial regression have been incorporated into spatial statistical software applications largely through two ways: (1) tightly coupled systems where spatial techniques are fully integrated in GIS software, for

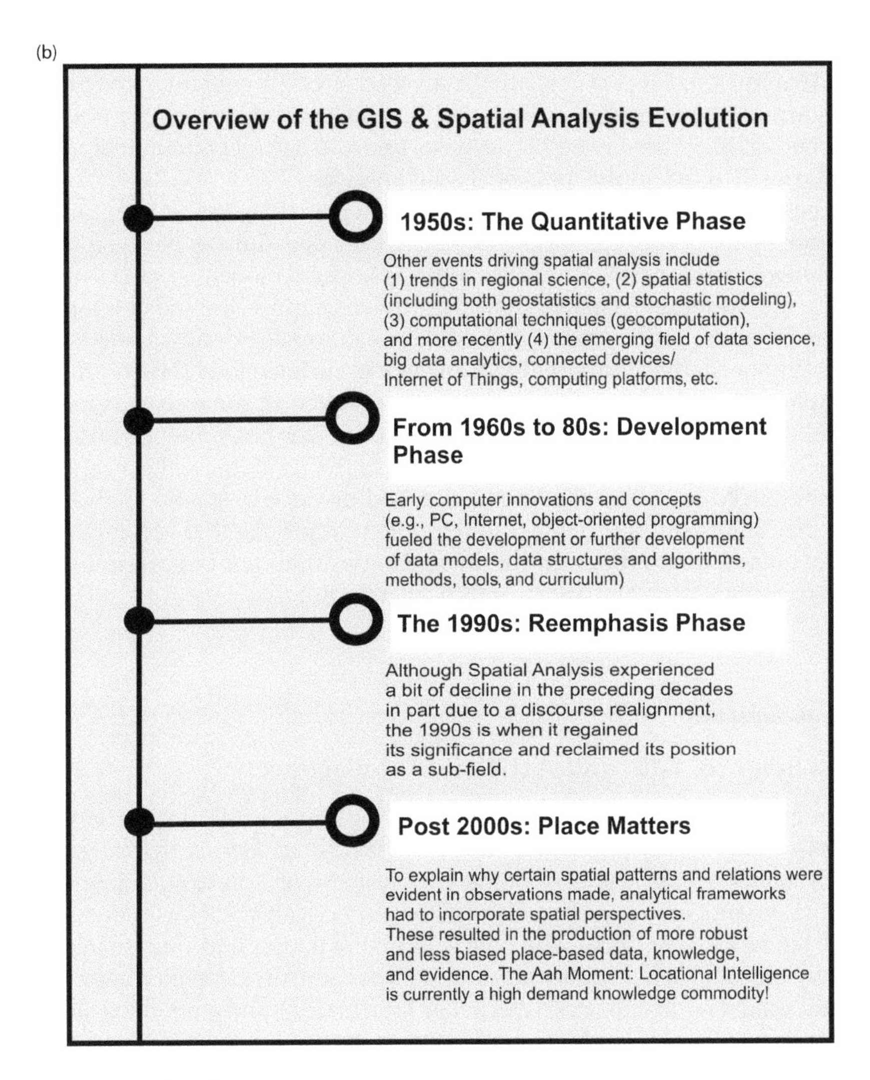

FIGURE 1.2 (Continued)
(b) A timeline showing an overview of the evolution of spatial analysis and GIS from the 1950s to the present.

example, ArcGIS, IDRISI, and MapInfo; and (2) loosely coupled systems where an open-source or off-the-shelf commercial software is loosely integrated with statistical tools or models. Recent developments are predominately in the following areas: (1) spatial data mining and predictive analytics; (2) new computing platforms and methods for analyzing very large-scale spatial datasets; (3) geocomputation, algorithm design, and development; and (4) bioinformatics, gene sequencing, and visual and spatial analytic methods. Recent changes are also reflected in the publication of several articles and availability of sophisticated computer software and computing platforms. The rapid expansion of topics in spatial analysis is strongly evident in the premier GIS journal, the *International Journal of Geographical Information Science*. As of October 25, 2019, there were 2213 articles that had a direct mention of or were primarily focused on the topic of spatial analysis.

Other notable developments in spatial analysis include the integration of commonly used concepts and methods developed during the quantitative revolution, such as network analysis functions, spatial modeling, spatial metrics; impact of computing on spatial analysis; software engineering and development; methodological developments and advances in topology and geometry; data visualization; and new computing frameworks such as cloud (Nelson 2012).

Implications of such rapid changes have resulted in some confusion and a clear lack of understanding of spatial methods. This book intends to address these concerns. We hope to offer our readers the fundamental concepts and tools required to master the knowledge and practice in spatial analysis. We also plan to expose our readers to the potential pitfalls that accompany the use of spatial methods given the increasing availability of easy-to-use and user-friendly spatial statistical analytical software.

Data Science, GIS, and Artificial Intelligence

In an effort to look into the future, we should get some inspiration from a well-cited report, published in the *International Journal of Geographical Information Science* on the GARP modeling systems, on how to automate spatial prediction and spatial analysis and GIS, respectively (Stockwell and Peters 1999; Janowicz et al. 2020). The data science era is upon us: punctuated by a rapidly growing field of data science and CyberGIS, virtual reality, augmented reality, mixed reality, drones, Internet of Things (IoT), and embedded systems and sensor technologies. Numerous connected devices deployed through the use of Precision Positioning and Autonomous Systems and Global Navigation Satellite Systems send data to a geospatial cloud, which bring together key elements of historical data analytics, streaming data analytics, location intelligence, and artificial intelligence with unmatched performance to explore, visualize, and analyze more than 8 million data sensors. GPS units or GPS-enabled sensor devices are now commonplace. They are now available on almost every sensor device, from navigation systems to smartphones, fitness devices, IoT, gaming, or smart hybrid watches. They are used for tracking movements and measuring travel distances and directions. As a result, we have a very large amount of location-based data and information that is being generated every second for use with other datasets.

In the current era of big data and data-driven decision-making, data science is a game-changing paradigm and one that could potentially open up new possibilities and exciting avenues for spatial data analysis. This field of data science has evolved largely as an amalgam of data-intensive disciplines, notably mathematics, statistics, computer science, operations research, geomatics, physics, business intelligence, and more. Terms such as data deluge, *data tsunami,* or *tidal waves* are being used to describe the large-scale databases that are now readily on hand to capture these activities in real time and address complex societal and scientific research questions. Further, given the critical role of individualization in big data, particularly when it comes to managing the location and pattern of movements, behaviors, and interaction of individuals in space, GIS, GPS, and spatial analysis lie at the heart of these recent developments. Traditional technologies and analytical approaches are still valid and reliable, but there is growing consensus that these are no longer efficient or effective enough to harness the massive and valuable storehouse of information. The sheer volume and rapidity at which the data are being generated these days, the urgent need to preprocess and integrate the different sources, types, and formats of data, and the fast turnaround time required to analyze and present the results now require the development of new computational tools and techniques. These emerging trends also call for a new cadre of data scientists with a broader set of skills and competencies including the ability to work in a collaborative environment with scholars from different disciplines and analytical domains.

The emerging field of data science is central, for example, to our understanding of the impact of social media networks on human activities. The social media landscape features local to global content that is normally published and shared through a large network of users. Social media users utilize the platform to connect, communicate, play, or engage in e-commerce. However, knowledge gaps exist in fully understanding human activities over social media landscapes. This is further compounded by the rapid growth being experienced in information technology (IT) and the increasing number of users since 2007. IT services are rapidly diffusing into urban communities at an increased rate, and many people rely on these services for information seeking and networking purposes. Also, the scale and volume at which data are being generated from IT data centers on a daily basis on the social media is both extensive and intensive. The IT data centers offer unique opportunities for data scientists to study human activities and behavior in never before imagined ways. If successfully exploited, the analysis of large-scale datasets generated from CyberGIS systems/data centers will increase our knowledge and understanding of human activities over social media networks. Although scholarly debates continue over ways to characterize, store, and process the big data, geospatial data scientists are more interested in the geospatial attributes of such information, and the analytical possibilities that lay ahead. Several challenges have been identified including the best practices for accessing, storing, and integrating the big geospatial data; maintaining consistency in geographic metadata; standards and protocols for maintaining privacy and security; data curation and quality control; data processing; visualization;

and analysis and presentation of results. In Chapter 8, we elaborate on these issues and discuss the geospatial data science approaches that are being developed to meet the grand challenges.

Geographic Data: Properties, Strengths, and Analytical Challenges

Geographic (or spatial) datasets, whether "big," large scale, or small scale, consist of quantifiable/qualitative information drawn from objects, people, events, and other observational units that have a spatial reference. This spatial reference may be explicit, as in an address or a grid reference, or it may be implicit, as in a pixel in the middle of a satellite image. In the context of a GIS, we typically have spatial objects and fields. Spatial objects refer to geographic features that can be represented using a vector model, and fields are geographic features that can be represented using a raster model (Figure 1.3).

A major strength that comes with using spatial data is that the data representation can take the form of many levels. At the conceptual level, we can take a philosophical view that considers representation of the world through spatial reasoning, spatiotemporal reasoning, and temporal reasoning. We can also reason beyond the two-dimensional (2D) perspective by thinking about representation in terms of three-dimensional (3D) or multiple dimensions. At the logical level, we have a GIS data model. This enables us to utilize a set of mathematical constructs to describe, formalize, and represent selected aspects of the real world in a computer. Simply, these are the means through which we formalize the real-world/geographic features into an abstract computer model. As indicated earlier, the two commonly encountered models for representing spatial data in a computer are the vector and raster models. These models are formalized in a computer using mathematical models. Indeed, the formalization of continuous space (field-like geographic feature) is typically encoded by approximations based on tessellations (Samet 1990, 1995; Egenhofer et al. 1999), whereas noncontinuous space (object-like geographic feature) is typically encoded with appropriate vector data structures. The GIS data model will encode interactions and relationships through a set of constructs between spatial and attribute information based on relations. These relations could be topological (e.g., meet, intersect, near, contain), directional (e.g., left, top, bottom, right or west, east, south, north), or metric spaces (e.g., distance function). In general, the data models have certain fundamental characteristics or functional relationships that allow them to support vector and raster data structures, geometric properties, algorithms, database structure, and maps and coordinate systems. Fundamental topics and knowledge in *topology, geometry, algebra,* and *cognitive science* guide the process of data representation in a GIS.

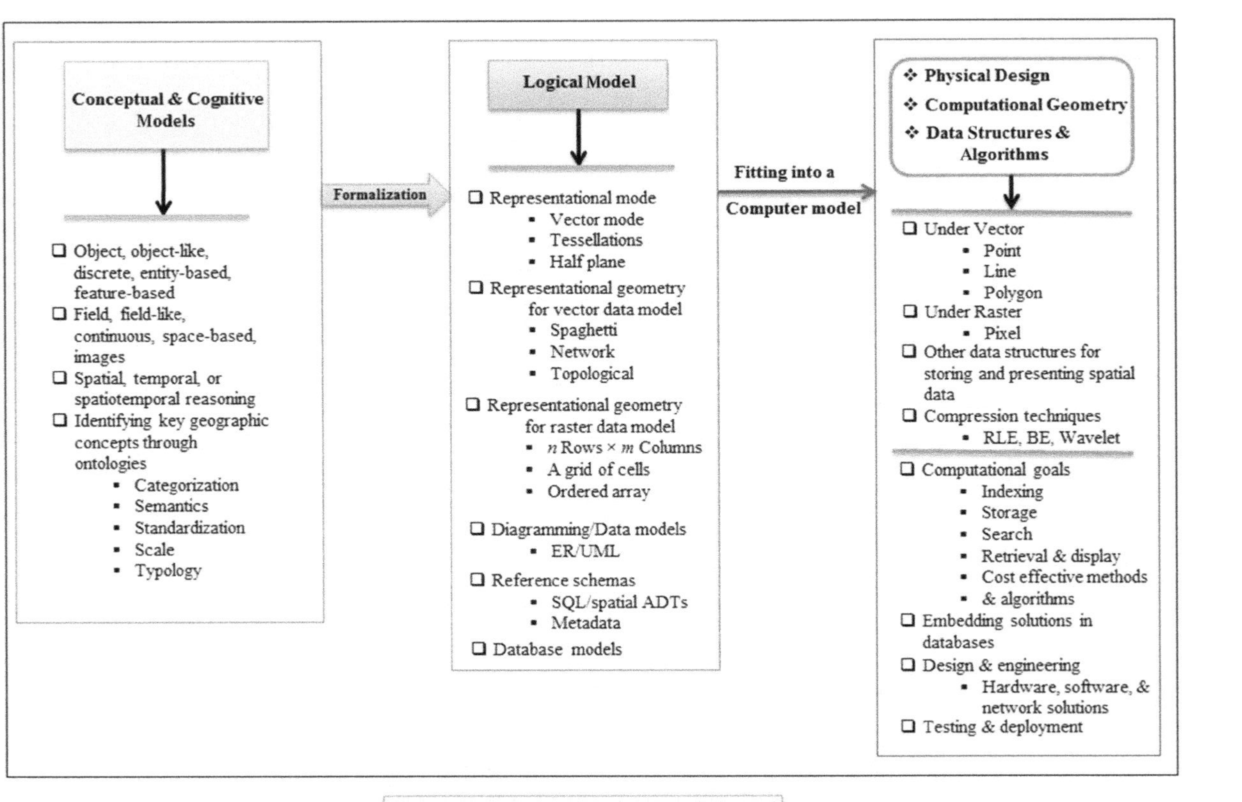

FIGURE 1.3
The process of representing data in a GIS. Different conceptual, logical, and computational geometries for the GIS data representation process.

It must be noted that data structures are simply encodings that work well in a computer setting. For the vector model, we have three basic types: points, lines, or polygons, whereas in the raster model, we have an array of pixels, grids/lattices organized in a matrix format of rows and columns. At the computational level, we can implement the spatial data structure by using features, feature classes, and raster datasets. In a spatial database, we can represent both objects and fields; describe relationships between object types and apply reference schemas; and specify data structures, algorithms, storage, retrieval, and search operations.

Key characteristics of spatial features are that they are irregular in shape; and there is a scale effect in all observed and measured spatial phenomena. The types of representations, including their geometrical shapes and properties, have a strong bearing on the conceptualization and formulation of study hypotheses, and analysis of the spatial features. Specifically, observed and measured spatial data have the following basic characteristics: (1) variations in measured values, that is, large-scale variations change slowly, whereas small-scale variations change quickly/normally uneven; and (2) similarities of measured observations at locations close together. The variations in a given spatial distribution exist at different scales and may depict a low or high degree of spatial variation.

When starting out with the analysis of spatial data, there are key concepts to take into consideration, such as spatial scale, dependency, and proximity. In addition, there are certain attributes that are unique to spatial data and could single-handedly derail a study by influencing the accurate estimation of the statistical parameters. These include the boundary problem (impact of artificial or natural border lines on spatial distributions), the scale problem, the pattern problem (spatial autocorrelation), and the modifiable areal unit problem (MAUP). When encountered in spatial analysis, these problems may confound the underlying relationships within the spatial data structures and could result in systematic uncertainties in the derived estimates. Following are descriptions of these concepts, the unique spatial data challenges, and their impacts on research findings.

Concept of Scale

Spatial and temporal scales are central to better decision-making and scientific research because nature is so complex, with many processes occurring at different scales. To capture these processes, it not only requires a scale that is representative, but also an optimal scale through which different measurements can be taken (Oyana et al. 2014). Scale can refer to any of the following: cartographic/map scale (small, medium, large), magnitude of study (amount of detail), geographic extent, accuracy (positional and attribute accuracy), and measurement, process, and time scale. In this book, *scale* will be conceptualized as the geographic extent and amount of detail in a study, while accounting for spatial and temporal variability. It involves understanding all sorts of geographic divisions in your study area, and the accompanying parameters in terms of both socioeconomic and biophysical aspects. In ecological settings, living organisms—human beings included—organize themselves in patches or in some form of spatial structure. The interaction among living organisms and the ecological processes depends on their immediate environment. These processes may be nonrandom or random. Therefore, a proper understanding of an appropriate scale for studying parameters in such populations is significant if accurate results are to be obtained. Also, the variation behind these processes could be due to numerous factors including (1) time of day—morning, afternoon, evening, or night; (2) climate—winter, autumn, spring or summer, rainy or dry season, temperature; (3) slope gradient—low, medium, or high; (4) feeding and reproductive habits; (5) altitude—low, medium, or high; (6) directional influences—angle of the sun/the intensity of the sun and wind; and (7) presence of food or water bodies. Knowledge of these factors and related processes may help the data scientist decide on the most appropriate scale for a given study.

Concept of Spatial Dependency

The statistical conditions governing the parameter of a sample population under investigation are normally based on two assumptions, namely, the degree of independence of the parameter and whether the variance is identically distributed (Griffith and Amrhein 1997; Kleinbuam et al. 1998). Due to natural variability, these assumptions only hold for a population with a high degree of certainty, and where the population is able to maintain the same variance. In the real world, however, these assumptions may not hold. Natural processes are normally dependent and occur in a random manner, or they may occur simultaneously. For instance, as noted in an earlier example, living organisms in ecological settings are organized in patches or some kind of spatial structure that could result in spatial dependency (Legendre and Fortin 1989; Legendre 1993; Fotheringham and Rogerson 1994; Oyana et al. 2014). This notion of spatial dependency is best captured in Tobler's first law of geography, which states that "everything is related to everything else, but near things are more related than distant things." Also noteworthy are two key aspects of spatial dependency: (1) spatial variable dependency (spatial autocorrelation and spatial correlation) and (2) spatial relations dependency (spatial homogeneity and spatial heterogeneity). These underlying spatial structures have major implications on how research problems are formulated, data sampling, measurement, and how hypotheses are tested.

Concept of Spatial Proximity

The concept of spatial proximity is different from spatial dependency though in most instances geographic features that are proximal to one another are more likely to exhibit similarities and therefore spatial dependency.

Notwithstanding this, it is important to clarify the differences between the two concepts. In spatial data analysis, spatial proximity provides valuable knowledge and topological information regarding the relative location of points, lines, and areal features in the database. It is a function of distance and the degree of connectivity between objects, people, or places in the geospatial database. There are several algorithms for computing spatial proximity with measures based on linear distance, costs, time, and networks within the system. Spatial proximity is an important concept that is woven into virtually every geographic analysis that deals with spatial pattern, mobility, interaction, association, and diffusion of people, objects, ideas, events, and processes. For example, in economic/retail geography, the concept is used to study the agglomeration of firms, conduct site selection or trade area

TASK 1.1 SPATIAL PROXIMITY

We now review three examples that show how we can measure and analyze the concept of spatial proximity.

1. Suppose we are analyzing the effects of planes taking off or landing on a runway in a residential neighborhood. What would be the impact on individuals living close to the runway?

   A likely response to this problem would be those residents living close by will likely complain about ambient noise exposure from aircraft take-off and landing. Specifically, the noise distance–decay model may show that aircraft noise levels decompose between 243 and 250 m from the runways, and the day or night sound levels will have a directional bias.
2. Suppose we are analyzing the effects of bird nesting in a habitat near a river. What would be the impact on nests located near the river?

   A likely response to this problem would be that nests located in a bird habitat that are in close proximity to the river may have an increased source of nest-building materials and access to water and food resources.
3. Suppose we are analyzing the locational advantage of restaurants located near the main street. What would be the impact on the restaurants located nearby?

   A likely response to this problem would be the impact of a restaurant closer to a main street may be a change in the need for parking spaces, increased human traffic, and increased patronage and use of the restaurant. If the main street is a busy roadway with human traffic, then it would likely result in more business and higher profits.

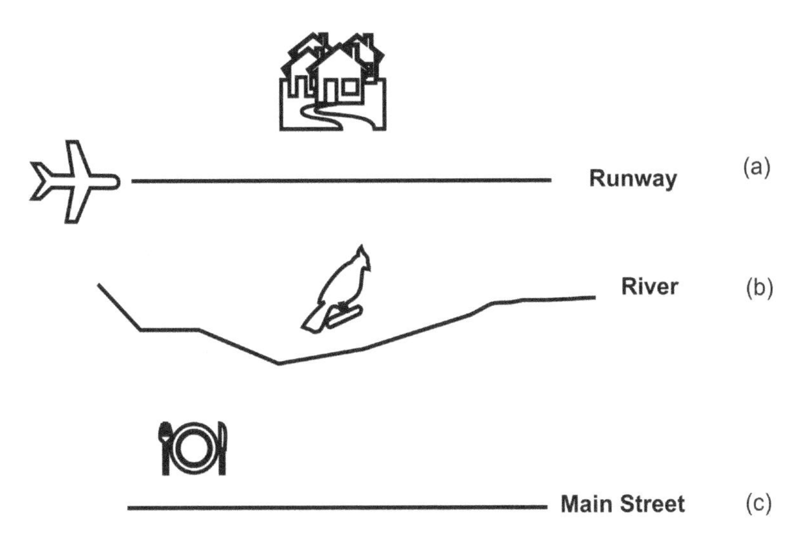

FIGURE 1.4
Spatial proximity.

analysis, evaluate consumer behavior, delineate activity spaces, and assess the diffusion of innovations or interchange of ideas and knowledge. In medical geography, spatial proximity is used to study disease transmission patterns, model atmospheric dispersion of air pollutants, or develop chemical plumes or footprints over which residents may be exposed to environmental hazards. Examine the illustrations in Figure 1.4 and indicate how you would apply the concept of spatial proximity.

Modifiable Areal Unit Problem

MAUP is a form of ecological fallacy associated with the aggregation of data into areal units (Figure 1.5). It identifies problems associated with the partitioning of spatial data (the "zoning problem") or the size of the spatial units on which the data are mapped (the "aggregation problem"). Both of these spatial configurations can influence the statistical models, correlations, and other statistical estimates generated from the data. Specifically, there are two effects that could arise from MAUP, or the system of modifiable areal units. First, MAUP could have a scale effect, which is the tendency for different statistical results to be obtained from the same set of data when the information is grouped at different levels of spatial resolution (e.g., enumeration areas, census tracts, cities, regions). Normally, the larger the unit of aggregation, the better, on average, is the correlation between two variables. A second MAUP effect, the aggregation effect, could result from different areal arrangements of the same data to produce different statistical findings. Given these MAUP effects, as geospatial data scientists, we cannot categorically state that the results of our analytical studies are independent of the spatial units being used; rather our results could be influenced by the configuration and size of the spatial units. For example, in Figure 1.5, using the same variables in the study of Chicago, the results data gathered from census block groups are likely to be different from those produced at higher levels of aggregation, such as the census tract level, the community district level,

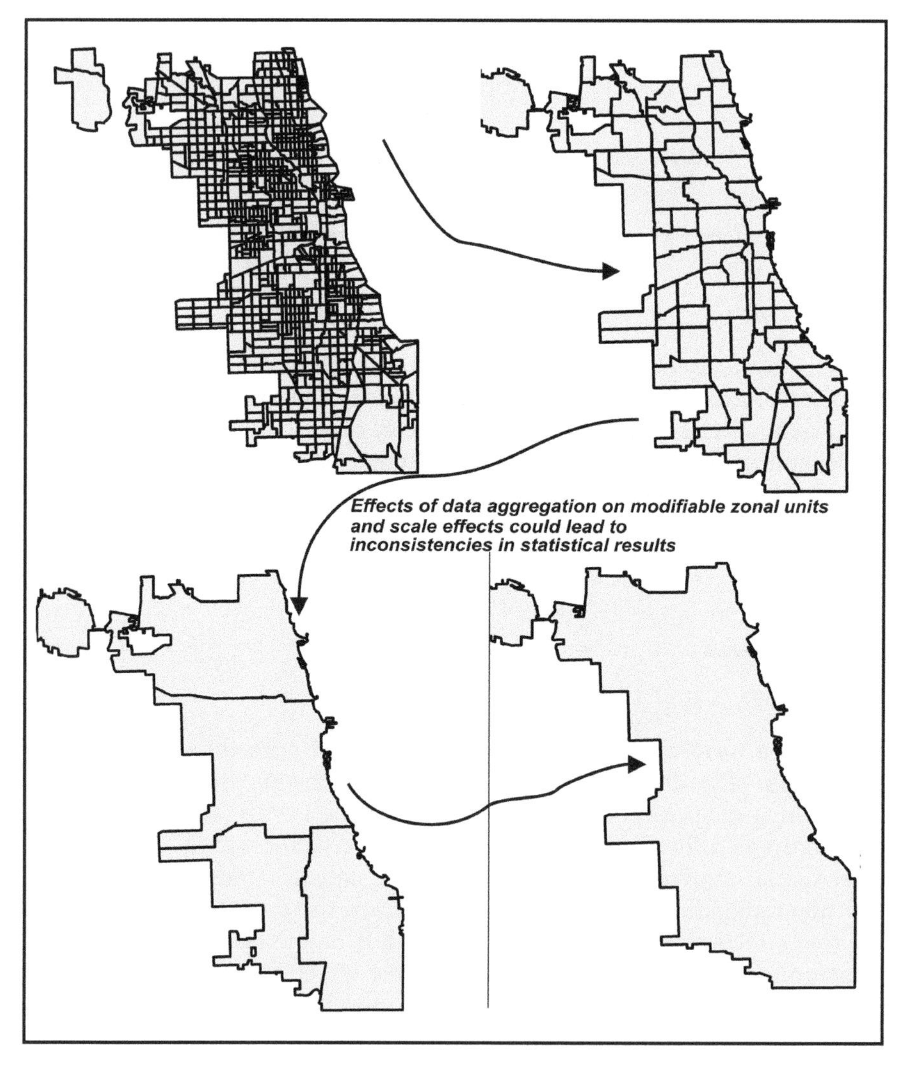

FIGURE 1.5
Modifiable areal unit problem that is a result of effects of data aggregation.

Spatial Units of Analysis—Hierarchical Model

Socioeconomic Units/Administrative

- Individual Level—11th
- Household Level—10th
- Sub-parish Level (Enumeration Areas, Blocks, Block Groups, ZIP Codes, Village etc.)—9th
- Parish Level (Census Tracts)—8th
- Sub County Level—7th
- County Level—6th
- State Level (District)—5th
- Sub National Level—4th
- National Level—3rd
- Continent Level—2nd
- Global Level—1st

Environmental Units

- Watershed Level—4th
- Sub Catchment—3rd
- Catchment—2nd
- Basin—1st

Other Spatial Units: Pixel/Grid Size at Different Spatial Resolutions, Voronoi, etc.

FIGURE 1.6
Commonly used spatial units of analysis.

or higher. As such, the task of obtaining valid generalizations or comparable results from multiple studies is extraordinarily difficult. Figure 1.6 presents commonly used spatial units of analysis. When using modifiable areal units, it is important to study their effects.

TASK 1.2 MEASURING AND ANALYZING THE IMPACTS OF GEOGRAPHIC SCALE

To further illustrate the MAUP concept, let us work through the following questions. Figure 1.7a depicts three options, or levels in which the data derived for the same region can be aggregated.

a. Using Figure 1.7a, list which level is divided into the smallest areal units. From the illustration, Level 1 is divided into the smallest areal units.
b. Using Figure 1.7a, list which level is divided into the largest areal units. From the illustration, Level 3 has the largest divisions.
c. Using the drawing tools in MS Word, redraw two different zonal configurations using the Level 1 image as a base. Level 1 can be redrawn and modified in many forms. Example solutions are illustrated in Figure 1.7b.

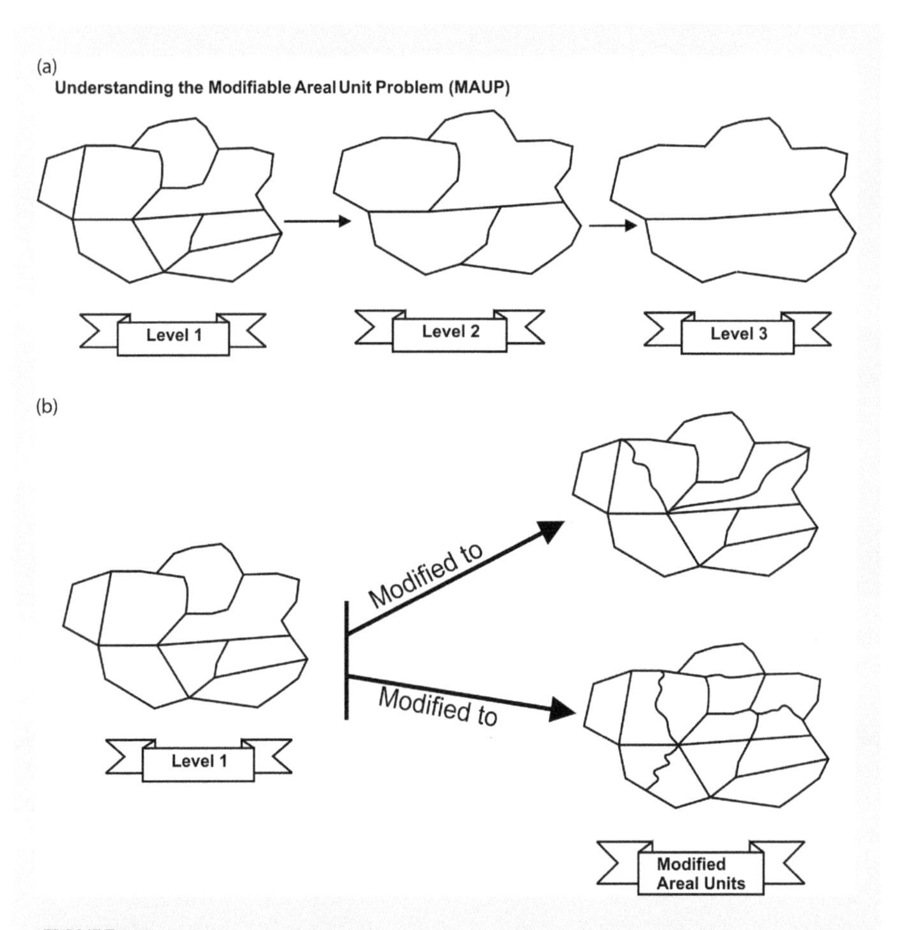

FIGURE 1.7
(a) Aggregation of modifiable spatial units from Level 1 to Level 3. (b) An example of an aggregated spatial unit. A solution for Task 1.2c.

d. Using Table 1.2, briefly describe the correlation results based on the different spatial units.

   The correlations between renters and owners are inconsistent across different spatial units due to either a scale effect or a zone effect. We can observe that the value correlation is low at the block group level (0.075), whereas it is high at the county level (0.984). We can use this information to study the aggregation effects.

e. Based on questions 1 through 4, define the modifiable unit areal problem.

   Recalling our earlier discussion, MAUP is a potential source of error in spatial studies that use data aggregated into zones.

TABLE 1.2

Statistical Relationships Between Homeowners and Renters in the State of Illinois Using the 2000 U.S. Census Data

Partitioning Levels	Spatial Units	N (# of Observations)	Renter versus Owner[a]
Level 1	Block Groups	9843	0.075
Level 2	Census Tracts	2966	0.104
Level 3	County	102	0.984

Note: Analysis depicts the impact of geographic scale common in spatial units that are created in arbitrary manner.

[a] r is the Pearson's correlation coefficient.

Delineated/aggregated zones are often done arbitrarily, which will yield different correlation results; this is known as the zoning effect. In addition, when data tabulated at multiple levels of spatial resolution or multiple geographic scales in a nested hierarchy are analyzed, they may produce results that are inconsistent across the various spatial scales. This is known as the scale effect.

Concept of Spatial Autocorrelation

The detection of spatial autocorrelation is very useful in spatial analysis, identifying underlying data structures, the degree of spatial randomness, or clustering in the data. For a given variable, spatial autocorrelation entails the assessment of that variable in reference to the spatial location of the observational units. It measures the level, nature, and strength of interdependencies among the data points (or observational units) within the variable in terms of both space and the attribute under consideration. Point values over space or time are described as autocorrelated variables if there is a systematic spatial/temporal variation in the variable when analyzing for a spatial/temporal pattern; this phenomenon is said to be exhibiting spatial/temporal autocorrelation.

There are different levels of spatial autocorrelation (Figure 1.8). For example, when a like value is adjacent to another, these values are described as depicting a positive spatial autocorrelation; when dissimilar values are adjacent to each other, they are described as depicting a negative spatial autocorrelation; and when there is a realization of a genuinely independent random process, then this exhibition has no significant spatial autocorrelation (neutral). The measures of spatial autocorrelation are primarily aimed at testing whether a variable in one position is significantly dependent on that same variable in other nearby positions.

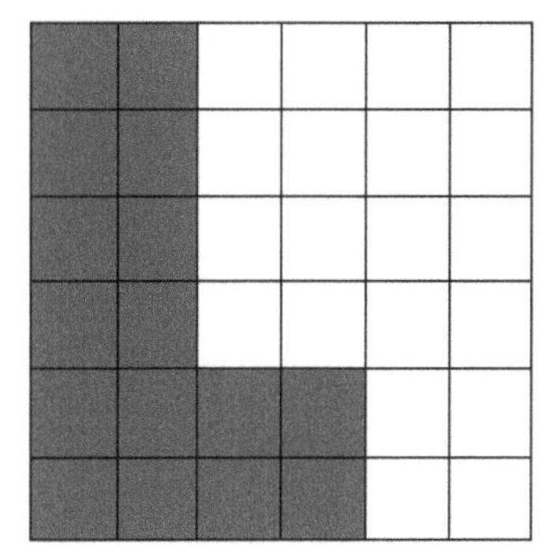
Positive spatial autocorrelation

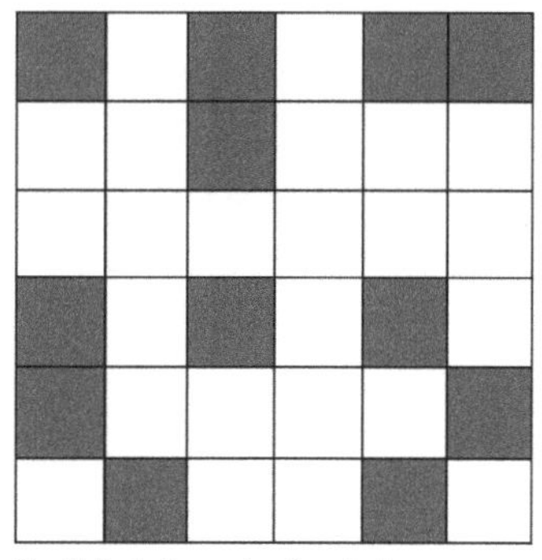
Spatially-independent/neutral

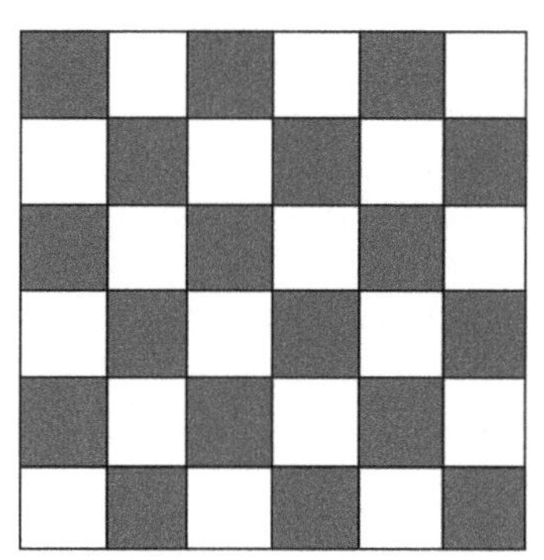
Negative spatial autocorrelation

FIGURE 1.8
Different illustrations of the concept of spatial autocorrelation.

TASK 1.3 SPATIAL AUTOCORRELATION

Let us now apply our knowledge of spatial autocorrelation to the problem set that follows. We use Figure 1.9 to learn about this concept (understanding spatial autocorrelation with Moran's scatterplot). This figure depicts different patterns of spatial autocorrelation.

Based on the illustrations, we can match each concept with its corresponding letter:

a. *Positively autocorrelated/clustered*: It is "c"—areal unit patterns are most tightly clustered, and Moran's I is close to or equal to +1.
b. *Negatively autocorrelated/dispersed*: It is "a"—areal unit patterns are the points mostly dispersed, and Moran's I is close to or equal to 0.

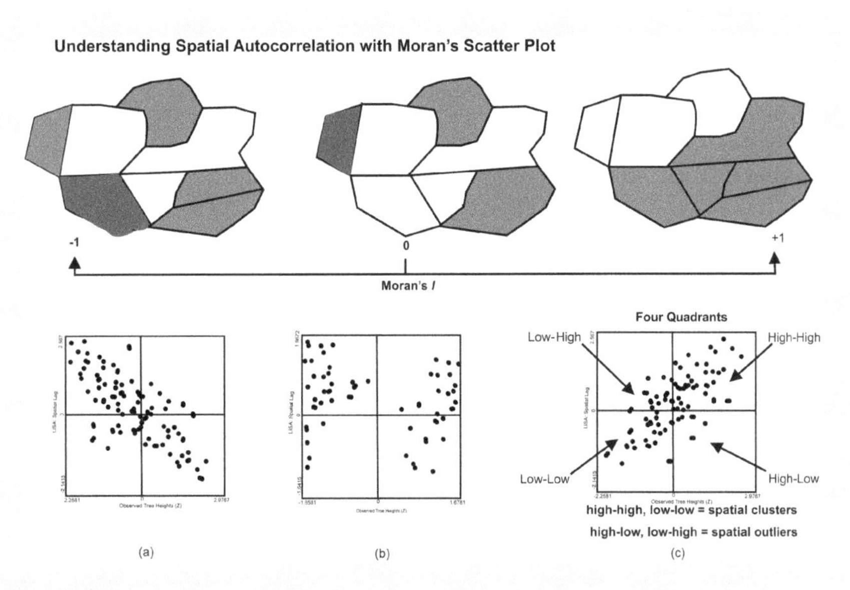

FIGURE 1.9
Different patterns of spatial autocorrelation: (a) negatively autocorrelated, (b) neither positively nor negatively autocorrelated, (c) positively autocorrelated.

c. *Neither positively nor negatively/neutral/independent/random*: It is "b"—areal unit patterns are randomly distributed, and Moran's I is equal to 0.

d. Based on your own research interests, identify a variable that is likely to be spatially autocorrelated. Will it be positively or negatively autocorrelated? Explain.

In medical geography, one example of spatial autocorrelation can be drawn from the distribution of a contagious disease such as the Ebola virus that spreads through direct transmission of bodily fluids. A localized outbreak that began in a small West African country (Guinea) in December 2013 gradually spread to neighboring countries of Sierra Leone and Liberia. By the end of July 2014, nearly 1100 cases had been reported resulting in 729 deaths, one of the deadliest outbreaks in the history of the disease. The current global 2020 pandemic of Severe Acute Respiratory Syndrome Coronavirus 2 (SARS-CoV-2) and COVID-19 in Wuhan, China, that was first detected by the BlueDot algorithm using foreign news reports and airline ticketing data, can be studied using this similar spatial autocorrelation concept. A spatial analysis of the disease patterns would reveal a stronger positive spatial autocorrelation with the communities close to the regions of these countries reporting the highest incidence and fatality rates, than with those that are farther away.

Conclusion

The field of spatial analysis has been dramatically transformed over the last two decades as new applications have been added to the suite of tools and technologies used to analyze geographic features. In this chapter, we explored trends in the evolution of the field and examined the types and properties of spatial data, and the inherent challenges that accompany their use. A set of challenge exercises that draw from key concepts and themes introduced in this chapter follows. Also included is a glossary of the key terms used in this chapter.

Worked Examples in R and Stay One Step Ahead with Challenge Assignments

The R Sample Code examples for this chapter are provided in an online downloadable link. Note that all of the activities that we undertake in spatial analysis frequently draw from the core concepts covered in Chapter 1.

Working with R

For R installation and administration instructions, I recommend you spend some time reviewing this very useful link (https://cran.r-project.org/doc/manuals/r-release/R-admin.html). Before you begin to work on your projects and exercises using R, make sure you have installed R-packages. R is vital for your success in statistical computing and production of graphics. Graphics are important visuals for data exploration and presentation, as we seek to acquire a 360° perspective.

Getting Started

Upon the installation of R, it can be launched from the Applications folder (MacOS) or Desktop Icon (Windows). Just type in the following commands in the console to check if RStudio has been properly installed and running.

Always, start any new session by loading external data first by setting the workspace directory. In this example `("../ChapterData/Chapter1 _ Data _ Folder")`, this is the workspace directory or simply the path of the folder where the file you are working on is located.

Ensure that all relevant packages are installed by attempting to load them each time you start a session using the following command: `library()`. If the package is unavailable, then ensure that it is either installed or updated.

Try to load data as follows:

```
setwd("..ChapterData/K24901 _ Data _ Folders/Chapter1 _
Data _ Folder")
airquality=read.csv('airquality.csv',header=TRUE, sep=",")
```

##The above code reads the file `airquality.csv` into a data frame airquality. `Header=TRUE` specifies that the data includes a header, and `sep=","` specifies that the values in data are separated by commas. This short tutorial is adopted from Pandey P. A Comprehensive Guide to Data Visualization in R for Beginners. An overview of the `R` visualization capabilities. https://towardsdatascience.com/a-guide-to-data-visualisation-in-r-for-beginners-ef6d41a34174

```
str(airquality)
```

This will display the internal structure of an R object and gives an overview of the rows and columns of the airquality dataset.

```
head(data,n) and tail(data,n)
```

The head outputs the top n elements in the dataset, while the tail method outputs the bottom n.

```
summary(airquality)
```

The summary method displays descriptive statistics for every variable in the dataset, depending on the type of the variable.

Working with Spatial Data

There are two ways of loading data in R, either it can be built-in or can be loaded from external sources. In this book, we have taken the approach of loading it from external sources. The link for downloading external datasets for this book is https://www.taylorfrancis.com/books/9781003021643

We use the following command to load necessary R packages, particularly the Geospatial Data Abstraction Library extensions that we work with throughout this book.

#load the necessary packages

```
library(rgdal)
```

```
library(dplyr)
```

set your workspace directory:

```
setwd("..ChapterData/K24901 _ Data _ Folders/Chapter1 _
Data _ Folder")
```

list and view the files in the workspace directory (It should list the files):

```
list.files(path=".")
```

The sample R codes for illustrating two spatial analysis concepts (Modifiable Areal Unit Problem and Spatial Autocorrelation) are provided in Chapter 1 R-Code file. Please run more codes and continue to practice on your own. This will get you ready to work with R.

Tips for Working with R

- Set your session by ensuring the correct path to the work directory
- Ensure required packages are installed or updated
- Note the six common R commands (note input files have been assumed into the command lines)

```
list.files(path=".")
```

Read MS Excel File (.xls)

```
library("readxl")
llinios _ cnty _ agri    <-    read _ excel("llinios _ cnty _
agricultural _ statistics.xls")
```

Read CSV file (.csv)

```
airquality=read.csv('airquality.csv',header=TRUE, sep=",")
```

Read Shapefile (.shp)

```
library(rgdal)
library(sp)
TCEQ   <-   readOGR('TCEQ _ Air _ Monitoring _ Sites _ PR _
studyregion.shp',  layer='TCEQ _ Air _ Monitoring _ Sites _
PR _ studyregion')
```

Read geodatabase (.gdb) file

```
library(rgdal)
list.files(path.expand("Noise _ OHare _ Geodatabase.gdb"))
```

```
[1]  Study _ Area _ Census _ BLGP.lyr"   "Noise _ OHare _
Geodatabase.gdb" > fc _ list=ogrListLayers("Noise _ OHare _
Geodatabase.gdb")
print(fc _ list)
```

Read a specific layer in the Noise_OHare_Geodatabase.gdb

```
library(sf)
luca <- st _ read("Noise _ OHare _ Geodatabase.gdb", layer=
"Study _ Area _ Census _ BLGP")
read _ sf("Noise _ OHare _ Geodatabase.gdb", layer="Study _
Area _ Census _ BLGP")
library(sf)
luca <- st _ read("Noise _ OHare _ Geodatabase.gdb", layer=
"Study _ Area _ Census _ BLGP")
```

Stay One Step Ahead with Challenge Assignments

Concept: In this chapter, we discussed five principal concepts and practical examples including data represent, spatial scale, spatial proximity, spatial autocorrelation, and modifiable areal unit problem. In spatial analysis, we often visualize, analyze, and model subsets of data patterns at different spatial, temporal, or spatiotemporal scales; we determine whether different subsets of data patterns are similar, dissimilar, interconnected, or not by examining their underlying spatial structure; and we use different spatial units to make or record observations.

Task: We use concepts from this chapter to describe spatial features around places where we live, play, work, worship, shop, eat, or learn.

1. Define start and end point addresses between two locations/places using the Get Direction's tool. Identify how far it is to your best friend's neighborhood, how far it is to your favorite grocery store or restaurant, how far it is to your workplace or friend's gym, and how far it is to your school or park. Compile these distances in a table, and rate them as close or far.
2. Use your local knowledge of these places or the online aerial/satellite image. Is the neighborhood that you live in similar or dissimilar to surrounding areas in terms of types of residential patterns? Are the trees that are in your compound, community, or neighborhood similar or dissimilar to surrounding areas in terms of height or age? Are the visual patterns of housing designs similar or dissimilar to the surrounding areas? Are the crime, disease, or unemployment rates in your place similar or dissimilar to the surrounding areas? We use

a specific attribute from a place to describe whether observed events are similar or dissimilar to one another. One of these measures is the distance between the events, time of event occurrence, or any other attribute of interest. For example, if we wish to determine the closeness of the observed patterns using the distance or time measures, then we must evaluate their similarity or dissimilarity in comparison to an established theoretical pattern. Also, we may investigate whether spatial patterns that lie in close proximity are similar or dissimilar in values for any measured attribute of interest.

3. We normally describe human or natural features using measurement or observational units such as census blocks/tracts, neighborhoods, watershed, catchment, basin, patch, physical features that surround them, or place names, etc. We record the time of occurrence of events to assess spatial processes. Units are applied to qualitatively or quantitatively describe towns or cities; soil or vegetation patterns near the rivers, lakes, or mountains; and grazing patterns of wild animals in an open African Savanna grassland or type of human settlement near the rivers, mountains, or lakes.

Approach/strategy for solving task: Use Google Maps or MapQuest tools to locate places of interest through your computer or mobile phone internet. Complete the tasks described previously through the use of local knowledge, online maps, and aerial or satellite layers. Compile your findings/observations into a meaningful short report.

Possible solutions:

1. In Task 1.1, we measured the spatial proximity of features using distance metrics.
2. In Task 1.2, we determined whether observed patterns are similar or dissimilar in space and in attribute. We examined the spatial/temporal patterns of places using the concept of spatial/temporal autocorrelation.
3. In Task 1.3, we learned how to record observations through a spatial unit. We use units, marks, or cartographic symbols to represent human or natural features at different spatial/temporal scales in a map. Recall, we learned that observational units/geographies are modifiable, so they will yield inconsistent analytical results across different spatial scales. They are also spatially autocorrelated due to the existence of spatial dependency (nonindependence). This spatial dependency therefore must be considered at all levels in an analysis. Spatial analytical methods are designed to successfully incorporate and account for the underlying spatial structure of an observed pattern. Additional tasks to help learners to further understand the nature of geographic or spatial data are provided under Tasks 1.4 and 1.5.

TASK 1.4 WORKING WITH GEOGRAPHIC DATA AT MULTIPLE RESOLUTIONS AND FORMATS

1. Suppose we are studying the impacts of land use and land cover changes in the city of Chicago over the last hundred years.
2. Suppose we are conducting a study of residential and commercial usage of broadband technologies in the city of Chicago over the last 12 years.
   a. List spatial datasets necessary to analyze studies (1) and (2) in two separate tables. Include format, scale/resolution, possible source of data, date of data collection, and sampling framework used to collect the observations or type of instrument/sensor used to record the data.
   b. How would you standardize the spatial datasets to complete the two studies?

TASK 1.5 IDENTIFYING PROBLEMS THAT ARE UNIQUE TO SPATIAL DATA

1. Earlier in the chapter, we presented a case scenario involving work with a local food bank agency to investigate the soil properties in the community prior to implementing an urban community garden initiative. Based on what you have learned in this chapter:
   a. How would you characterize the geographic features used to collect the soil samples?
   b. Describe the unique challenges that you are likely to encounter when analyzing the soil data.

Review and Study Questions

1. Differentiate between statistical and spatial analysis. Describe the significant milestones in the development of spatial analysis.
2. Since the quantitative revolution, what are the major activities and innovations that have spurred the development of spatial analysis?
3. Describe the properties of spatial data. What are the strengths associated with the use of spatial data?

4. Choose one of the following geographic concepts, and explain its role/impact on potential results derived from spatial analysis:
    - Spatial scale
    - Spatial proximity
    - Spatial autocorrelation
    - MAUP

Glossary of Key Terms

geographic information science: The basic science that cultivates the use of theoretical concepts and principles that inform and facilitate the development of GIS technology (Goodchild 1997).

geographic information system: Refers to the use of a computational framework to present, store, manipulate, manage, visualize, analyze, and optimize spatial data. GIS technology can primarily be identified using three well-known streams, which include location, the use of computer-based technology, and application-driven/functional aspects.

***geographic* versus *spatial*:** The term *geographic* is typically used to refer to the earth, its two-dimensional surface, and its three-dimensional atmosphere, oceans, and subsurface, whereas the term *spatial* refers to the multidimensional frame that references data. For instance, medical images are referenced to the human body, engineering drawings are referenced to a mechanical object, and architectural drawings are referenced to a building (Goodchild 1997).

geographic visualization: Involves the use of computers to make sense of spatial data by employing different graph encodings. It includes three activities: exploration, analysis, and synthesis and presentation. It also entails the use of the cognitive domain to assess expressiveness and effectiveness of any data encoding and decoding processes.

logical positivism: A way of thinking that evaluates the truth or falsity of empirical knowledge/cause-and-effect statements; must be verifiable.

paradigm: A set of assumptions, norms, thoughts, concepts, and values that govern scientific work and process.

spatial analysis: Entails an examination of data that is associated with location. It is a crucial analytical component of GIS. We can describe and analyze the distribution of features or spatial patterns across the study region. Through spatial analysis, we can understand the distribution of certain characteristics associated with those spatial patterns.

statistics: Helps with the collection and measurement of observations; provides an analytical framework for explaining distributions, providing estimates, and generating random numbers. When representative observations are made, they may provide supporting evidence about these events.

References

Andrienko, N., and G. Andrienko. 2010. Spatial generalization and aggregation of massive movement data. *IEEE Transactions on Visualization and Computer Graphics* 17(2): 205–219.

Bailey, T.C., and A.C. Gatrell. 1995. *Interactive Spatial Data Analysis*. London: Longman Scientific and Technical.

Barnes, N., R.A. Cartwright, C. O'Brien, B. Roberts, I.D.G. Richards, and C.C. Bird. 1987. Spatial patterns in electoral wards with high lymphoma incidence in the Yorkshire Health Region. *British Journal of Cancer* 56: 169–172.

Besag, J., and P.J. Green. 1993. Spatial statistics and Bayesian computation. *Journal of the Royal Statistical Society Series B-Methodological* 55(1): 25–37.

Besag, J., P. Green, D. Higdon, and K. Mengersen. 1995. Bayesian computation and stochastic-systems. *Statistical Science* 10(1): 3–41.

Besag, J., and J. Newell. 1991. The detection of clusters in rare diseases. *Journal Royal Statistical Society, Series A (Statistics in Society)* 53: 127–128.

Besag, J., J. York, and A. Mollie. 1991. Bayesian image-restoration, with 2 applications in spatial statistics. *Annals of the Institute of Statistical Mathematics* 43(1): 1–20.

Bithell, J.F. 1995. The choice of test for detecting raised disease risk near a point source. *Statistics in Medicine* 14: 2309–2322.

Bithell, J.F. 1999. Disease mapping using the relative risk function estimated from areal data. In Lawson, A.B., A. Biggeri, D. Bohning, E. Lasaffre, J.-F. Viel, and R. Bertollini (eds.), *Disease Mapping and Risk Assessment for Public Health*. New York: John Wiley & Sons, pp. 247–255.

Chen J., R.E. Roth, A.T. Naito, E.J. Lengerich, and A.M. MacEachren. 2008. Geovisual analytics to enhance spatial scan statistic interpretation: An analysis of US cervical cancer mortality. *International Journal of Health Geographics* 7: 57.

Clark, P.J., and F.C. Evans. 1954. Distance to nearest neighbor as a measure of spatial relationships in populations. *Ecology* 35: 445–453.

Clark, P.J., and F.C. Evans. 1955. On some aspects of spatial pattern in biological populations. *Science* 121: 397–398.

Conley J., M. Gahegan, and J. Macgill. 2005. A genetic approach to detecting clusters in point data sets. *Geographical Analysis* 37(3): 286–314.

Cuzick, J., and R. Edwards. 1990. Spatial clustering for inhomogeneous populations. *Journal of the Royal Statistical Society, Series B* 52(1): 73–104.

Diggle, P.J. 1990. A point process modeling approach to raised incidence of a rare phenomenon in the vicinity of a prespecified point. *Journal Royal Statistical Society, Series A (Statistics in Society)* 153: 349–362.

Diggle, P.J., and B.S. Rowlingson. 1994. A conditional approach to point process modeling of elevated risk. *Journal Royal Statistical Society, Series A (Statistics in Society)* 157: 433–440.

Diggle, P.J., J.A. Tawn, and R.A. Moyeed. 1998. Model-based geostatistics. *Journal of the Royal Statistical Society Series C (Applied Statistics)* 47(Part 3): 299–326.

Efron, B. and C. Morris. 1975. Data analysis using Stein's estimation and its generalization. *Journal of the American Statistical Association* 70: 311–319.

Efron, B. and C. Morris. 1976. Multivariate empirical bayes and estimation of covariance matrices. *Annals of Statistics* 4(1): 22–32.

Egenhofer, M.J., J. Glasgow, O. Gunther, J.R. Herring, and D.J. Peuquet. 1999. Progress in computational methods for representing geographical concepts. *International Journal of Geographical Information Science* 13(8): 775–796.

Excoffier, L., G. Laval, and S. Schneider. 2005. Arlequin (version 3.0): An integrated software package for population genetics data analysis. *Evolutionary Bioinformatics* 1: 47–50.

Fotheringham, A.S., and P. Rogerson. 1994. *Spatial Analysis and GIS*. London: Taylor and Francis Group.

Fotheringham, A.S., C. Brunsdon, and M. Charlton. 2000. *Quantitative Geography: Perspectives on Spatial Data Analysis*. London: SAGE.

Garzon, M.B., R. Blazek, M. Neteler, R.S. de Dios, H.S. Ollero, and C. Furlanello. 2006. Predicting habitat suitability with machine learning models: The potential area of *Pinus sylvestris L.* in the Iberian Peninsula. *Ecological Modelling* 197(3–4): 383–393.

Gatrell, A.C., T.B. Bailey, P.J. Diggle, and B.S. Rowlingson. 1996. Spatial point pattern analysis and its application in geographical epidemiology. *Transactions of the Institute of British Geographers* 21(1): 256–274.

Glantz, S. 2012. *Primer of Biostatistics*, 7th ed. New York, NY: McGraw-Hill.

Goodchild, M.F. 1997. What is geographic information science? *NCGIA Core Curriculum in GIScience*. University of California, Santa Barbara.

Griffith, D., and C. Amrhein 1997. *Multivariate Statistical Analysis for Geographers*. Englewood Cliffs, NJ: Prentice Hall.

Guillot, G., A. Estoup, F. Mortier, and J.F. Cosson. 2005. A spatial statistical model for landscape genetics. *Genetics* 170(3): 1261–1280.

Guo, D.S., J. Chen, A.M. MacEachren, and K. Liao. 2006. A visualization system for space-time and multivariate patterns (VIS-STAMP). *IEEE Transactions on Visualization and Computer Graphics* 12(6): 1461–1474.

Haggett, P., A.D. Cliff, and A. Frey. 1977. *Locational Analysis in Human Geography*, 2nd ed. New York, NY: John Wiley and Sons.

Hall, A., P. Ahonen-Rainio, and K. Virrantaus. 2014. Knowledge and reasoning in spatial analysis. *Transactions in GIS* 18(3): 464–476.

Janowicz, K., S. Gao, G. McKenzie, Y. Hu, and B. Bhaduri. 2020. GeoAI: Spatially explicit artificial intelligence techniques for geographic knowledge discovery and beyond. *International Journal of Geographical Information Science* 34(4): 625–636.

Kleinbuam, D.G., L.L. Kupper, K.E. Muller, and A. Nizam. 1998. *Applied Regression Analysis and Other Multivariable Methods*. Belmont, CA: Duxbury Press.

Kothur, P., M. Sips, A. Unger, J. Kuhlmann, and D. Dransch. 2014. Interactive visual summaries for detection and assessment of spatiotemporal patterns in geospatial time series. *Information Visualization* 13(3): 283–298.

Lawson, A.B. 1989. *Score Tests for Detection of Spatial Trend in Morbidity Data*. Dundee: Dundee Institute of Technology.
Legendre, P. 1993. Spatial autocorrelation: Trouble or new paradigm? *Ecology* 74(6): 1659–1673.
Legendre, P., and M.J. Fortin. 1989. Spatial pattern and ecological analysis. *Vegetatio* 80: 107–138.
Longford, N.T. 1993. *Random Coefficient Models*. Oxford: Clarendon Press.
Lewis, M.S. 1980. Spatial clustering in childhood leukemia. *Journal of Chronic Diseases* 33(11–12): 703–712.
Marmion, M., M. Luoto, R.K. Heikkinen, and W. Thuiller. 2009. The performance of state-of-the-art modelling techniques depends on geographical distribution of species. *Ecological Modelling* 220(24): 3512–3520.
Marshall, R. 1991. A review of methods for the statistical analysis of spatial patterns of disease. *Journal of the Royal Statistical Society. Series A (Statistics in Society)* 154(3): 421–441.
Matheron, G. 1963. Principles of geostatistics. *Economic Geology* 58(8): 1246–1266.
Miller, H.J. 2004. Tobler's first law and spatial analysis. *Annals of the Association of American Geographers* 94(2): 284–289.
Moran, P.A.P. 1948. The interpretation of statistical maps. *Journal of Royal Statistical Society, Series B* 10: 243–251.
Munasinghe, R.L., and R.D. Morris. 1996. Localization of disease clusters using regional measures of spatial autocorrelation. *Statistics in Medicine* 15: 893–905.
Myers, D.E. 1994. Spatial interpolation: An overview. *Geoderma* 62: 17–28.
Myers, D.E. 1997. Statistical models for multiple-scaled analysis. In Quattrochi, D.A., and M.F. Goodchild (eds.), *Scale in Remote Sensing and GIS*. New York, NY: CRC Lewis Publishers, pp. 273–293.
Nelson, T.A. 2012. Trends in spatial statistics. *The Professional Geographer* 64(1): 83–94.
Ohno, Y., K. Aoki, and N. Aoki. 1979. A test of significance of geographical clusters of disease. *International Journal of Epidemiology* 8: 273–280.
Openshaw, S., M. Charlton, C. Wymer, and A. Craft. 1987. A mark 1 geographical analysis machine for the automated analysis of point data sets. *International Journal of Geographical Information Systems* 1(4): 335–358.
Openshaw, S., A.W. Craft, M. Charlton, and J.M. Birch. 1988. Investigation of leukemia clusters by use of a Geographical Analysis Machine. *Lancet* 1(8580): 272–273.
O'Sullivan, D., and D.J. Unwin. 2010. *Geographic Information Analysis*, 2nd ed. New York, NY: John Wiley & Sons.
Oyana, T.J., S.J. Johnson, and G. Wang. 2014. Landscape metrics and change analysis of a national wildlife refuge at different spatial resolutions. *International Journal of Remote Sensing* 35(9): 3109–3134.
Rogerson, P.A. 2005. *Statistical Methods for Geography*. London: SAGE.
Rossi, R.E., D.J. Mulla, A.G. Journel, and E.H. Franz. 1992. Geostatistical tools for modeling and interpreting ecological spatial dependence. *Ecological Monographs* 62(2): 277–314.
Ripley, B.D. 1981. *Spatial Statistics*. New York, NY: John Wiley and Sons.
Sabel, C.E., A.C. Gatrell, M. Loytonen, P. Maasilta, and M. Jokelainen. 2000. Modelling exposure opportunities: Estimating relative risk for motor neuron disease in Finland. *Social Science and Medicine* 50(7–8): 1121–1137.
Samet, H. 1990. *Applications of Spatial Data Structures: Computer Graphics, Image Processing and GIS*? Reading, MA: Addison-Wesley/ACM.

Samet, H. 1995. Spatial data structures. In: W. Kim (ed.), *Modern Database Systems: The Object Model, Interoperability and Beyond*. New York: ACM Press and Addison-Wesley, pp. 361–385.
Stockwell, D.R.B., and D.P. Peters. 1999. The GARP modeling system: Problems and solutions to automated spatial prediction. *International Journal of Geographic Information Systems* 13: 143–158.
Wallet, F., and C. Dussent. 1997. Multifactorial comparative study of spatial point pattern analysis methods. *Journal of Theoretical Biology* 187(3): 437–447.
Wallet, F., and C. Dussent. 1998. Comparison of spatial point patterns and processes characterization methods. *Europhysics Letter* 42(5): 493–498.
Waller, L.A., and G.M. Jacquez. 1995. Disease models implicit in statistical tests of disease clustering. *Epidemiology* 6: 584–590.
Ye, X.Y., B. She, L. Wu, X.Y. Zhu, and Y.Q. Cheng. 2014. An open source toolkit for identifying comparative space-time research questions. *Chinese Geographical Science* 24(3): 348–361.
Zurita, G.A. and M.I. Bellocq. 2010. Spatial patterns of bird community similarity: Bird responses to landscape composition and configuration in the Atlantic forest. *Landscape Ecology* 25(1): 147–158.

2

Making Scientific Observations and Measurements in Spatial Analysis

Learning Objectives

1. Define successful strategies for spatial data collection.
2. Identify potential sources of spatial datasets.
3. Apply a sampling framework to collect spatial observations/events.
4. Successfully process and prepare spatial datasets for analysis.

Introduction

The process of making scientific observations starts with an important realization that naturally occurring phenomena and processes are very complex, and as data scientists, we must come up with simple and creative ways to effectively measure and represent them. In spatial analysis, the strategies for collecting and processing data are the keys to scientific success, and many of these analytical strategies have been inspired by several schools of thought. Chief among them are the logical positivists who recommend the use of research designs that rely on direct observations with the help of our senses, established protocols, artificial sensors, or instrumentation to validate research hypotheses. While data generated from primary sources are the most ideal in such designs, the increasing availability of secondary data sources has made it possible for a variety of spatial analyses to be done using computer programs and without necessarily conducting any taxing experiments. The purpose of this chapter is to underscore the relevance of data collection, how and why data are collected, potential gaps in the data collection, and the accompanying processing needed to ensure quality and accuracy in the observations. Studies that are carefully designed with the appropriate mix of data and analytical strategies used for execution, analysis, and interpretation will yield meaningful scientific conclusions and recommendations. Studies drawn from reliable and scientifically valid measures are often the ones that are easily verifiable and replicable, yielding a solid body of evidence and new knowledge for use in policy formulation and scientific decision-making.

Scales of Measurement

In both traditional statistics and spatial analysis, the choice of analytical methods used to address our research questions largely depends on the nature and characteristics of the variables that are used to calibrate the naturally occurring phenomena. Variables may be characterized by continuous or discrete data values, quantitative or qualitative measures. The means by which we systematically observe and assign data values to these variables are referred to as the scales of measurement. There are four commonly used scales of measurement: nominal, ordinal, interval, and ratio. The first two (nominal and ordinal) are qualitative scales, and the last two (interval and ratio) are quantitative scales of measurement. In a statistical context, measures that are recorded on a qualitative scale are evaluated using nonparametric statistics, whereas the measures recorded on a quantitative scale are evaluated using parametric statistics.

Nominal Scale

This is the simplest means of assigning data values to a variable. Most raw datasets or ungrouped categories that are still in their original format fit this description. A nominal scale describes the means by which ungrouped categories of data are evident without numerical reference. Descriptive or qualitative statements can be employed to identify such observations (Figure 2.1). For example, people in Chicago may be classified in categories such as young, adult, or elderly; gender may be classified as male or female. Several geographic objects are measured using this scale, for example, land use categories, types of zoning, vegetation types or biomes, and types of settlements. Nominal scale variables may also have observed values that are classified as dichotomous/binary using only two categories, for example, yes or no, either one or zero, and true or false.

Ordinal Scale

This is the second form of recording values for a qualitative variable that involves the ordering of observations in rank order. Any organization of observed values that is applied through some ordering or ranking normally fits this description. This scale has both identity and magnitude properties, and at times its use can result in strongly or weakly ordered observations (Figure 2.2). Strong ordering refers to a situation in which the

FIFA World Cup 2014

Group A	Group B
Brazil	Netherlands
Mexico	Chile
Croatia	Australia
Cameroon	Spain

Gender

Male
Female

Direction

North
East
South
West

Types of Roads

Street
Highway
Lane
Major Roads
Freeway
Avenues
Interstate
Super Highways
Interchange

Land Use and Land Cover Categories

Water
Barren
Shrub land
Vegetation
Wetlands
Developed
Forest

Busiest Ports

Shanghai, China
Singapore, Singapore
Hong Kong, China
Shenzhen, China
Busan, South Korea
Ningbo-Zhoushan, China
Guangzhou Harbor, China
Jebel Ali, Dubai, UAE
Rotterdam, Netherlands
Hamburg, Germany
Los Angeles

FIGURE 2.1
Examples of data recorded using the nominal scale.

ranks are assigned to observed values, whereas weak ordering occurs when individual observations are grouped into unique categories. For example, observed values of household income can be grouped together as low, medium, or high, or the weather can be described as being mild, moderate, or severe. In weakly ordered observations, it is easy to differentiate

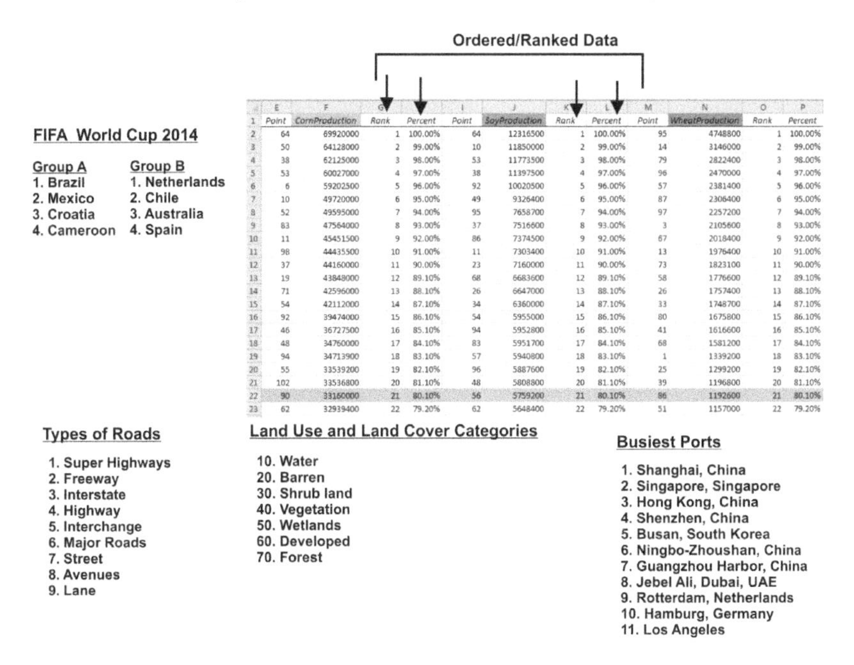

	E	F	G	H	I	J	K	L	M	N	O	P
1	Point	CornProduction	Rank	Percent	Point	SoyProduction	Rank	Percent	Point	WheatProduction	Rank	Percent
2	64	69920000	1	100.00%	64	12316500	1	100.00%	95	4748800	1	100.00%
3	50	64128000	2	99.00%	10	11850000	2	99.00%	14	3146000	2	99.00%
4	38	62125000	3	98.00%	53	11773500	3	98.00%	79	2822400	3	98.00%
5	53	60027000	4	97.00%	38	11397500	4	97.00%	96	2470000	4	97.00%
6	6	59202500	5	96.00%	92	10020500	5	96.00%	57	2381400	5	96.00%
7	10	49720000	6	95.00%	49	9326400	6	95.00%	87	2306400	6	95.00%
8	52	49595000	7	94.00%	95	7658700	7	94.00%	97	2257200	7	94.00%
9	83	47564000	8	93.00%	37	7516600	8	93.00%	3	2105600	8	93.00%
10	11	45451500	9	92.00%	86	7374500	9	92.00%	67	2018400	9	92.00%
11	98	44435500	10	91.00%	11	7303400	10	91.00%	13	1976400	10	91.00%
12	37	44160000	11	90.00%	23	7160000	11	90.00%	73	1823100	11	90.00%
13	19	43848000	12	89.10%	68	6683600	12	89.10%	58	1776600	12	89.10%
14	71	42596000	13	88.10%	26	6647000	13	88.10%	26	1757400	13	88.10%
15	54	42112000	14	87.10%	34	6360000	14	87.10%	33	1748700	14	87.10%
16	92	39474000	15	86.10%	54	5955000	15	86.10%	80	1675800	15	86.10%
17	46	36727500	16	85.10%	94	5952800	16	85.10%	41	1616600	16	85.10%
18	48	34760000	17	84.10%	83	5951700	17	84.10%	68	1581200	17	84.10%
19	94	34713900	18	83.10%	57	5940800	18	83.10%	1	1339200	18	83.10%
20	55	33539200	19	82.10%	96	5887600	19	82.10%	25	1299200	19	82.10%
21	102	33536800	20	81.10%	48	5808800	20	81.10%	39	1196800	20	81.10%
22	90	33160000	21	80.10%	56	5759200	21	80.10%	86	1192600	21	80.10%
23	62	32939400	22	79.20%	62	5648400	22	79.20%	51	1157000	22	79.20%

FIGURE 2.2
Examples of data recorded using the ordinal scale. All the groups have been assigned ranks or ordered.

between categories but not within categories, whereas in strongly ordered observations, it is easy to differentiate between individual observations using assigned ranks.

Interval Scale

This is the third form of assigning data values to a variable using equal intervals or defined intervals, for example, temperature or time. As a quantitative scale, it provides a more precise measurement of individual observations than the nominal or ordinal scales. It also has the properties of identity, magnitude, and equal intervals. We can classify measurements of length and height of buildings, the stem width and height of trees, height of the terrain, road width and length, width and length of rivers, and age of individuals using equal intervals (Figure 2.3). We can describe all of the measureable attributes of a variable on this scale. An interval can also be established between values measured for the buildings, trees, terrains, roads, and rivers. This scale has a defined interval, for example, temperature or time. It is an ordered, constant scale, but without a natural zero.

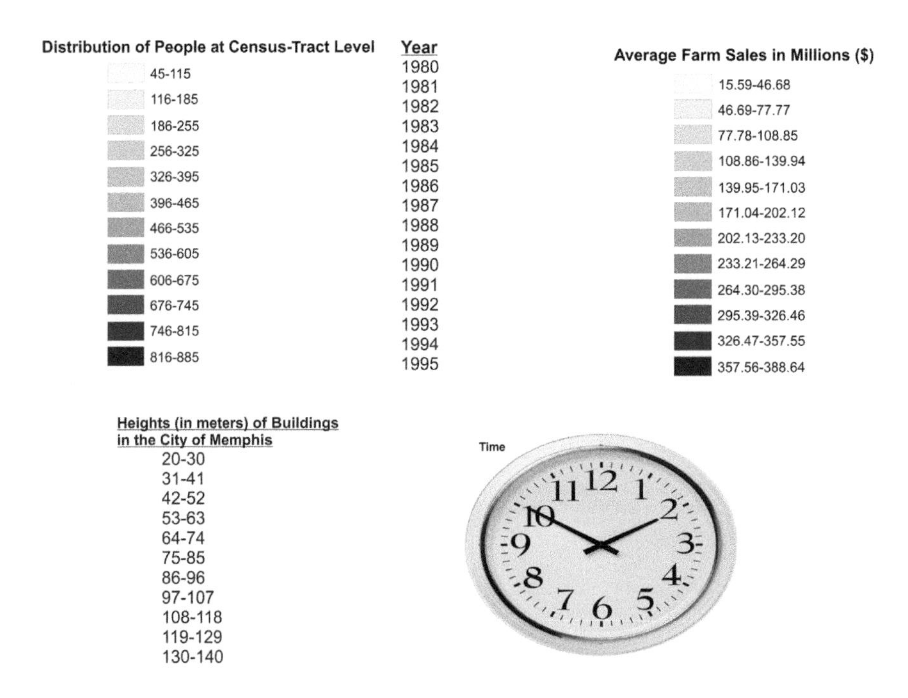

FIGURE 2.3
Examples of data recorded using the interval scale.

Ratio Scale

This is the fourth means by which one can record data values. The ratio scale has all the qualities of nominal, ordinal, and interval scales plus the advantage of zero having a precise meaning when it is assigned to an observation. So, the value of its origin or zero position indicates the absence of the quantity being measured for a given object. This scale has identity, magnitude, equal intervals, and absolute zero properties. One can provide precise measurements of length and height of a building, stem width and height of a tree, height of the terrain, road width and length, width and length of a river, weight of an object, and individual age. All of these measureable attributes are attainable using this scale (Figure 2.4).

To further demonstrate the types of measurement scales and other data considerations, in Task 2.1 let us review some of the best practices for spatial data collection. We also learn how to construct deductive and inductive hypothetical arguments using the spatial dataset as an example. Having these practical skills will ensure greater success in data collection and contribute further to our understanding of scales of measurement.

Ratio Scale

Table

agricul_ILL_stats

POP2010	POP10_SQMI	WHITE	BLACK	AMERI_ES	ASIAN	HAWN_PI	OTHER	MULT_RACE	HISPANIC	MALES	FEMALES	AGE_UNDER5
4125	11	4117	166	35	12	0	20	63	40	2235	2178	211
4418	24.3	4580	132	2	24	6	23	33	51	2407	2393	263
6566	32.3	4888	2278	10	68	0	20	84	107	3516	3832	450
5159	18.2	5023	2	16	9	0	8	26	32	2552	2532	270
5179	20.5	5507	2	8	7	0	2	11	10	2672	2865	351
6622	21.6	5580	1265	6	9	5	47	38	273	4420	2530	280
6133	35.6	5941	38	21	16	0	38	32	171	3007	3079	361
8001	31.7	6040	3347	27	35	2	52	87	138	4760	4830	600
6195	21.5	6245	4	12	12	0	9	50	54	3054	3278	399
6057	18.4	6340	17	46	4	2	6	30	56	3125	3320	337
6638	29.8	6892	10	6	28	3	6	26	32	3374	3597	394
6910	15.7	7103	16	11	8	1	15	35	39	3565	3624	414
7574	19.2	8090	21	9	8	3	15	67	72	4059	4154	469
8249	18.9	8470	58	22	11	1	12	47	55	4160	4461	512
9817	19.7	10031	8	7	19	2	19	31	48	4998	5119	579
13886	39.8	10756	1825	35	17	5	138	102	368	7706	5172	604
10962	31.6	11123	12	22	17	2	26	51	68	5508	5745	712
12633	40	12310	48	27	21	0	31	49	94	6122	6364	722
12112	53.2	12660	51	22	58	6	33	107	95	6314	6623	737
13070	32.8	12941	46	29	33	1	33	97	138	6454	6726	722
13628	35.5	13000	61	23	38	4	457	112	1162	6803	6892	934
14061	28.9	13982	35	14	46	0	57	107	176	6819	7422	911

4 (0 out of 102 Selected)

agricul_ILL_stats

FIGURE 2.4
Examples of data recorded using the ratio scale.

TASK 2.1 BEST PRACTICES FOR SPATIAL DATA COLLECTION

- Determine whether your analysis requires the use of primary / or secondary data sources. In a geographic information system (GIS), most data are available in digital format.
- Understand the approaches to data collection.
- Identify appropriate methodologies and resources required for data collection.
- Develop a solid data management plan (processing, manipulation, sharing, access and storage, quality control measures).

Two Main Approaches for Data Collection That Involve Deductive and Inductive Reasoning

Data collection approaches are guided by both deductive and inductive reasoning. Deductive reasoning entails evaluating the validity or soundness of an argument that is logically derived from a set of generalized principles or statements to arrive at a conclusion (i.e., general to specific). This is accomplished in spatial analysis when we use theory or theoretical foundation to guide our research and derive a set of hypotheses. Deduction logic is applied in the classical view of probability; in this type of reasoning, if the two premises are valid and sound, then the conclusion is considered to be true. An example of a deductive argument is as follows:

All wetlands have bird habitats

The city of Carbondale has a wetland

Therefore, the city of Carbondale has a bird habitat

Inductive reasoning entails making or evaluating generalized statements based on specific statements (i.e., specific to general). In spatial analysis, we can derive a set of general principles or a set of hypotheses from a sample of specific observations and use that information to develop a generalized set of empirical conclusions. Inductive logic is applied in the relative frequency interpretation view of probability. An example of an inductive argument is as follows:

A bird habitat existence was confirmed in 90% of the wetlands in North America

The city of Cairo has a wetland in the south

Therefore, the city of Cairo has a bird habitat

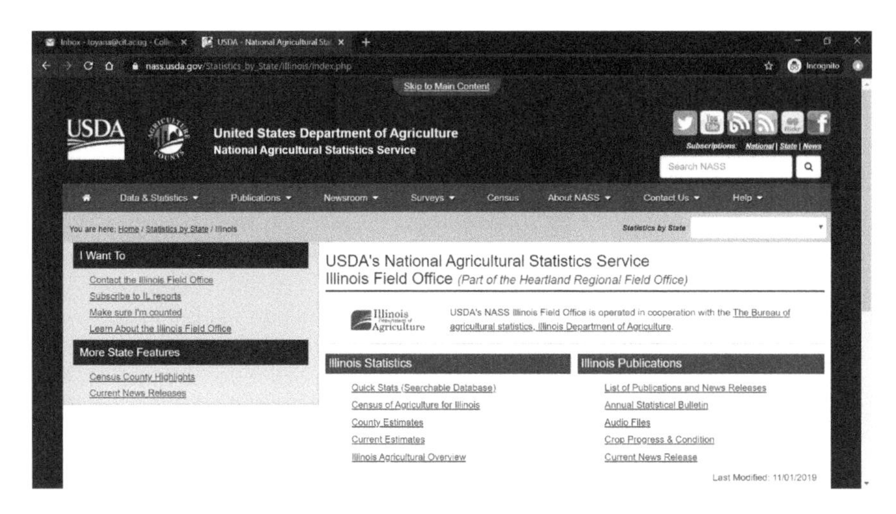

FIGURE 2.5
A screenshot of the website for downloading National Agricultural Statistics Data.

To demonstrate the two approaches, let us process an agricultural dataset from the state of Illinois at county level (Figure 2.5). Our goal is to apply these concepts to formulate data-driven hypotheses. To do so, first, open your internet to access the database, and paste the following link https://www.nass.usda.gov/Statistics_by_State/Illinois/index.php. Review the metadata for this dataset, and then click on County Estimates to review Illinois County Statistics by year.

Question: How many years of data are posted on this website? Knowing this information helps us to better understand the temporal attributes of the dataset. On review, we observe that numerous county-level agricultural datasets have been posted on this website since 2005. We will explore the 2008 dataset, partly because this is when the U.S. economy experienced significant recession, so it would be interesting to review its effects on the agricultural sector, specifically focusing on Illinois. This dataset is provided in Chapter2DataFolder (file name: *Illinois_cnty_ agricultural_statistics.xlsx*).

Question: In Table 2.1, fill in the missing information for the measurement scales on which the factors/variables were recorded and field data types for the attributes.

TABLE 2.1
Scale of Measurements and Database Data Types for Each Variable/Factor

Factor/Variable	Scale of Measurements	Data Type
WheatAcre		Integer
SoyAcre		
CornYield		
GroupArea	Nominal	
RankCornProd		

TASK 2.2 GENERATING DESCRIPTIVE STATISTICS

1. Using the *Illinois_census_county.dbf* file, sort each of the four variables listed in Table 2.2 in ascending or descending order. The results of a sorted information table would look like Table 2.2. Now, derive the mean, standard deviation, and confidence intervals for these variables. We cover this in detail in Chapter 3, but the mean shows the spread and distribution of observation around their center point, whereas the standard deviation provides information about the variations of observations in each variable. The confidence interval captures the class width of the observations in each variable.
2. Make a statement regarding the agricultural data using the *deductive logic approach.*

   Here is a sample hypothesis of deductive logic approach: All counties in Illinois have farms (since the minimum number of farms is 73). Jackson County is located in the state of Illinois. We can therefore conclude that there are farms in Jackson County.
3. Is the following statement an example of deductive logic? "All farms have more than zero acres of crops. Tina's Apples is an apple farm. Therefore, crop acreage at Tina's Apples is greater than zero." *This is a deductive logic argument because it uses a set of generalized statements that all farms have more than zero acres of crops to arrive at the conclusion.*
4. Make a statement regarding the agricultural data that is an example of inductive logic.

   Here is a sample hypothesis of inductive logic approach: All counties have between 73 and 1622 farms (a mean 753 farms). If we select 20 counties randomly, we can expect half to have more than 753 farms.

TABLE 2.2

Basic Descriptive Data for Selected Agricultural Variables in Illinois from the 2010 U.S. Census Data (Illinois_census_county.dbf)

Variable	Minimum	Maximum	Range
NO_FARMS07 (Number of Farms)	73	1,622	1,549
AVG_SIZE07 (Average Size of a Farm)	45	885	840
CROP_ACR07 (Acres of Crops)	6,388	647,350	640,962
AVG_SALE07 (Average Sales per Farm '000)	15.59	388.64	373.05

5. Is the following statement an example of inductive logic? "Ten percent of Illinois counties have more than 376,178 acres of crops. Therefore, if we randomly selected 10 counties from Illinois's 102 counties, we would expect nine of these counties to have less than 376,178 acres of crops."

   The statement is an example of inductive logic because it uses specific statements to arrive at a general conclusion. The statement relies on the assumption that our sample is representative of the study population (farms per county), and induces the result of the sample demonstrating the same patterns as the population. Although the statement is an example of inductive reasoning, it is likely that random sampling error may not support the stated conclusion.

Now, open the *Illinois_census_county.dbf* either in MS Excel, R, or any statistical software. Use this attribute table to statistically evaluate any of the four variables presented in Table 2.1. The goal is for you to understand the quantitative data well enough to make deductive and inductive logical statements using this information. Review and provide a list of scale of measurements for any four selected variables of your choice. You should also evaluate the minimum, maximum, mean, or median values of the field(s) of the four variables you have selected. It is acceptable to evaluate different statistical attributes of the field (e.g., percentile, and 95% confidence limits). Be creative.

Population and Sample

The previous problem set raises an important question that often comes up when designing an analytical project, namely, the distinction between a population and a sample. In both deductive and inductive reasoning approaches, differentiating between these two is critical for drawing the appropriate conclusions and inferences from the data. A population consists of the entire collection of events, objects, or subjects that are being studied, whereas a sample consists of a representative portion or subset of those events, objects, or subjects in a population of interest. Simply put, a sample is a mirror image subset of a parent population. The central purpose of a sample is to use it to make inferences about the population from which it was drawn. A population of interest, denoted by the denominator in the standard deviation as N, must be clearly and properly defined so that observations can be obtained for the purpose of statistical analysis. But when computing the standard deviation for a sample, the denominator is given by $n - 1$, as the sample represents a subset of the larger population.

When collecting information for statistical purposes, we could use the entire population or a sample. Due to exorbitant costs and the feasibility of conducting a large-scale study, at times, a sample is most appropriate. If a sample is properly drawn from a population, then it will contain the same characteristics from it. However, for a sample to be valid, each event stands the same chance of being selected and is independent of the selection of another event in that population; thus, this strategy minimizes a selection bias. The descriptive measures that explain a population are called *parameters*, whereas the ones that explain a sample are called *statistics*.

Sampling: This is the act of drawing a representative portion of a population. It is primarily concerned with the selection of a subset of observational units within a population with the intention of estimating its characteristics. In geography, our primary goal is to sample across space, or in some studies, we collect samples across space and time. Due to spatial and spatiotemporal variations, an effective spatial sampling strategy requires that these factors be taken into account. Given that sampling must meet the classical assumptions of randomness and independence in observations, we must take into consideration the nature of geographic data on data collection when performing analysis. It is, therefore, imperative to have a list of the elements and characteristics of a population before a sample is drawn. A representative sample should capture the essence of these elements and its characteristics. In having this knowledge of all elements and characteristics of a population, we normally want to produce unbiased estimates, so we can use any of the three types of random sampling methods: (1) simple random sampling, (2) stratified sampling, and (3) sequential sampling.

A simple random sampling is most appropriate when each event within a population has an equal chance of being selected and no subgroups are evident. However, when dealing with subgroups and a strong element of homogeneity within those groups is identified, a stratified sampling approach is recommended. This type of sampling entails splitting the population into subgroups of interest and sampling each of the subgroups either sequentially or randomly. Sequential sampling entails selecting observational units in a population based on a specified interval. However, to minimize a selection bias in the sample, the first unit must be selected randomly before the sequence is established.

Spatial Sampling

Spatial sampling refers to obtaining a representative sample of a study region that reflects the spatial structure. When designing a scheme for spatial sampling, several considerations must be made regarding the spatial dependency, spatial pattern, temporal pattern, or spatiotemporal pattern of the data. In addition to the three sampling designs mentioned earlier, we can draw samples using cluster sampling, transect sampling, or contour sampling. There are four types of sampling units for spatial sampling: (1) point sampling, (2) area sampling, (3) linear sampling (transect across the landscape), and (4) plotless sampling (common in forest vegetation survey). A detailed example of spatial sampling that includes point, area, and linear sampling follows. Two specific examples of spatial sampling are offered: one for collecting physical attributes of land cover and the other for supporting a health study.

Spatial Sampling Example 1: Suppose the central objective of our spatial sampling design is to assess the variation of leaf area index (LAI) and photosynthetically active radiation (PAR) in the fragile mountain ecosystem of Mt. Elgon located in eastern Uganda (Oyana et al. 2015, 2017). To accomplish this objective, there is a justified need to collect representative field measurements of LAI and PAR. We can design sampling protocols based on a systematic grid framework to collect LAI and PAR sample data based on a high-resolution sensor LP-80 AccuPAR Ceptometer. This instrument is a lightweight optical and portable PAR sensor and consists of 80 sensors, spaced 1 cm apart with a data storage capacity of 1 MB RAM (over 2000 measurements) and minimum spatial resolution of 1 cm. By measuring light intensity above and below the vegetation canopy, it assesses PAR interception of canopy and calculates LAI.

The LAI sample data can be combined with Landsat Thematic Mapper images and any other high-resolution images to generate LAI maps. This sensor measures PAR and LAI from the crops in the field and structural diversity of canopies in sample agricultural plots, which drive both the within- and the below-canopy microclimate; determines and controls canopy water interception, radiation extinction, and water and carbon gas exchange; and is therefore a key component of the biogeochemical cycles in mountain ecosystems.

The study area is located within an area that can be divided into 1566 blocks of area 1 km × 1 km in Manafwa watershed (Figure 2.6). The study area consists of 663 grids (the area sampling unit is based on a latitude/longitude 1000 m × 1000 m grid), so we can plan to have 28 sample agricultural plots—11 in the lower catchment and another 17 in the upper catchment. For the purpose of standardizing the sampling approach, the southwest corner of each grid will be taken to correspond to the intersection of the latitude/longitude lines. In each grid, a sampling unit will be composed of four subsample plots; that is, each sample plot/cluster will be composed of 36 microplots placed within a sampling unit of 1 km^2. Each sampling plot is expected to have one or more crop classes/plant species due to the heterogeneous nature of cultivated crops and vegetation. The plots are designed to cross the maximum possible variations within and between the classes and to monitor the crop and vegetation dynamics. Each subplot measures 30 m × 30 m and is located at 250 m apart within the sampling tract. Within each subplot, nine, 1 m transect lines distanced 7 m apart will be selected and used to measure LAI and PAR

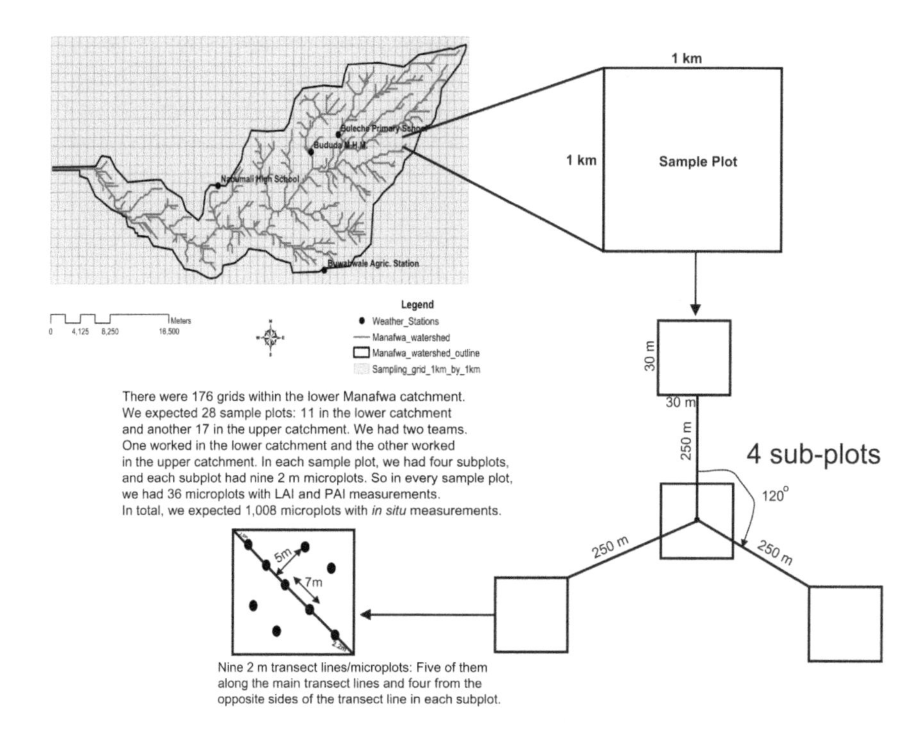

FIGURE 2.6
A schematic overview of the sampling framework for *in situ* measurements.

values. For all the subplots, we can sample a total of 1008 data points. This sample design makes it possible to combine the LAI field data with remotely sensed images at a range of spatial resolutions from 1×1 m to 1×1 km. Using this sampling design, we can make *in situ* measurements of LAI and PAI values at different spatial resolutions, two times (in the morning and afternoon) on a daily basis over a 10-day period.

In the study region, we can expect to find the following:

1. Differences in LAI and PAR (below and above canopy) estimates between crops in the upper and lower catchments
2. Differences in LAI and PAR (below and above canopy) estimates between efficiently managed agricultural fields and intensively cultivated agricultural fields
3. Differences in PAR and LAI (below and above canopy) estimates in agricultural fields close to stream flows, stable slopes, and forest stands
4. Differences in LAI and PAR (below and above canopy) estimates in different biomes
5. Temporal differences in LAI and PAR (below and above canopy) estimates

The dataset can be linked to Landsat images and Normalized Difference Vegetation Index (NDVI) vegetation profiles to assess the sensitivity of different crops over a period in the study area. The design can help advance our understanding of the functionality of mountain ecosystems and biogeochemical cycles.

Spatial Sampling Example 2: Another illustration of spatial sampling can be drawn from a health disparities project. Suppose we want to determine whether physical environmental factors play a significant role in influencing health outcomes. We can employ GIS and advanced computational tools to integrate, identify, and analyze spatial clusters of environmental stressors and lifestyle risk factors that influence the prevalence and distribution of obesity and type 2 diabetes over time. We can hypothesize that the prevalence of obesity and diabetes is a spatiotemporal phenomenon, with clustering resulting from the underlying spatial structure of the physical environment, together with socioeconomic and demographic factors. Given that unsatisfactory health outcomes have been reported in ten states located in the southeastern United States, a typical sampling strategy can be devised to investigate this hypothesis. The sampling strategy could entail a benchmark model for two states (Mississippi and Florida) at census-tract level that will be replicated in eight other states in this region. Mississippi has been identified with exceedingly higher burden of obesity and type 2 diabetes than Florida, so this can be used to build our benchmark model. We will create a comprehensive case control study design with Mississippi and Florida serving as our cases and controls, respectively, to accomplish our epidemiological study objectives.

The study area for this project includes the following states: Alabama, Arkansas, Florida, Georgia, Louisiana, Missouri, Mississippi, North Carolina, Virginia, and South Carolina (Figure 2.7). Together these ten states have 1316 counties and a total population of 71,221,706 (U.S. Census Bureau). The race/ethnicity composition within these states shows a very strong concentration of African Americans and Hispanics (41% of the population compared to 29% of the total U.S. population). The study area provides a diverse population and unique neighborhood characteristics that will be used to create baseline health information that will be used to explain differences in adverse health outcomes. We can randomly sample 10% of the households in Mississippi and Florida at census-tract level to be our target population. Using the target population, a 45-minute questionnaire survey instrument can be administered to support this epidemiological study. The questionnaire will be systematically administered to collect self-reported measures of health and physical activities from the sample population. The sample population data can be supplemented with five sets of existing secondary datasets outlined as follows.

Data description of five sets of relevant data for the epidemiological study:

1. Neighborhood demographic data from the U.S. Census Bureau.
2. Boundary data and other relevant layers from Florida Geographic Data Library and Mississippi Geospatial Clearinghouse, respectively.

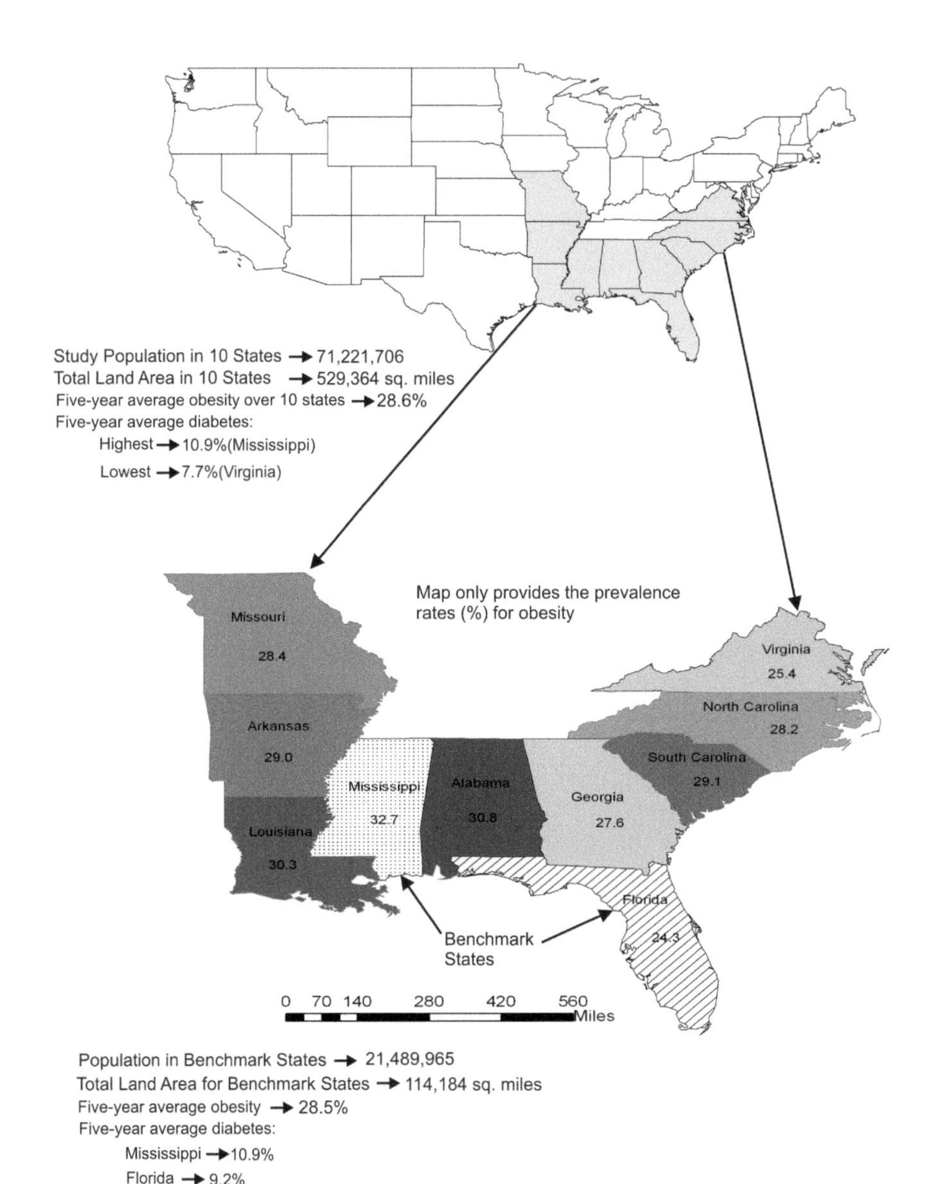

FIGURE 2.7
Study area map of case-control study design for obesity and type 2 diabetes.

3. Individual-level data on health outcome and physical activity: The data on self-reported anthropometric measures (height and weight) and other health outcomes are from two national surveys: Behavioral Risk Factor Surveillance System (BRFSS) and National Health and Nutrition Examination Survey (NHANES). Both surveys have adequate information that can be used for calculation of the prevalence

of obesity and type 2 diabetes. The BRFSS collects data on health risk behavior, preventive health practice, and health-care access for adults aged 18 years and older in the United States. The census tract-level benchmark model can be built based on this dataset. The NHANES provides the health and nutritional status of adults and children in the United States. The physical examination data consist of medical, dental, physiological measurements, and laboratory tests. The body mass index (BMI) measure will be created and categorized as "normal weight" if an individual's BMI is between 18 and 24.9, "overweight" if the BMI is between 25 and 29.9, and "obese" if the BMI is greater than 30. Individuals will be considered to have diabetes if they responded "Yes" to the question "Have you ever been told by a doctor that you have diabetes?" We assess physical activity using BRFSS/NHANES questions, for example, type of physical activity, distance in miles, and how long. Individuals who reported that they did not engage in any of these activities over the past month will be considered inactive.

4. Built environment data: Food environment data from the InfoUSA, Inc. (Omaha, Nebraska) and Dun and Bradstreet, Inc. (Short Hills, New Jersey).
5. Remote sensing data and existing products: Remotely sensed data in the form of aerial photographs and satellite images can be compiled, including (1) *orthoimagery* which will provide an aerial view and spatial perspective of neighborhoods/National Agriculture Imagery program with 1 m spatial resolution; (2) *elevation*, 3 and 10 m resolution from the U.S. Geological Survey; (3) the *1992 and 2001 National Land Cover Dataset and Light Detection and Ranging* (LiDAR) for land cover and elevation, which are useful for 3D/4D visualizations; and (4) combined *high-resolution multispectral data* (IKONOS, QuickBird, LiDAR, hyperspectral image) to extract sidewalk inventory data.

Having introduced the two examples of sampling strategies earlier, let us now examine how to process a specific spatial dataset and learn how to describe the statistical or spatial distributions in a sample or a population. We also learn how to draw different types of samples that have been described in previous sections.

1. To find out the top/largest and bottom/smallest producers of corn, soybeans, and wheat in Illinois state, we need to summarize the agriculture statistics (i.e., average sales, acreage, yields, and production) in MS Excel using the nine reporting agricultural statistics districts based on the Group or GroupArea field in the attribute table. You will see that the top producers are the following: for corn, it is Mclean County; for soybean, it is Mclean County; and for wheat, it is Washington County. Hardin, DuPage, and Rock Island counties

produced the least corn, soybean, and wheat, respectively, statewide. Corn production is highest in the northern half of Illinois, with the exception of the greater Chicago area. Soybean production seems to be higher in the west-central portion of the state, with low production areas in the south and northwest. Wheat production is highest in the southern half of the state. The reason why Illinois is a corn and not a soybean state is due to favorable growing conditions, the use of corn for the production of ethanol gas, and a comparative advantage. If we review the summary table of key agricultural statistics, we would be able to comment on the distribution of agricultural production across the state of Illinois (Table 2.3).

TASK 2.3 PROCESSING A SPATIAL DATASET

Attribute processing is a common task when preparing a spatial dataset for analysis. Most spatial datasets are available in unstructured, semistructured, or structured formats. Thus, a lot of time is normally spent in preprocessing the dataset. Once the dataset has been processed, one has to review or check for accuracy. Due to the availability of modern computing systems and citizen sensors, spatial datasets are constantly being generated and archived in large data warehousing.

As an example of processing archived spatial datasets, let us analyze the 2008 Illinois Agricultural Statistics obtained earlier from the Illinois Department of Agriculture, U.S. Department of Agriculture https://www.nass.usda.gov/Statistics_by_State/Illinois/index.php). The dataset was tabulated at county level and made available in a PDF format and posted on this website. Several processed datasets related to this chapter are stored in Chapter2_Data_folder (Illinois_census_county, Agricultural_Exported_GISdataset, llinois_cnty_agricultural_statistics, and agricultural regions).

Feel free to explore these datasets on your own. The preprocessing of the dataset was undertaken as follows:

The data were downloaded; the tables were extracted from the PDF document, these were then converted to MS Excel. The columns and rows were all cleaned up and formatted to a database format. The key units for the agricultural data were acreage planted for all purposes, acreage harvested for grain, yield per acre in bushels, and production in bushels. Other relevant demographic and boundary data were obtained from the U.S. Census Bureau. Two files that were created contain the spatial information (Illinois_census_county), and the other file has countylevel attribute information on Illinois agricultural statistics (Illinois_cnty_agricultural_statistics). We will use this information to perform spatial analysis after doing a bit of attribute data processing as outlined later.

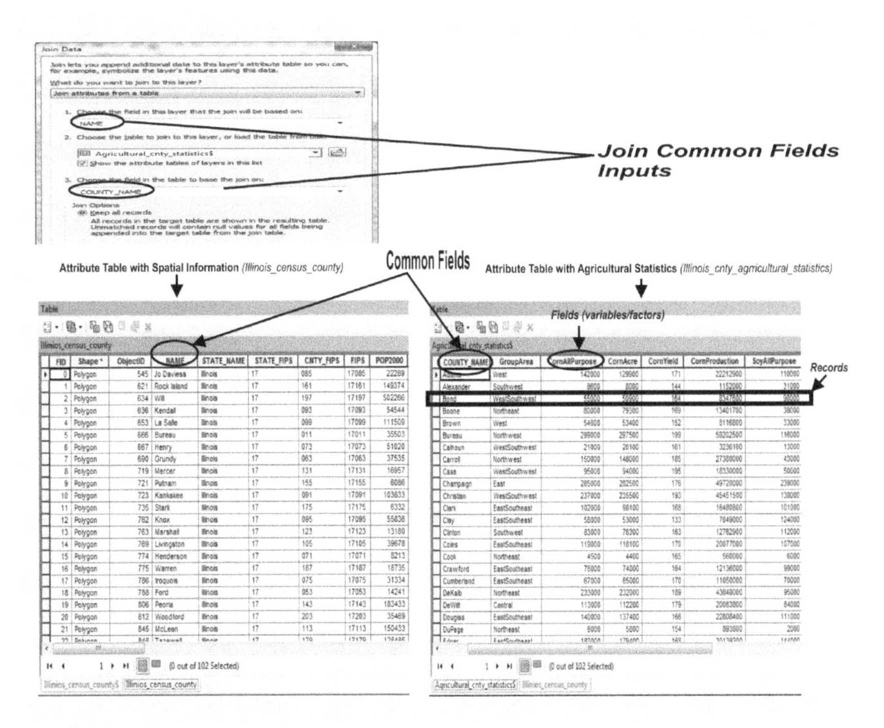

FIGURE 2.8
Attribute table concepts and what is required to complete the join table task. There are 102 county-level records in the two tables.

First, we should join the attribute table of Illinois_cnty_agricultural_statistics to that of Illinois_census_county so that the information for agricultural statistics is attached. (Use COUNTY_NAME and NAME as common fields to join the tables.) A visual illustration of common fields and other attribute table concepts is provided in Figure 2.8.

We can explore the same dataset using this sample R-Code as follows:

```
#load the necessary packages
#Type each command in sequence and on its own line and wait
for a response, review what is on screen before typing in the
next one.

library(rgdal)
library(dplyr)
setwd("../ChapterData/Chapter2_Data_Folder")
list.files(path=".")
# Question: How many files are listed? You should have 75 as
shown previously.
```

TABLE 2.3

A Summary of Agricultural Statistics by Group Area

Group Area	Average Sale	Acreage	Corn Yield	Soy Yield	Wheat Yield	Corn Production	Soy Production	Wheat Production
Central	211.81	268,748.91	194	51	32	310,902,91	492,850,0	137,682
East	267.35	443,525.86	176	48	40	429,853,14	874,021,4	355,543
East Southeast	151.57	217,203.00	161	46	55	159,280,27	505,920,0	895,120
Northeast	187.98	190,633.73	172	43	38	194,664,00	273,587,3	207,555
Northwest	217.65	277,936.17	183	47	46	312,176,92	334,265,8	178,825
Southeast	109.49	141,437.75	146	40	48	595,066,7	280,771,7	948,183
Southwest	92.38	155,774.33	152	43	54	601,923,3	318,970,8	160,1667
West	172.88	232,268.22	189	49	21	249,612,78	431,665,6	298,444
West Southwest	167.01	253,813.15	170	47	53	224,728,62	426,0969	760,431

```
# Explore the spatial data:

Illinois_census_county <- readOGR('Illinois_census_county.
shp', layer='Illinois_census_county')
plot(Illinois_census_county)

# Now let's read in the nonspatial data file; "Illinois_cnty_
agricultural_statistics.xls". Since it is an Excel file, we
need to load a package that enables us to read in Excel files.
If it's not installed yet use; install.packages("readxl") and
load it using the library function.
#install.packages("readxl")
library("readxl")
## Warning message:
## package "readxl" was built under R version 3.4.4
Illinois_cnty_agri <- read_excel("Illinois_cnty_agricultural_
statistics.xls")
head(Illinois_cnty_agri)
#Your attribute should like this
# extract first few (default 6) parts of the data shown next
# A tibble: 6 × 16
CNTY_FIPS COUNTY_NAME Group GroupArea CornAllPurpose CornAcre
CornYield
 <chr> <chr>       <dbl> <chr>      <dbl>   <dbl>   <dbl>
1 001  Adams       30    West       142000  129900  171
2 003  Alexander   80    Southwest  8600    8000    144
3 005  Bond        60    WestSout~  55000   50900   164
4 007  Boone       20    Northeast  80000   79300   169
5 009  Brown       30    West       54000   53400   152
6 011  Bureau      10    Northwest  299000  297500  199
#… with nine more variables: CornProduction <dbl>,
SoyAllPurpose <dbl>,
# SoyAcre <dbl>, SoyYield <dbl>, SoyProduction <dbl>,
WheatAllPurpose <dbl>,
# WheatAcre <dbl>, WheatYield <dbl>, WheatProduction <dbl>
#In case you want to view the entire attribute table, use
View(Illinois_cnty_agri)
```

2. From the Excel spreadsheet, using Rank and Percentile, we can sort corn, soybean, and wheat production in the state of Illinois (see screenshots in Figures 2.9 through 2.12). Using this approach, we can describe their distribution as follows:
   a. There are 21 records in the 80th percentile and above for corn, soybean, and wheat production (Figure 2.9).
   b. There are 30 records between the 50th and 80th percentiles for corn, soybean, and wheat production (Figure 2.10).
   c. There are 30 records between the 20th and 50th percentiles for corn and soybean, and 19 records for wheat production (Figure 2.11).

	E	F	G	H	I	J	K	L	M	N	O	P
1	Point	CornProduction	Rank	Percent	Point	SoyProduction	Rank	Percent	Point	WheatProduction	Rank	Percent
2	64	69920000	1	100.00%	64	12316500	1	100.00%	95	4748800	1	100.00%
3	50	64128000	2	99.00%	10	11850000	2	99.00%	14	3146000	2	99.00%
4	38	62125000	3	98.00%	53	11773500	3	98.00%	79	2822400	3	98.00%
5	53	60027000	4	97.00%	38	11397500	4	97.00%	96	2470000	4	97.00%
6	6	59202500	5	96.00%	92	10020500	5	96.00%	57	2381400	5	96.00%
7	10	49720000	6	95.00%	49	9326400	6	95.00%	87	2306400	6	95.00%
8	52	49595000	7	94.00%	95	7658700	7	94.00%	97	2257200	7	94.00%
9	83	47564000	8	93.00%	37	7516600	8	93.00%	3	2105600	8	93.00%
10	11	45451500	9	92.00%	86	7374500	9	92.00%	67	2018400	9	92.00%
11	98	44435500	10	91.00%	11	7303400	10	91.00%	13	1976400	10	91.00%
12	37	44160000	11	90.00%	23	7160000	11	90.00%	73	1823100	11	90.00%
13	19	43848000	12	89.10%	68	6683600	12	89.10%	58	1776600	12	89.10%
14	71	42596000	13	88.10%	26	6647000	13	88.10%	26	1757400	13	88.10%
15	54	42112000	14	87.10%	34	6360000	14	87.10%	33	1748700	14	87.10%
16	92	39474000	15	86.10%	54	5955000	15	86.10%	80	1675800	15	86.10%
17	46	36727500	16	85.10%	94	5952800	16	85.10%	41	1616600	16	85.10%
18	48	34760000	17	84.10%	83	5951700	17	84.10%	68	1581200	17	84.10%
19	94	34713900	18	83.10%	57	5940800	18	83.10%	1	1339200	18	83.10%
20	55	33539200	19	82.10%	96	5887600	19	82.10%	25	1299200	19	82.10%
21	102	33536800	20	81.10%	48	5808800	20	81.10%	39	1196800	20	81.10%
22	90	33160000	21	80.10%	56	5759200	21	80.10%	86	1192600	21	80.10%
23	62	32939400	22	79.20%	62	5648400	22	79.20%	51	1157000	22	79.20%

FIGURE 2.9
Screenshot showing the 80th percentile and above for corn, soybean, and wheat production.

	E	F	G	H	I	J	K	L	M	N	O	P
1	Point	CornProduction	Rank	Percent	Point	SoyProduction	Rank	Percent	Point	WheatProduction	Rank	Percent
22	90	33160000	21	80.10%	56	5759200	21	80.10%	86	1192600	21	80.10%
23	62	32939400	22	79.20%	62	5648400	22	79.20%	51	1157000	22	79.20%
24	86	32703300	23	78.20%	6	5635000	23	78.20%	28	1152900	23	78.20%
25	56	32106500	24	77.20%	21	5594700	24	77.20%	75	1079700	24	77.20%
26	34	31559000	25	76.20%	102	5432400	25	76.20%	53	924600	25	76.20%
27	68	31240600	26	75.20%	46	5428500	26	75.20%	38	903000	26	75.20%
28	23	30139200	27	74.20%	40	5425200	27	74.20%	56	894000	27	74.20%
29	29	28911400	28	73.20%	74	5415000	28	73.20%	40	885600	28	73.20%
30	69	28880000	29	72.20%	15	5305000	29	72.20%	34	882000	29	72.20%
31	89	28315000	30	71.20%	27	5296500	30	71.20%	99	869400	30	71.20%
32	66	27878400	31	70.20%	87	5280800	31	70.20%	30	841800	31	70.20%
33	8	27380000	32	69.30%	69	5231200	32	69.30%	60	742400	32	69.30%
34	74	26591200	33	68.30%	55	5217300	33	68.30%	24	689000	33	68.30%
35	27	26232500	34	67.30%	97	5203800	34	67.30%	17	628300	34	67.30%
36	60	24770200	35	66.30%	90	5179200	35	66.30%	63	591600	35	66.30%
37	75	23777000	36	65.30%	13	5067600	36	65.30%	77	518700	36	65.30%
38	32	23218000	37	64.30%	14	4995000	37	64.30%	31	501500	37	64.30%
39	72	22833600	38	63.30%	29	4926600	38	63.30%	71	483000	38	63.30%
40	21	22808400	39	62.30%	79	4783500	39	62.30%	84	464800	39	62.30%
41	88	22451800	40	61.30%	80	4780600	40	61.30%	101	448000	40	61.30%
42	59	22300400	41	60.30%	58	4752000	41	60.30%	27	439200	41	60.30%
43	1	22212900	42	59.40%	25	4627600	42	59.40%	82	416000	42	59.40%
44	20	20083800	43	58.40%	12	4600000	43	58.40%	12	394400	43	58.40%
45	15	20077000	44	57.40%	19	4540800	44	56.40%	42	384000	44	57.40%
46	9	18330000	45	56.40%	66	4540800	44	56.40%	65	379600	45	56.40%
47	36	18208500	46	55.40%	99	4429800	46	55.40%	18	371700	46	55.40%
48	70	17748000	47	54.40%	17	4312000	47	54.40%	11	338100	47	54.40%
49	31	17601200	48	53.40%	1	4169700	48	53.40%	19	335800	48	53.40%
50	45	17427600	49	52.40%	20	4037600	49	52.40%	89	330000	49	52.40%
51	99	17399100	50	51.40%	3	4005000	50	51.40%	45	326800	50	51.40%
52	65	17285400	51	50.40%	52	3960600	51	50.40%	6	303400	51	50.40%
53	12	16480800	52	49.50%	73	3956000	52	49.50%	85	292500	52	49.50%

FIGURE 2.10
Screenshot showing between the 50th and 80th percentiles for corn, soybean, and wheat production.

d. There are 21 records at or below the 20th percentile for corn and soybean production, and 32 records for wheat production (Figure 2.12). You do realize that there are more than 30 counties with a zero value for wheat production. Your turn: List the counties with zero values for wheat production.

	E	F	G	H	I	J	K	L	M	N	O	P
1	Point	CornProduction	Rank	Percent	Point	SoyProduction	Rank	Percent	Point	WheatProduction	Rank	Percent
52	65	17285400	51	50.40%	52	3960600	51	50.40%	6	303400	51	50.40%
53	12	16480800	52	49.50%	73	3956000	52	49.50%	85	292500	52	49.50%
54	63	16274000	53	48.50%	33	3912000	53	48.50%	2	291400	53	48.50%
55	47	15752000	54	47.50%	31	3807000	54	47.50%	98	259200	54	47.50%
56	101	15640800	55	46.50%	51	3751500	55	46.50%	9	226800	55	46.50%
57	26	15184000	56	45.50%	41	3748000	56	45.50%	74	222000	56	45.50%
58	57	15174500	57	44.50%	98	3601800	57	44.50%	21	199800	57	44.50%
59	43	15126400	58	43.50%	72	3572400	58	43.50%	61	186000	58	43.50%
60	25	14586300	59	42.50%	32	3539100	59	42.50%	8	179400	59	42.50%
61	40	14128000	60	41.50%	75	3526000	60	41.50%	100	178400	60	41.50%
62	95	13497000	61	40.50%	70	3432300	61	40.50%	91	169600	61	40.50%
63	4	13401700	62	39.60%	71	3271500	62	39.60%	55	162400	62	39.60%
64	84	13229600	63	38.60%	88	3264800	63	38.60%	59	149100	63	38.60%
65	14	12762900	64	37.60%	67	3220000	64	37.60%	70	112000	64	37.60%
66	87	12167500	65	36.60%	59	3192600	65	36.60%	83	100800	65	36.60%
67	81	12151200	66	35.60%	28	3188000	66	35.60%	50	94500	66	35.60%
68	17	12136000	67	34.60%	60	3180100	67	34.60%	88	81000	67	34.60%
69	42	11556700	68	33.60%	39	3147600	68	33.60%	78	78000	68	33.60%
70	96	11294200	69	32.60%	89	3087000	69	32.60%	32	65000	69	32.60%
71	18	11050000	70	31.60%	18	3058000	70	31.60%	81	64900	70	31.60%
72	30	10475400	71	30.60%	65	2865600	71	30.60%	4	0	71	0.00%
73	97	10320000	72	29.70%	30	2857700	72	29.70%	5	0	71	0.00%
74	85	8881500	73	28.70%	93	2692800	73	28.70%	7	0	71	0.00%
75	80	8580000	74	27.70%	36	2514300	74	27.70%	10	0	71	0.00%
76	3	8347600	75	26.70%	9	2430400	75	26.70%	15	0	71	0.00%
77	51	8184400	76	25.70%	82	2392000	76	25.70%	16	0	71	0.00%
78	78	8131500	77	24.70%	42	2373500	77	24.70%	20	0	71	0.00%
79	5	8116800	78	23.70%	47	2335900	78	23.70%	22	0	71	0.00%
80	58	8066000	79	22.70%	84	2281600	79	22.70%	23	0	71	0.00%
81	79	7919200	80	21.70%	8	2082500	80	21.70%	29	0	71	0.00%
82	93	7276500	81	20.70%	81	2054400	81	20.70%	35	0	71	0.00%
83	13	7049000	82	19.80%	45	1922800	82	19.80%	36	0	71	0.00%

FIGURE 2.11
Screenshot showing between the 20th and 50th percentiles for corn, soybean, and wheat production.

	E	F	G	H	I	J	K	L	M	N	O	P
1	Point	CornProduction	Rank	Percent	Point	SoyProduction	Rank	Percent	Point	WheatProduction	Rank	Percent
72	30	10475400	71	30.60%	65	2865600	71	30.60%	4	0	71	0.00%
73	97	10320000	72	29.70%	30	2857700	72	29.70%	5	0	71	0.00%
74	85	8881500	73	28.70%	93	2692800	73	28.70%	7	0	71	0.00%
75	80	8580000	74	27.70%	36	2514300	74	27.70%	10	0	71	0.00%
76	3	8347600	75	26.70%	9	2430400	75	26.70%	15	0	71	0.00%
77	51	8184400	76	25.70%	82	2392000	76	25.70%	16	0	71	0.00%
78	78	8131500	77	24.70%	42	2373500	77	24.70%	20	0	71	0.00%
79	5	8116800	78	23.70%	47	2335900	78	23.70%	22	0	71	0.00%
80	58	8066000	79	22.70%	84	2281600	79	22.70%	23	0	71	0.00%
81	79	7919200	80	21.70%	8	2082500	80	21.70%	29	0	71	0.00%
82	93	7276500	81	20.70%	81	2054400	81	20.70%	35	0	71	0.00%
83	13	7049000	82	19.80%	45	1922800	82	19.80%	36	0	71	0.00%
84	67	6884700	83	18.80%	63	1821300	83	18.80%	37	0	71	0.00%
85	33	6816000	84	17.80%	85	1804800	84	17.80%	43	0	71	0.00%
86	73	6555400	85	16.80%	101	1747200	85	16.80%	44	0	71	0.00%
87	82	6058800	86	15.80%	24	1717800	86	15.80%	46	0	71	0.00%
88	28	4870600	87	14.80%	4	1587600	87	14.80%	47	0	71	0.00%
89	41	4755300	88	13.80%	43	1569500	88	13.80%	48	0	71	0.00%
90	24	4498200	89	12.80%	77	1457400	89	12.80%	49	0	71	0.00%
91	39	3877600	90	11.80%	61	1396000	90	11.80%	52	0	71	0.00%
92	61	3685200	91	10.80%	2	1192000	91	10.80%	54	0	71	0.00%
93	7	3236100	92	9.90%	5	1187700	92	9.90%	62	0	71	0.00%
94	77	2622400	93	8.90%	100	1068600	93	8.90%	64	0	71	0.00%
95	91	1955800	94	7.90%	78	1045000	94	7.90%	66	0	71	0.00%
96	100	1750000	95	6.90%	91	896800	95	6.90%	69	0	71	0.00%
97	49	1228200	96	5.90%	76	624200	96	5.90%	72	0	71	0.00%
98	2	1152000	97	4.90%	44	620100	97	4.90%	76	0	71	0.00%
99	44	1086300	98	3.90%	7	576000	98	3.90%	90	0	71	0.00%
100	76	1026800	99	2.90%	50	268600	99	2.90%	92	0	71	0.00%
101	22	893800	100	1.90%	16	259600	100	1.90%	93	0	71	0.00%
102	16	560000	101	0.90%	35	72700	101	0.90%	94	0	71	0.00%
103	35	331000	102	0.00%	22	62700	102	0.00%	102	0	71	0.00%

FIGURE 2.12
Screenshot showing less than the 20th percentile for corn, soybean, and wheat production.

3. We can select 34 samples using a simple random sampling strategy (use CO_FIPS as your field). Here is a list of 34 County FIPS generated using a simple random sampling strategy: 099, 081, 195, 061, 063, 111, 031, 165, 029, 025, 153, 031, 137, 139, 143, 093, 069, 173, 127, 149, 087, 109, 055, 091, 093, 143, 163, 053, 117, 119, 043, 079, 099, and 149.

4. We can select 34 samples using systematic sampling with a random strategy. This can be accomplished either in MS Excel or R-Code. We need to carefully select the first sample to minimize any potential selection bias. We have selected CO_FIPS #63 as our first one, so we can now select every third county after this.
   a. If you wish to generate your own starting point, you can randomly generate 102 integers using MS Excel (formula: "= RAND ()*(MAX-MIN)+MIN," that is, "RAND ()*(102-1)+1," hit ENTER key to refresh and generate a new number) or simply use this website http://www.random.org/integers/. To be truly random, use every third count in the list of numbers generated at this website as representative of the CO_FIPS. Here is a list of 34 County FIPS generated using systematic sampling with a random strategy: 063, 113, 001, 147, 123, 187, 203, 037, 093, 129, 169, 041, 035, 083, 027, 189, 145, 165, 003, 103, 201, 199, 069, 183, 045, 139, 173, 023, 005, 163, 081, 181, 007, and 099.

Assuming each of the points represents farm locations in each of the counties, they can be used to collect additional data for spatial analysis. We can constrain each polygon (county) and randomly assign 4 individual-level points to have a total of 408 sampling point (Figure 2.13).

Finally, there are nine spatial regions/subpopulations used for reporting agricultural statistics districts. These will be used to stratify Illinois, and in each stratum a sample will be drawn randomly using a two-stage sampling design process. First, we determine the number of observations in each subpopulation/stratum (Table 2.4). Second, we randomly draw 34 samples from each of the nine spatial regions using sample size percentage (Table 2.4).

TASK 2.4 DERIVING A SAMPLE FROM A SPATIAL DATASET

As noted earlier, the adoption of an effective sampling strategy can help in achieving a cost-effective, representative population sample for spatial analysis. Spatial sampling requires that one covers space or time periods that accurately represent the population. When a sample is representative, conclusions can be generalized to the population, and unbiased estimators with confidence intervals with known precisions can be derived. In this task, we examine simple random, stratified, systematic, and two-stage sampling designs. We use these designs to create or draw a representative sample.

To complete the sampling task, we need agricul_ILL_stats.shp located in the Chapter 2 data folder. We use the agricul_ILL_stats (.dbf) attribute table to draw 34 samples from 102 counties based on different sampling designs. We will compile and save each of the samples for further analysis in Chapter 3.

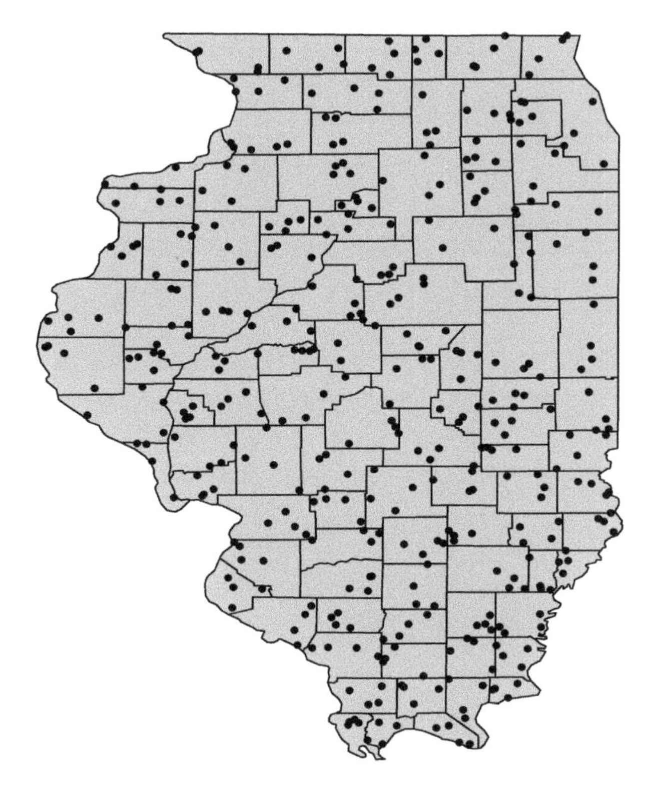

FIGURE 2.13
Randomly created individual-level four points per county. There are 408 sampling points that could be used to collect additional data for spatial analysis.

TABLE 2.4
Pre-Knowledge Information for Selecting a Representative Subpopulation

Region	Average Farm Size	SD/ AVG_ SIZE[a]	Average Sale	Ave_ CornPr[a]	SD_ CornPro[a]	Ave_ SoyPro[a]	Sample Size[b]	*n*
Southeast	355.25	211.27	109.49	595,066,7	352,005,3	280,771,7	6%	2
Southwest	288.9167	86.36	92.38	601,923,3	469,746,8	318,970,8	6%	2
East Southeast	332.6667	87.99	151.57	159,280,27	779,153,8	505,9200	8%	3
Northeast	261.5455	143.45	187.98	194,664,00	192,546,62	273,587,3	9%	3
West Southwest	372.0769	83.94	167.01	224,728,62	139,572,01	426,096,9	12%	4
West	386.3333	47.16	172.88	249,612,78	992,745,4	431,665,6	12%	4
Central	411	87.13	211.81	310,902,91	148,903,01	492,850,0	13%	4
Northwest	334.3333	69.44	217.65	312,176,92	165,829,36	334,265,8	17%	6
East	474.2857	51.10	267.35	429,853,14	147,366,66	874,0214	17%	6

[a] SD/AVG_SIZE/standard deviation average farm size, Ave_CornPr/average corn production, SD_CornPro/standard deviation average corn production, Ave_SoyPro/average soybean production.

[b] Sample size (*n*) for each stratum must total 34 observations, and it is derived in the last column.

	A	B	C	D	E	F	G
1	ObjectID	NAME	STATE_NAME	STATE_FIPS	CNTY_FIPS	rand()	Spatial regions
2	1520	Saline	Illinois	17	165	0.4461	Southeast
3	1361	Edwards	Illinois	17	047	0.2529	Southeast
4	1378	Washingt	Illinois	17	189	0.5969	Southwest
5	1665	Alexande	Illinois	17	003	0.8330	Southwest
6	1202	Effingham	Illinois	17	049	0.5282	Eastsoutheast
7	1087	Coles	Illinois	17	029	0.1504	Eastsoutheast
8	1160	Cumberla	Illinois	17	035	0.3558	Eastsoutheast
9	3080	Cook	Illinois	17	031	0.5022	Northeast
10	690	Grundy	Illinois	17	063	0.9613	Northeast
11	3081	Kane	Illinois	17	089	0.4188	Northeast
12	990	Cass	Illinois	17	017	0.1553	Westsouthwest
13	1024	Sangamor	Illinois	17	167	0.6557	Westsouthwest
14	1044	Morgan	Illinois	17	137	0.5399	Westsouthwest
15	1048	Pike	Illinois	17	149	0.0780	Westsouthwest
16	775	Warren	Illinois	17	187	0.8233	West
17	973	Adams	Illinois	17	001	0.1263	West
18	958	Schuyler	Illinois	17	169	0.4498	West
19	993	Brown	Illinois	17	009	0.9562	West
20	763	Marshall	Illinois	17	123	0.6751	Central
21	812	Woodforc	Illinois	17	203	0.2412	Central
22	922	Mason	Illinois	17	125	0.6238	Central
23	806	Peoria	Illinois	17	143	0.9407	Central
24	721	Putnam	Illinois	17	155	0.8678	Northwest
25	545	Jo Daviess	Illinois	17	085	0.0766	Northwest
26	3062	Stephens	Illinois	17	177	0.1494	Northwest
27	3078	Ogle	Illinois	17	141	0.6848	Northwest
28	3098	Lee	Illinois	17	103	0.4514	Northwest
29	3063	Winnebag	Illinois	17	201	0.9212	Northwest
30	788	Ford	Illinois	17	053	0.8869	East
31	955	Piatt	Illinois	17	147	0.1924	East
32	929	Champaig	Illinois	17	019	0.5085	East
33	911	Vermilion	Illinois	17	183	0.8846	East
34	786	Iroquois	Illinois	17	075	0.1409	East
35	723	Kankakee	Illinois	17	091	0.7277	East

FIGURE 2.14
Screenshot of the sampled areas using systematic sampling approach.

Based on the information presented in Table 2.4, we can draw a sample that will meet the sampling requirements for the nine stratified regions of agriculture in Illinois. Sampled results would look like the results in Figure 2.14.

Conclusion

In this chapter, we explored the key fundamentals in spatial data design including the measurement scales of variables, the distinction between population and sample, the types of sampling strategies, and steps toward processing the spatial data once they have been secured through primary or secondary sources. Having gained the practical skills in these areas through the sample exercises given earlier, it is now your turn to complete the challenge exercise that follows.

Worked Examples in R and Stay One Step Ahead with Challenge Assignments

Step I. View Data Structure

Before you do anything else, it is important to understand the structure of your data and that of any objects derived from it.

```
#Type each command on its own line and wait for a response
before typing in the next one.
class(Illinois_cnty_agri)
sapply(Illinois_cnty_agri, class) # show classes of all columns
names(Illinois_cnty_agri) # show list components
dim(Illinois_cnty_agri) # dimensions of the object, if any
```

Step II. Basic Data Summaries

Use a loop to generate and summarize attribute values of the Illinois_cnty_agri nonspatial dataset.

```
# Using the code that is given below
for (i in 1:dim(Illinois_cnty_agri)[[2]])
{
a <- names(Illinois_cnty_agri)[[i]]
b <- Illinois_cnty_agri [[i]]
if (is.numeric(b) == TRUE)
  {
  cat("Attribute Number", i)
  print(a)
  print(summary(b)) # Gives the Min, 1st quantile, Median,
  Mean, 3rd quantile & Max
  print("")
  }
}
```

Step III. Exploring the Spatial Data

Let's summarize our spatial data, Illinois_census_county.shp using the following commands

```
#Type each command on its own line and wait for a response,
review what is on screen before typing in the next one.

summary(Illinois_census_county)
str(Illinois_census_county [1,])

## The sample R codes for data processing are located in
Chapter2-R-Code and the dataset that is used to implement the
code is stored in Chapter2_Data_folder. The data folder has two
files. One file, which contains the spatial information, is
named Illinois_census_country, and the other file that contains
county-level attribute information on Illinois agricultural
statistics is named Illinois_cnty_agricultural_statistics. The
trick is to identify a common primary key to link the two
datasets together. Please run the codes and continue to
practice on your own. This will set you up, and you will gain
greater exposure to R. Remember practice makes perfect!
```

Stay One Step Ahead with Challenge Assignments

Concept: In this chapter, we learned about making scientific observations and measurements, which entails a systematic process of acquiring knowledge about phenomena of interest. We can do this either directly through the use of our human senses or by recording information using tools and instruments. Science has become a way of life whereby every day we are using mobile devices or sensors (e.g., citizen sensor) that record our locations, daily activities, building height, or body weight. The data is recorded using a specific metric that may consist of nonnumeric or numeric attribute with a unit. Our curiosity to learn and gain knowledge leads us to ask specific research questions, formulate hypotheses, design experiments and ensure proper collection of primary data, and validate or reject the hypothesis using observations. In this chapter, we also discussed these scientific concepts and especially focused on a hypothesis-driven process through which we collect and analyze spatial data. We presented the four scales of measurement, two main scientific approaches for data collection, sampling, and data processing, since most data is currently available through secondary sources.

Task: We use concepts from this chapter to collect scientific observations around places where we live, play, work, worship, shop, eat, or learn.

1. Choose one of any of three topics given as follows:
    - Understanding the urban food environment in low- and high-income settings
    - Understanding access of primary health-care services among racial and ethnic groups
    - Understanding activity patterns among young working adults
2. Outline a research approach to collect observations to study the selected problem.

Approach/strategy for solving task: Design a robust sampling data collection mechanism for studying this problem. Use a high-grade Global Positioning System (GPS) unit or GPS-enabled mobile devices with submeter accuracy (it should have less than a 3-m locational accuracy) to collect at least 30 data points and any relevant attribute within a 10-mile radius around places where you live, play, work, worship, shop, eat, or learn. Compile the data in a MS Excel Spreadsheet. Also, find and collect secondary data for the problem you are studying. Compare the two datasets. Are there any significant differences in terms of error for the two datasets? Compile your findings/observations into a meaningful short report.

Possible solutions: Answers will vary, but here is a possible solution for understanding the urban food environment in low- and high-income settings. Below is a specific research outline for studying this problem:

1. *Hypothesis*: There are no significant differences in food access between low- and high-income settings in the city of Chicago, Illinois.
2. *Define*: Define food access, low-income, and high-income measures.
3. *Design*: Use either a prospective or respective cohort study. Develop a solid spatial sampling framework that will capture the variation of key factors—such as demographics, social stressors, socioeconomic status, family/household characteristics, location of food stores, type of food stores (e.g., groceries, farmer's market, restaurants), operating hours of food stores, individual/neighborhood characteristics, income, education, diet intake and requirements/preferences, etc.
4. *Analysis plan*: Determine the most appropriate food access measures, including distance to nearest food store or fixed distance (spatial access), food stores per unit area or per 1000 people (spatial access), temporal access (operation hours), and spatiotemporal access (spatial and temporal considerations). Conduct spatial statistical tests to determine whether there are differences in the food environment between low- and high-income settings in Chicago.
5. *Challenges, alternative approaches, and ideas for future direction*: Develop a strategy to deal with and minimize errors between the primary and secondary datasets, incorporate a community-engaged research component to understand dietary practices, and plan a follow-up study. Conceptual challenges exist in terms of clearly differentiating between low-income and high-income settings relative to 5As (accessibility, availability, affordability, acceptability, and accommodation). The ideas presented here can be replicated to other urban areas. This research outline template can be adopted to design a study for other towns, cities, or places nearest to where you live.

TASK 2.5 KEY STEPS IN SPATIAL STATISTICAL DESIGN

The keys to successful design and use of geographic data in a research project are as follows:

- Knowledge of observations/phenomena/events.
- Review of data collection and sampling strategy.
- Review of scales of measurement.
- Knowledge of geographic scales and map projection.
- Knowledge of analytical frameworks to facilitate data analysis. This includes the ability to explore, detect, and explain spatial patterns plus a thorough grounding in the knowledge and relevant skills of methods, tools, and systems.

TASK 2.6 QUESTS FOR SPATIAL DATASETS

1. Search for two separate spatial datasets on the internet that can be used for spatial analysis in a specified application of particular interest to your work. The datasets must be spatially explicit with at least six variables measured across the different measurement scales. Once you have the appropriate spatial datasets, make two deductive and inductive statements/arguments.
2. Due to different data reporting systems, inconsistencies in records, and other sources of uncertainty, there are always gaps in a dataset. Suggest two ways to address this common problem.
3. Suppose we are asked to design a study to investigate the commuting patterns of young working adults in the city of Chicago. Outline a point-by-point research plan that covers the five points mentioned in Task 2.5.

Review and Study Questions

1. Using examples of variables in your area of interest, describe the four scales of measurement. What are the unique properties of each scale?
2. Distinguish between a population and a sample. In a spatial statistical data design, what are the benefits of compiling sample data (if any) over the entire population data?
3. What is spatial sampling? Using examples from your research area, explain how you would go about conducting point, linear, or areal sampling.
4. What are the merits and demerits of simple random, stratified, or sequential sampling?
5. Distinguish between cluster sampling, transect sampling, and contour sampling in a spatial sampling strategy.

Glossary of Key Terms

deductive reasoning: The making of or the evaluation of the validity or soundness of an argument that logically derives from a set of generalized principles to arrive at a conclusion.

hypothesis: This is simply the process of induction and deduction. A theory is actually the basis for suggesting lots of testable hypotheses. It is a prediction that expresses the expected outcome in any given situation; for example, there is a spatial association between surrounding pollution source(s) and persons with respiratory illness living within a radius of 1000 m, or persons with respiratory illnesses living within a radius of 1000 m are geographically associated with nearby pollution source(s). In experimental research, the hypothesis is usually a prediction of how the manipulation of the independent variable will influence the behavior of a dependent variable. There are two types of hypotheses, the Null and Alternative, denoted as H_0 and H_1, respectively.

inductive reasoning: The making of or the evaluation of generalized statements based on specific statements.

law: A verified statement with universal application or a generalized body of observations.

measurement scale: The systematic means of defining variables by assigning data values to the observations. The four scales (ratio, interval, ordinal, and nominal) have unique properties that influence the uses and applications of different statistical techniques.

model: A simplified/an abstract representation of reality or an object or system. It can be conceptual, statistical, or mathematical.

theory: A coherent and replicable system of tested ideas or hypotheses or evidence that explains a phenomenon.

References

Illinois Department of Agriculture, U.S. Department of Agriculture, Available at https://www.nass.usda.gov/Statistics_by_State/Illinois/index.php (accessed on May 3, 2020).

Oyana, T.J., E., Kayendeke, and S. Adu-Prah. 2017. Assessing performance of Leaf Area Index in a monitored mountain ecosystem of the Manafwa Catchment on Mount Elgon, Uganda. *International Journal of Applied Geospatial Research* 8(1): 65–81.

Oyana, T.J., E. Kayendeke, Y. Bamutaze, and D. Kisanga. 2015. A field assessment of land use systems and soil properties at varied landscape positions in a fragile ecosystem of Mount Elgon, Uganda. *African Geographical Review* 34(1): 83–103.

3

Using Statistical Measures to Analyze Data Distributions

Learning Objectives

1. Understand basic statistical concepts and measures.
2. Generate and interpret descriptive statistics.
3. Generate and interpret descriptive spatial statistics.
4. Understand probability concepts and applications.

Introduction

In Chapter 1, we mentioned that the field of spatial statistics draws from statistics, mathematics, and related disciplines. Several of the techniques in spatial analysis are variants of traditional procedures used in these fields with the added dimensions and modifications made to cope with the unique properties of spatial data. The foundation for statistical measures and knowledge was laid through the work of well-known statisticians (Varberg 1963; David 1998), including Ronald Fisher (experimental design, analysis of variance, and likelihood-based methods), Karl Pearson (Pearson's chi-square test), Francis Galton (correlation and regression), Gertrude Cox (experimental design), Frank Yates (experimental design and Yates' algorithm), Kirstine Smith (optimal design theory), John Tukey (exploratory data analysis and graphical presentation of data), William Sealy Gosset (student's *t*-test), and George E. P. Box (experimental design, quality control, and time series analysis).

Knowledge of the means by which we organize spatial data using traditional statistical measures is therefore essential and useful for advanced analysis using geospatial techniques. Specifically, knowledge of key concepts and theories in statistics such as descriptive measures, sampling, and probability theories helps a geographer to (1) draw a representative sample, (2) assess the state of a distribution in a group of observations, (3) compare groups or observations, (4) explain observations, (5) identify and test explanatory variables, and (6) predict estimates and analyze uncertainty. Statistical approaches may be grouped into univariate or multivariate methods depending on the number of variables used to address the research questions. Techniques that focus on one variable at a time are univariate techniques, and those that examine the joint assessment of multiple variables are multivariate and are often the more advanced approaches. Statistical approaches can also be characterized as exploratory or confirmatory in nature, descriptive or inferential, predictive, and prescriptive. These terminologies are not confined to traditional statistics; they are commonly used to describe spatial statistical methods as well. This chapter provides an overview of statistical and spatial statistical methods that are commonly used in summarizing data. Understanding the statistical distribution of a dataset helps a data scientist gain fundamental knowledge to move the analysis forward. This chapter illustrates basic statistical methods using a number of datasets with nonspatial or spatial characteristics. The illustrations are based on a few sets of observations and will be used to deepen our knowledge and understanding of the basic statistical measures. Statistical summaries, plots, maps, or worktables shown in this chapter can be generated using MS Excel or any statistical software packages, such as R, SPSS, SAS, and ArcGIS.

Descriptive Statistics

All statistical approaches noted earlier typically begin with a comprehensive evaluation of the spectrum of data values obtained for each of the variables included in a dataset. These assessments rely on the use of descriptive measures that are presented in a numerical, tabular, or graphical format. Regardless of the format used, descriptive measures are generated to provide a fundamental understanding of the distribution of observations in a dataset. Using tabular summaries (such as frequency tables), graphical summaries (such as bar charts, line graphs, boxplots, stem and leaf, Normal QQ plots), and statistical summaries (mean, median, standard deviation), these statistics help us organize our data. They may also offer suggestive clues about the patterns and trends present in the data, and possibly help generate new research hypotheses.

Descriptive statistics differ from inferential statistics in the sense that the latter are used in the estimation of population parameters and the testing of hypotheses using information drawn from sample data. Descriptive statistics often provide the preliminary information about the sample characteristics, which could then be used for undertaking inferential statistics so that a specific hypothesis can be confirmed or rejected. Both approaches support efforts through which inferences about a population can be made that could be helpful in quantifying statistical relationships, making generalizations and statistical predictions. Two of the most commonly used sets of descriptive measures are the measures of center and the measures of dispersion. These measures are described along with their geographic counterparts in spatial analysis.

Measures of Central Tendency

Mode refers to the value that occurs most frequently in a specific set of ungrouped observations. For example, in Table 3.1, the mode is 43. If the observations are grouped, then one has to select the class with the most frequency as the modal class. The midpoint value for this class is referred to as the crude mode.

Median refers to the middle value in a specific set of ranked observations, or the centermost value in a ranked list of observations. If one has an odd number of observations in the dataset, the middlemost value in the set of ranked observations defines the median. However, if the number of observations is even, the median is defined by the midpoint of the two values. In Table 3.1, the median value (7 in rank) is 37, and we have an odd number of observations (11) in this set. The median can also be viewed as the 50th percentile in a data distribution.

Mean, also known as the arithmetic mean or simply an average, refers to the sum of a specific set of observations divided by the number of observations in the set. Simply put, it is the average value in a specific set of observations. It is great for interval- or ratio-scaled variables. Unlike the median, the mean is sensitive to the presence of outliers in a distribution.

The mean is statistically defined as follows:

$$\bar{X} = \frac{\sum_{i=1}^{n} X_i}{n} \text{ or simply } \frac{X_1 + X_2 + X_3 + .. + X_m}{n}$$

where $\bar{X}$ is the mean of variable X, X_i is the value of the observation i, Σ is a Greek summation symbol, and n is the number of observations in a given set.

Using data from Table 3.1, the mean for this set of observations can be derived as follows:

$$\bar{X} = \frac{16 + 18 + 21 + 32 + 34 + 37 + 43 + 43 + 45 + 60 + 72}{11}$$

Therefore, $\bar{X} = 38.27$.

TABLE 3.1

Eleven Sampled Tree Heights Near a Residential Area in a Chicago Neighborhood (in Meters)

16	43
18	43
21	45
32	60
34	72
37	

We can conclude that the average height for the tree heights derived for 11 samples near a residential area in a Chicago neighborhood is 38.27 meters (Figure 3.1).

We use n to derive the sample mean; however, for the population mean it is derived based on N. The two (*population* and *sample mean*) mainly differ because the degrees of freedom for a sample is based on the number of independent observations employed to calculate a statistic, which is reduced by one observation and denoted by $n - 1$.

Deriving a Weighted Mean Using the Frequency Distributions in a Set of Observations

There are certain applications that call for the use of weighted means over the traditional arithmetic means. The weights represent the magnitude

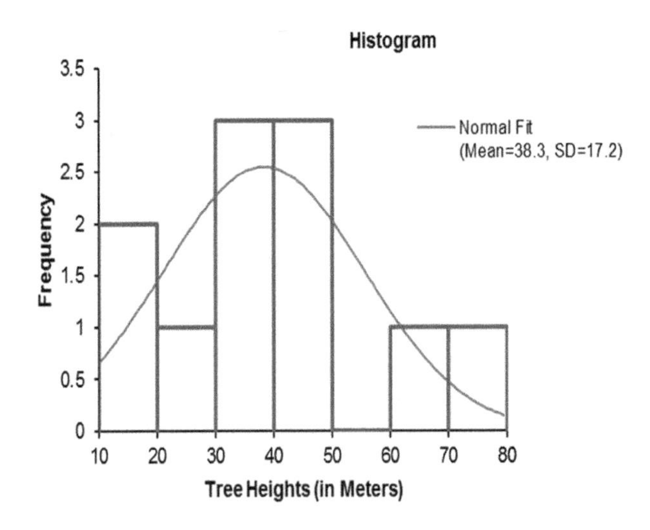

FIGURE 3.1
Frequency distribution of 11 sampled tree heights near a residential area in a Chicago neighborhood (in meters).

TABLE 3.2

Worktable for Deriving the Weighted Mean of Tree Heights Near a Residential Area in a Chicago Neighborhood (in Meters)

Class Interval (i)	Class Midpoint (X_i)	Class Frequency (f_j)	$X_i f_j$
16–25.99	21	3	63
26–35.99	31	2	62
36–45.99	41	4	164
46–55.99	51	0	0
56–65.99	61	1	61
66–75.99	71	1	71
Total		$\Sigma f_j = 11$	$\Sigma X_i f_j = 421$

or frequency (f_j) of the reported events, incidents, or attributes under investigation. The following example illustrates the computation of a weighted mean for tree heights in a residential area in a Chicago neighborhood. We group the data in Table 3.2 using the following steps:

1. Identify the largest and smallest values.
2. Derive the range.
3. Determine the number of classes.
4. Define the class interval.
5. Determine the frequency for each class.
6. Compile this information in a table, as has been done in Table 3.2.

$$\bar{X} = \frac{\sum X_i f_j}{\sum f_j} = \frac{421}{11} = 38.273$$

Measures of Dispersion

A measure of dispersion or variation is a descriptive statistic that quantifies the variability or the spread of a set of observations. Sources of errors in sampled estimates often consist of conceptual errors, sampling errors, measurement errors, or equipment operational errors, and measures of dispersion may help us quantify the extent to which sampled observations differ or vary from the true population values. A variety of statistical measures exist to quantify variability, including range, mean deviation, standard deviation, and variance. However, the most useful are standard deviation and variance that enable statisticians to assess the degree of dispersion in a set of observations.

TABLE 3.3

Summary Statistics for 11 Sampled Tree Heights Near a Residential Area in a Chicago Neighborhood (in Meters)

Mean	38.3%, 95% CI 26.7–49.8, SE 5.19	
Median	37.0, 98.8% CI 18.0–60.0	
Standard deviation	17.2%, 95% CI 12.0–30.2	
Variance	296.4	
Range	56	
Interquartile range	21.9	
Skewness	0.58	
Kurtosis	0.06	
Percentile	0th	16.0 (minimum)
	25th	22.8 (1st quartile)
	50th	37.0 (median)
	75th	44.7 (3rd quartile)
	100th	72.0 (maximum)

Abbreviations: CI, confidence interval; SE, standard error.

Range is a measure of dispersion that shows the difference between the highest (maximum) and lowest (minimum) values in a set of observations. In ungrouped data, it captures the difference between the maximum and minimum values. To obtain these values, one may wish to sort the data in order—in either ascending or descending order. For example, in Table 3.2, the range for 11 sampled tree heights is (72 – 16) = 56. In grouped data, it captures the difference between the upper value in the highest numbered midpoint and the lower value in the lowest numbered midpoint of a class interval. For example, in Table 3.2, the upper value is 71, and the lower value is 21; thus, the range is (71 – 21) = 50. You notice that there is a small difference of six in the range that is given by ungrouped and grouped data, which can be misleading. It is also possible to generate additional (e.g., quartile range) information from this dataset by dividing it further into equal portions or percentiles. An interquartile range (IQR) can be derived after dividing a ranked set of observations into four groups of equal size, followed by obtaining the interval between the 25th percentile (the lower quartile represented as *Q*1) and the 75th percentile (the upper quartile represented as *Q*3). In the dataset given in Table 3.3, *Q*1 is 22.8 and *Q*3 is 44.7, so IQR is (44.7 – 22.8) = 21.9.

Standard deviation is a summary statistic that measures the extent to which the data values are scattered around the mean (or center) of the distribution. Simply put, it quantifies the difference in the spread of a set of observations below and above the mean. It enables the statistician to determine whether observations in a set are tightly compact (a narrow standard deviation) or are spread out (a wide standard deviation). A narrow standard deviation indicates the observations are closely knit, and there is a low variation from the mean. A large standard deviation suggests that the observations are widely distributed, and there is a large variation from the mean. A large

TABLE 3.4

Worktable for Deriving Mean Deviation, Sample Variance, and Standard Deviation for 11 Sampled Tree Heights Near a Residential Area in a Chicago Neighborhood (in Meters)

Height (*m*)	$(X_i - \bar{x})$	$\bar{D} = \left(\sum\lvert x_i - \bar{x}\rvert\right)/n$	$(X_i - \bar{x})^2$		
16	−22.2727	22.2727	496.0744		
18	−20.2727	20.2727	410.9835		
21	−17.2727	17.2727	298.3471		
32	−6.27273	6.27273	39.34711		
34	−4.27273	4.27273	18.2562		
37	−1.27273	1.27273	1.619835		
43	4.727273	4.727273	22.34711		
43	4.727273	4.727273	22.34711		
45	6.727273	6.727273	45.2562		
60	21.72727	21.72727	472.0744	**Sample**	**Standard**
72	33.72727	33.72727	1137.529	**Variance**	**Deviation**
$\bar{x} = 38.3$		Σ=143.27	Σ=2964.18	2964.18/10	**SQRT (296.42)**
		$\bar{D}$=13.025		s^2=296.42	σ=17.22

variation is suggestive of a small sample size or the amount of uncertainty present in a set of observations. The standard deviation, which is denoted with a Greek letter "σ" for a population and "s" for a sample, is the value of the square root of the variance.

Variance is an important measure of dispersion or unevenness that indicates how a set of observations varies from the mean. If there is a wide variation from the mean, then the variance will be large; likewise, if it is small, then variation from the mean is narrow. It is a numerical value from the average of squared differences from the mean. The variance of a population is normally denoted by a Greek letter σ^2, whereas the variance of a sample is given by s^2.

In Table 3.4, the following equations have been used to derive mean deviation, standard deviation, and variance:

$$\text{Mean Deviation} \quad \bar{D} = \frac{\sum\lvert X_i - \bar{X}\rvert}{n}$$

$$\text{Sample Variance} \quad s^2 = \frac{\sum\left(X_i - \bar{X}\right)2}{n-1}$$

$$\text{Population Variance} \quad \sigma^2 = \frac{\sum(X_i - \mu)^2}{N}$$

Standard Deviation for a sample $\sigma = \sqrt{\frac{\sum (X_i - \mu)^2}{N}}$

Standard Deviation for a population $s = \sqrt{\frac{\sum (X_i - \bar{X})^2}{n-1}}$

Spatial Statistics: Measures for Describing Basic Characteristics of Spatial Data

The focus now shifts to spatial descriptive statistics. Unlike traditional descriptive statistics that deals with singletons, spatial statistics deals with observation recorded in pairs. Spatial descriptive statistics are used to measure the basic characteristics of spatial data. The foundation for spatial statistics was laid through the earlier work of Mercer and Hall (1911), Besag (1974), Besag et al. (1982), Cormack (1977), Fisher (1935), and Matheron (1963). Chapter 1 focused on some of these aspects. This chapter presents statistics that are applied to describe spatial data. Subsequent chapters cover more of these statistics and other advanced methods and strategies that are used to describe spatial data.

Given the uniqueness in spatial data, especially the need to understand the spatial structure, a number of spatial analytical statistics have been developed to deal with these data. Both theoretically and empirically, we know that spatial patterns or processes of a phenomenon offer fundamental clues about the nature of the spatial structure. Consequently, when studying spatial phenomena, we observe and measure specific events at different locations within a study region using a georeferenced system. The events are then uploaded into a geographic information system (GIS) for mapping and analysis. Once they are in a computer system, we can begin to quantify and understand any spatial distribution of phenomena. This is usually done by incorporating *X*- and *Y*-coordinates and the associated attributes into the spatial analysis framework. In a bid to understand the basic spatial characteristics, we apply spatial descriptive statistics. There are two common types of measures that can be undertaken: (1) one that measures centrality (spatial measures of central tendency) and (2) one that measures dispersions (spatial measures of dispersion) of events over space. These measures provide useful summaries of a spatial distribution.

We now illustrate spatial descriptive measures using an environmental quality dataset downloaded from the Texas Commission on Environmental Quality website (https://www.tceq.texas.gov/). The Texas Environmental Quality Database contains seven types of emissions: carbon monoxide (CO), nitrogen oxides (NO_x), volatile organic compounds (VOCs), particulate matter with an aerodynamic diameter of less than or equal to 10 μm (PM_{10}),

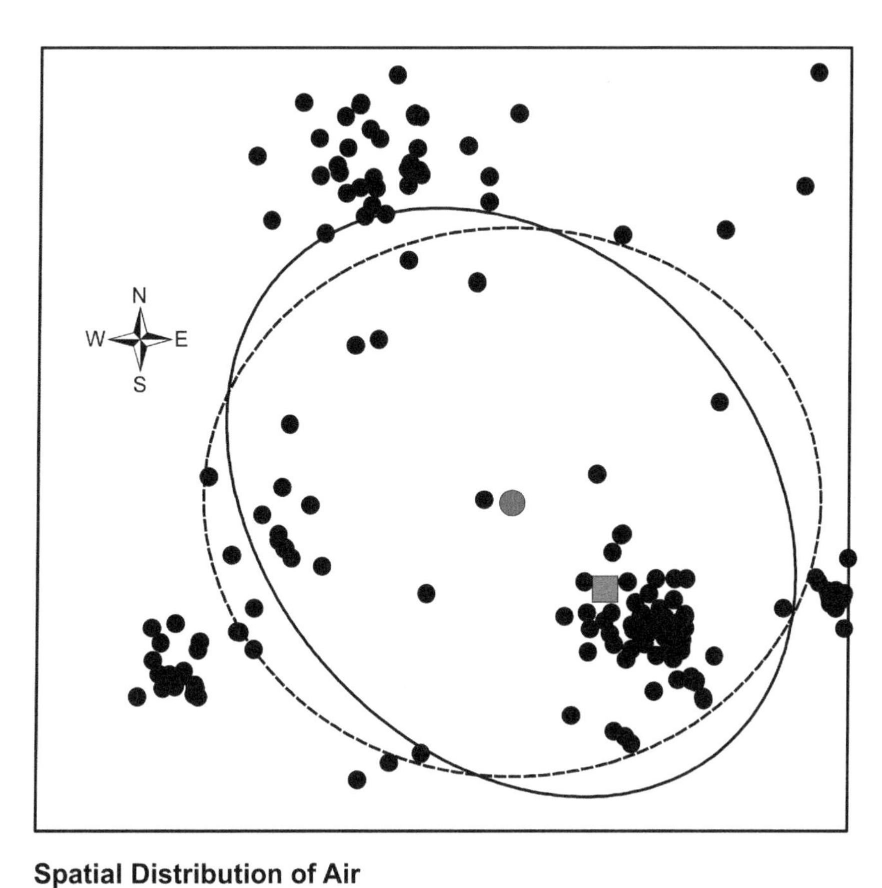

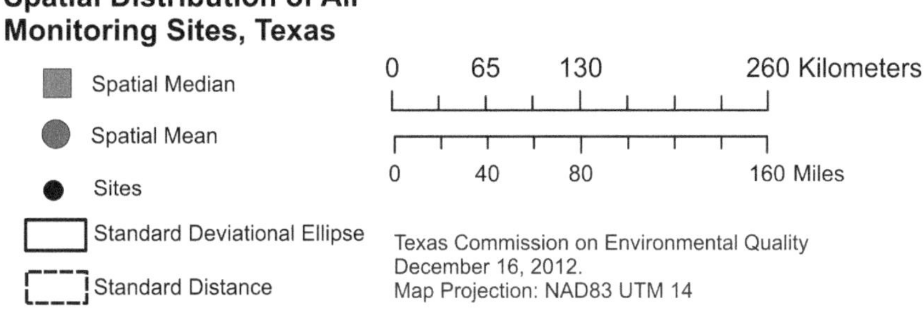

FIGURE 3.2
Spatial distribution of air monitoring sites in Texas.

particulate matter with an aerodynamic diameter of less than or equal to 2.5 μm ($PM_{2.5}$), sulfur dioxide (SO_2), and lead (Pb). It can be used to study the spatial distributions of emissions in Texas. Figure 3.2 shows spatial distributions of air monitoring sites. Figures 3.3 and 3.4 show the spatial distributions of CO, NO_x, PM_{10}, and SO_2 emissions in a three-dimensional (3D) perspective.

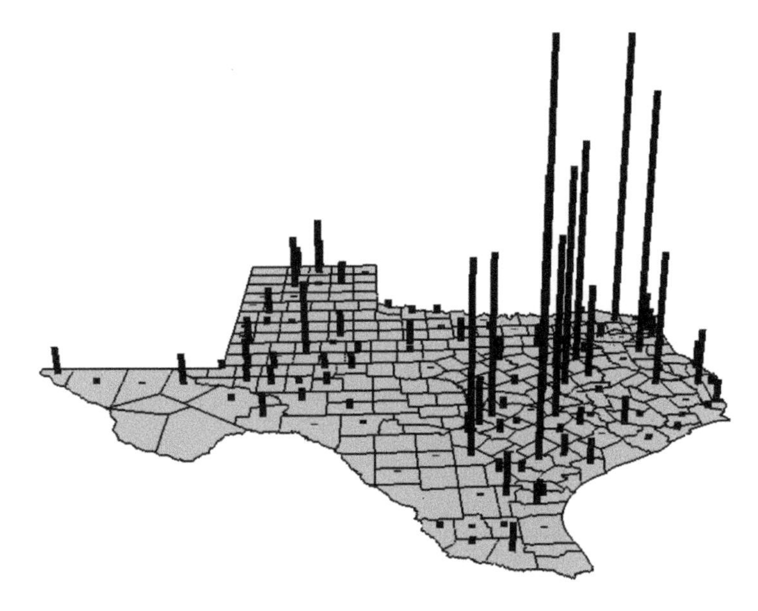

Average industrial carbon monoxide emissions in tons per year (tpy)
Lowest = 0, Highest (Titus County) = 21538.45
Mean = 219.20, Standard Deviation = 1390.02

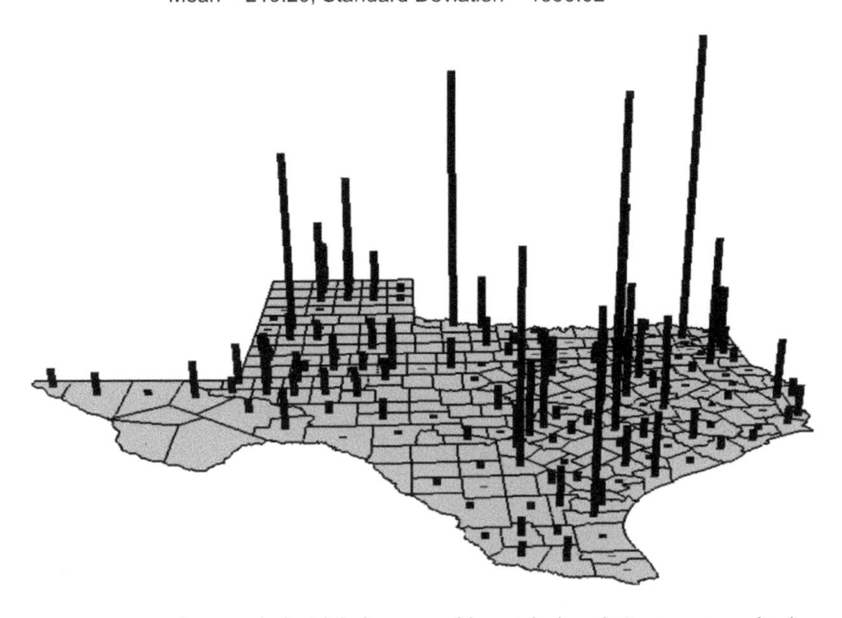

Average industrial nitrogen oxides emissions in tons per year (tpy)
Lowest = 0, Highest (Titus County) = 10169.31
Mean = 198.98, Standard Deviation = 708.15

FIGURE 3.3
Spatial distribution of carbon monoxide and nitrogen oxide emissions in Texas, presented in a three-dimensional perspective. (Data from Texas Commission on Environmental Quality.)

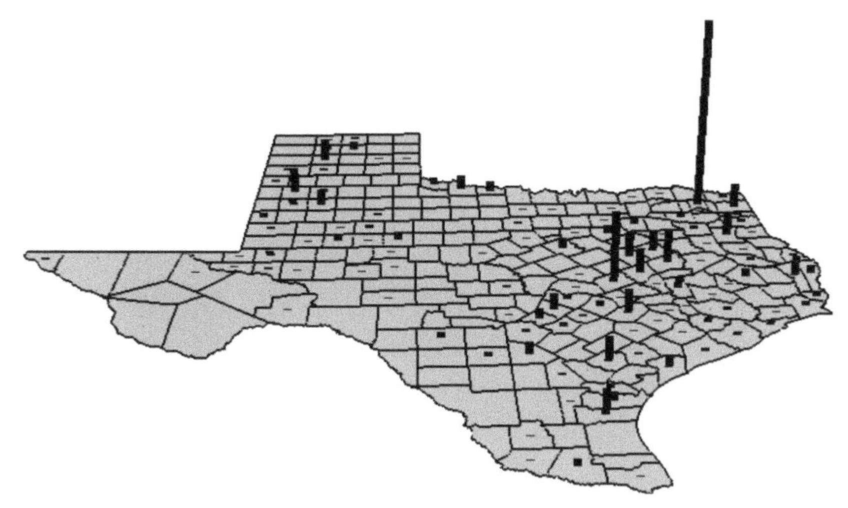

Average industrial particulate matter less than ten microns in diameter (PM10) emissions in tons per year (tpy)
Lowest = 0, Highest (Titus County) = 1432.12
Mean = 27.80, Standard Deviation = 102.56

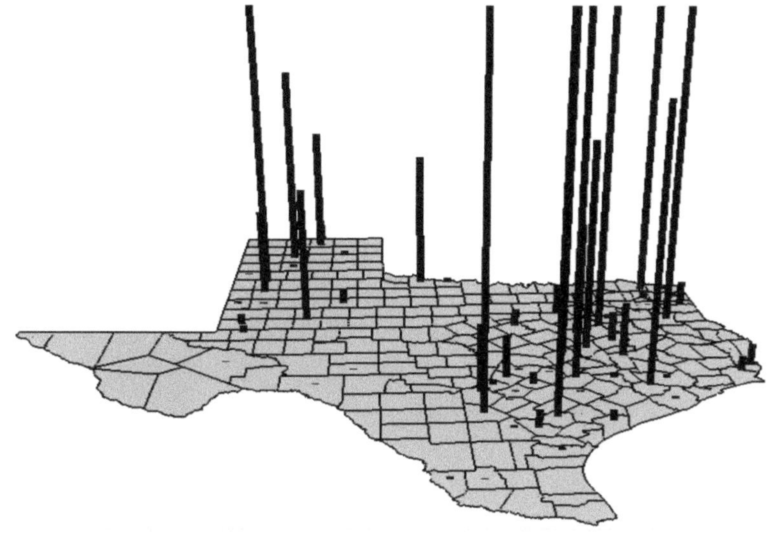

Average industrial sulfur dioxide emissions in tons per year (tpy)
Lowest = 0, Highest (Titus County) = 40030.82
Mean = 373.56, Standard Deviation = 2652.68

FIGURE 3.4
Spatial distribution of particulate matter and sulfur dioxide emissions in Texas, presented in a three-dimensional perspective. (Data from Texas Commission on Environmental Quality.)

Spatial Measures of Central Tendency

Spatial Mean/Mean Center: The spatial mean provides the average value of observed points for each of the X- and Y-coordinates. It shows the central point of spatial distributions of events. All the values for X and Y are separately summed up and divided by the total number of events/observations as follows:

$$\bar{X} = \frac{\sum_{i=1}^{n} X_i}{n}$$

$$\bar{Y} = \frac{\sum_{i=1}^{n} Y_i}{n}$$

where X_i and Y_i are the coordinates for feature i, and n is equal to the number of features.

These summary statistics provide the center of gravity of the spatial events being evaluated and are sensitive to outlying observations. The air monitoring sites presented in Figure 3.2 have a spatial mean of $\bar{X} = 772059$ and $\bar{Y} = 3385090$. In Table 3.5, for example, the spatial mean for the selected counties of Texas is $\bar{X}$=737059 and $\bar{Y}$=3401082.

Weighted Spatial Mean/Mean Center: As noted earlier, there are circumstances in which one may prefer to use the weighted mean. For spatial data, the weights represent the frequency or magnitude of the events observed at a given location. The summary statistic is produced by weighting each of the locational coordinates (X, Y) by the frequency values (or the variable that measures the magnitude of or characteristics observed in those locations). Unlike the spatial mean that assumes uniformity, the weighted spatial mean is able to capture the spatial variations and pulls toward the weighted points with the highest quantity. To derive this measure, we use the following formula:

$$\bar{X} = \frac{\sum_{i=1}^{n} X_i W_i}{\sum_{i=1}^{n} W_i}$$

$$\bar{Y} = \frac{\sum_{i=1}^{n} Y_i W_i}{\sum_{i=1}^{n} W_i}$$

where X_i and Y_i are the coordinates for feature i, and W_i is the weight at feature i.

Table 3.6 presents the weighted spatial mean for CO emissions in selected counties of Texas. In this example, we have weighted each of the coordinates

TABLE 3.5

Worktable for Deriving Spatial Mean for Industrial Emissions in Tons per Year (TPY) in Selected Counties of Texas

County	Number of Pollutant Sources	Average CO	Average NO_X	Average Pb	Average PM_{10}	Average $PM_{2.5}$	Average SO_2	*X*-Coordinate	*Y*-Coordinate
Milam	4	3381.62	651.17	0.0123	514.35	410.75	6538.11	693583	3407660
Hays	3	1309.44	802.01	0.0010	116.96	68.46	388.70	593410	3325620
Freestone	13	1866.30	531.36	0.0042	143.57	91.60	4955.47	770199	3511250
Fayette	4	1458.37	1802.01	0.1437	175.74	124.83	1347.52	700886	3306950
Rusk	14	2183.10	1127.20	0.0043	166.19	100.61	4933.45	900008	3560230
Titus	2	21538.45	10169.31	0.1424	1432.12	632.06	40030.82	876075	3682550
Atascosa	2	1500.79	1676.52	0.3595	83.60	33.78	5097.36	546120	3196270
Goliad	2	2081.51	1285.86	0.0219	169.16	89.28	6914.76	653773	3171000
Polk	5	1130.28	84.64	0.0032	54.08	47.36	0.98	899128	3414050
Robertson	6	1774.84	977.95	0.0117	182.71	128.30	1841.73	737407	3435240
								ΣX=7370589	**ΣY=34010820**
								$\bar{X}$=737059	$\bar{Y}$=3401082

TABLE 3.6

Worktable for Deriving Weighted Spatial Mean for Carbon Monoxide Emissions in Tons per Year (TPY) in Selected Counties of Texas

County	Average CO	X-Coordinate	Y-Coordinate	Weight CO*X	Weight CO*Y
Milam	3381.6182	693583	3407660	2345432896.01	11523405075.41
Hays	1309.4351	593410	3325620	777031882.69	4354683557.26
Freestone	1866.2958	770199	3511250	1437419158.86	6553031127.75
Fayette	1458.3652	700886	3306950	1022147751.57	4822740798.14
Rusk	2183.0947	900008	3560230	1964802694.76	7772319243.78
Titus	21538.4454	876075	3682550	18869293553.81	79316402107.77
Atascosa	1500.7864	546120	3196270	819609468.77	4796918546.73
Goliad	2081.5055	653773	3171000	1360832095.25	6600453940.50
Polk	1130.2816	899128	3414050	1016267834.44	3858837896.48
Robertson	1774.8402	737407	3435240	1308779587.36	6097002048.65
	Σ=38225			Σ= 30921885546.5	Σ= 135696972537.5
				$= \frac{30921885546.5}{38225}$	$= \frac{135696972537.5}{38225}$
				$\bar{X}$=808944.03	$\bar{Y}$=3549953.50

with CO emissions. We can also derive weighted spatial means for NO_x (789,033.44; 3,519,863.36), PM_{10} (791,093.56; 3,524,260.37), and SO_2 (800,846.69; 3,538,672.87).

Spatial Median/Median Center: The spatial median/median center provides an efficient way to estimate the location parameter of a statistical population. It is most effective when a distribution is spherical, and since the most preferred one is based on Euclidean space, the data must be projected to accurately measure distances. Suppose we have a projected set of finite (observation) points in space, the spatial median measure will minimize the sum of absolute distances toward the same points. It is less influenced by data outliers (so it can be applied to outlying observations) and serves as a popular estimator of the location of sparse data. The spatial median for air monitoring sites presented in Figure 3.2 is X = 834,275 m and Y = 3,323,400 m.

Spatial Measures of Dispersion

Spatial measures of dispersion measure the spatial variations or spread of observation points/events. Common methods that can be used to summarize the distribution of observation points include standard distance, weighted standard distance, and the standard deviational ellipse. These methods are extremely useful in situations where we seek to understand the centers of spatial distributions and the extent of dispersion of spatial events.

Standard Distance: The standard distance measures the extent to which observation points are dispersed around the spatial mean. It is a valuable statistic for understanding how compact observation points are distributed around their mean center. It is sensitive to outlying observation points. In Figure 3.2, the standard distance of air monitoring sites in Texas is 202,298 m. It is evident that this measure is large, implying that air monitoring sites are widely dispersed in the study region.

$$\mathrm{SD} = \sqrt{\frac{\sum_{i=1}^{n}(x_i - \bar{X})^2 + \sum_{i=1}^{n}(y_i - \bar{Y})^2}{n}}$$

where x_i and y_i are the coordinates for feature i, and n is equal to the number of features.

Weighted Standard Distance: This measure is produced by weighting the sum of the squared differences of x- and y-coordinates.

$$\mathrm{SD}_w = \sqrt{\frac{\sum_{i=1}^{n} w_i(x_i - \bar{X})^2 + \sum_{i=1}^{n} w_i(y_i - \bar{Y})^2}{\sum_{i=1}^{n} w_i}}$$

where x_i and y_i are the coordinates for feature i, w_i is the weight at feature i, and n is equal to the number of features. As with the weighted measures introduced earlier, this statistic has several applications in spatial analytics.

Standard Deviational Ellipse: This is a valuable measure of the dispersion of spatial events around the spatial mean. It gives the dispersion of observation points along the major and minor axes. It is a useful measure for summarizing data with a distributional directional bias. The measure can also be used in identifying distributional trends of geographic phenomena. This measure is able to account for both distance and orientation/directionality.

To derive the standard deviational ellipse, we must calculate three measures: spatial mean, angle of rotation from the point of origin (i.e., that is from spatial mean), and standard deviations along the x- and y-coordinates. The parameters are required for constructing a standard deviational ellipse for each type of observation point. The angle of rotation equation requires the mean center to be found so as to transform the coordinates in the region toward it. By rotating the coordinates clockwise about their new origin by a certain angle, we are able to determine the standard deviations along the x- and y-coordinates from the spatial mean. This helps in identifying the axes of the ellipse. It can be derived with or without the weight. However, the weighting provides a more realistic directional distribution because it adds the influence of weight field to the location. The size of the ellipse can be characterized using one, two, or three standard deviations.

Angle of rotation is given by

$$\tan\theta = \frac{\left(\sum_{i=1}^{n} x_i'^2 - \sum_{i=1}^{n} y_i'^2\right) + \sqrt{\left(\sum_{i=1}^{n} x_i'^2 - \sum_{i=1}^{n} y_i'^2\right) + 4\left(\sum_{i=1}^{n} x_i' y_i'\right)^2}}{2\sum_{i=1}^{n} x_i' y_i'}$$

where x_i' and y_i' are the deviations of x- and y-coordinates from the spatial mean.

Standard deviation along the x-axis is given by

$$\delta_x = \sqrt{\frac{\sum_{i=1}^{n} \left(x_i' \cos\theta - y_i' \sin\theta\right)^2}{n}}$$

Standard deviation along the y-axis is given by

$$\delta_y = \sqrt{\frac{\sum_{i=1}^{n} \left(x_i' \sin\theta - y_i' \cos\theta\right)^2}{n}}$$

The spatial deviational ellipse is shown by elliptical polygons in Figure 3.2. The standard deviation along the x-axis is 230,351.67 m and for the y-axis is 169,667.57 m, and the angle of rotation is 148.95. One can conclude that the air monitoring sites follow a northwest to southeast direction. There are more sites in the southeast, and the large standard deviations along the x- and y-axes suggest a wide dispersion (Figure 3.2).

We can explore the same dataset using this sample R-Code as follows:

Assume we have now practiced R and can recall earlier commands used in Chapters 1 and 2. We are now ready to proceed with exploring spatial statistics in R.

```
#Type each command in sequence and on its own line and wait
for a response before typing in the next one.
setwd("../ChapterData/Chapter3_Data_Folder")

# Select the Texas air monitoring sites data (TCEQ_Air_
Monitoring_Sites_PR_studyregion.shp) from the workspace
directory (../ChapterData/Chapter3_Data_Folder).
library(rgdal)
library(sp)
TCEQ <- readOGR('TCEQ_Air_Monitoring_Sites_PR_studyregion.
shp', layer='TCEQ_Air_Monitoring_Sites_PR_studyregion')
OGR data source with driver: ESRI Shapefile
```

```
Source: "TCEQ_Air_Monitoring_Sites_PR_studyregion.shp", layer:
"TCEQ_Air_Monitoring_Sites_PR_studyregion" with 175 features
It has 37 fields
Integer64 fields read as strings: OBJECTID SITE_ID AQS_SITE_C
ELEVATION ADDR_ZIP_C AIR_TOXICS METEOROLOG NITROGEN PM_25
OZONE CO_SO2_H2S PM10_OTHER LEAD CHROMIUM_V
plot(TCEQ, main="Spatial distribution of air monitoring sites
in Texas")
TCEQ_coords <- as.data.frame(coordinates(TCEQ))
View(TCEQ_coords)

# Now derive the spatial mean (x,y) using this command:
TCEQ_X_MEAN <- mean(TCEQ_coords$coords.x1); TCEQ_X_MEAN
TCEQ_Y_MEAN <- mean(TCEQ_coords$coords.x2); TCEQ_Y_MEAN

# Now derive the spatial median (x,y) using this command
TCEQ_X_MED <- median(TCEQ_coords$coords.x1); TCEQ_X_MED
TCEQ_Y_MED <- median(TCEQ_coords$coords.x2); TCEQ_Y_MED

# Now let us derive standard distance deviation (SDD) or the
standard deviational ellipse (SDE). # This is more complex
than the previous two. However, the aspace R library provides
functionalities for calculating SDD and SDE. This will make
our work much easier than a manual calculation.

#We start by installing it as follows:
install.packages("aspace")
library(aspace)
calc_sdd(id=1, filename="SDD_TCEQ_Output.txt", centre.xy
NULL, calccentre TRUE,
weighted FALSE, weights NULL, points= TCEQ_coords, verbose
FALSE)
calc_sde(id=1,points= TCEQ_coords)
plot_sdd(plotnew FALSE, plotcentre FALSE, centre.col="red",
centre.pch="1", sdd.col="red",sdd.lwd=1,titletxt="",
plotpoints TRUE,points.col="black")
plot_sde(plotnew FALSE, plotcentre FALSE, centre.col="red",
centre.pch="1", sde.col="red",sde.lwd=1,titletxt="",
plotpoints TRUE,points.col="black")
```

Random Variables and Probability Distribution

Along with the descriptive measures presented in the previous section, it is equally important for spatial data scientists to be conversant with the probability distributions, random variables, and formulation and testing of hypotheses. These concepts are introduced in the following sections.

Random Variable

This is a function/rule of the random process for assigning every outcome in the sample space of a random experiment a numerical value. Given the fact that random experimental results may yield nonnumerical values, the assignment of unique numbers to the outcome is done through a random process function. For example, suppose we hypothesized that "it will be cold tomorrow" in our neighborhood; the other option will be "it will not be cold." So, we can use a random variable to assign two unique numerical values to these two outcomes as follows:

$$X = \begin{cases} 1, \text{if it is cold} \\ 0, \text{if it is not cold} \end{cases}$$

This is typically achieved using a probability function, which assigns numerical values to a set of outcomes with an equally likely possibility for each member of a sample space. The two types of random variables are discrete and continuous. A random variable is typically associated with two mathematical functions: (1) a probability distribution (discrete random variable) takes on a finite value or any countable infinite set of values and (2) a probability density function (continuous random variable) takes on any infinite set of values that continuously varies within one or more intervals.

Probability and Theoretical Data Distributions

Concepts and Applications

The use of the term *probability* implies the possibility or likelihood of an event happening. In a statistical context, probability helps to advance our understanding of the science of uncertainty, chance, or likelihood. The probability function is a numerical function for describing a probability distribution. The numerical values of a probability normally range from 0 to 1, thus the value indicates whether the event will occur with each member of the sample space having an equal chance. A zero value indicates no chance that an event will occur, while a one value indicates a 100% chance that an event will occur. When we conduct an experiment, we obtain an outcome after observing or measuring a specific activity. It may simply mean "tossing a coin," "rolling a die," or "determining whether there will be a severe thunderstorm tomorrow." In general, such an experiment can be used to determine the probability of a given event B, in equally likely possibilities in the sample space as follows:

$$P(B) = \frac{\text{\# of possibilities that meet the set criteria in the sample space}}{\text{\# of equally likely possibilities in the sample space}}$$

Let us now focus on two of the examples noted above.

Experiment I: Undertaking a Tossing Coin Activity

In this experiment, there are only two possible outcomes when one tosses a coin once, it will be either a head or a tail. We can determine the likelihood of success of this experiment by calculating its probability. Let p represent the probability function, H represent heads, and T represent tails:

$$p(H) = \tfrac{1}{2} \text{ or } 50\% \text{ chance}$$

$$p(T) = \tfrac{1}{2} \text{ or } 50\% \text{ chance}$$

Suppose we tossed two coins or decided to toss this coin several times; the number of outcomes would definitely change. This is because there are many different ways to achieve the goal; also there are several combinations from which to choose the outcome. It gets even more complicated if we decide whether to allow any repetitions or no repetitions in this experiment. If we tossed two coins at once, there are four possible outcomes in the sample space: {H-H}, {H-T}, {T-L}, or {T-T}. Therefore, the probability of obtaining p (H-H or T-T) is one-quarter, the probability of obtaining p (H-T or T-H), or a match is one-half, and the probability of least one head or one tail is three-quarters.

We can use n^r to derive the combinations if repetitions and orders are allowed, where n is the number of possibilities to choose from, and r is the number of times. However, if repetitions and orders are not allowed, then we can use the following formula to derive all possible outcomes/combinations of a sample as follows:

$$C_r^n = \binom{n}{r} = \frac{n!}{(n-r)!(r!)}$$

Note that $r!$ is the factorial.

Experiment II: Undertaking the Rolling of a Die Activity

Rolling a die once has six possible outcomes (the numbers are 1, 2, 3, 4, 5, and 6). So, we can now work out the probability of the following:

$$p(1) = 1/6$$

$$p(1 \text{ or } 2) = 2/6 = 1/3$$

$$p(1 \text{ or } 2 \text{ or } 3) = 3/6 = 1/2$$

$p(1 \text{ and } 2) = 0/6$, this will yield what is termed as a mutually exclusive event.

Taking this a little further, we can roll a die twice. The two rolls will yield 36 possible outcomes (6^2). This is the sample space for deriving probability of a specific event in this experiment, and the batch of outcomes can be grouped to form a distribution. There are several types of theoretical distributions. One of the objectives of statistical analysis is to explore how well empirical distributions (observed from naturally occurring phenomena) match these theoretical distributions. The results can help us establish confidence bands in inferential statistics and could also serve as the basis for selecting the appropriate techniques in more advanced statistical analysis. Following are the most common theoretical distributions.

Binomial Distribution

A binomial distribution depicts the sequence of a fixed number of events ($x = 0, 1, 2, 3, n$) in a sample space that can be segregated into two outcomes, where x represents the number of times each event occurs in the experiment, and these events are independent of each other. The probability (p) of sampling each of the two outcomes in an event is the same [$p(X) = 0.5$]; the probability of sampling the occurrence or nonoccurrence of the event in a single experiment is given by p and q, respectively. A binomial distribution can be expressed mathematically as follows:

$$p(X) = \frac{n!p^x q^{n-x}}{X!(n-X)!}$$

$p(X)$ gives the probability of occurrence or nonoccurrence in n binomial experiments, whereas $X!$ represents the factorial. It involves the examination of the probability of discrete events and is evident when there are two mutually exclusive outcomes; for example, yes–no, success–failure, male–female, head–tail, or absence–presence events. This is typical of geographic applications that can be expressed using a binary framework, including the absence or presence of vegetation/animal species in a defined geographic location; whether people residing in a neighborhood have a college-level education or not; and whether the application of pesticides to an agricultural field improves crop yield or not. Let us now consider several examples of binomial distribution to help our understanding further. In MS Excel, the $p(X)$ formula would look like this: = ((FACT (n) × ((p)∧x) × ((q)∧(n–x)))/(FACT (X) × (FACT (n–X)))). One could use the binomial distribution function.

Experiment I: If a traveler from the city of St. Louis on the way to Chicago randomly stops at five convenience stores, find the probability that the traveler stops at exactly three stores. Each stop has six possible choices.

$$p(3) = \frac{5!\left(\frac{1}{6}\right)^3\left(\frac{5}{6}\right)^{5-3}}{3!(5-3)!} = \frac{120 \times 0.00462963 \times 0.694444}{3 \times 2 \times 2} = 0.032$$

Experiment II: In the last 3 years, the U.S. Transportation Security Administration found that 40% of the passengers passing through Chicago's O'Hare International Airport had banned liquids exceeding 100 mL. If ten passengers are selected randomly, find the probability that at least six of them have banned liquids. To find this probability, we have to calculate individual probabilities for either 5, 6, 7, 8, 9, or 10 and then add them up to get the answer.

$$p(5)=\frac{10!(0.4)^5(0.6)^{10-5}}{5!(10-5)!}=\frac{3628800\times 0.01024\times 0.0776}{120\times 120}=0.2007$$

$$p(6)=\frac{10!(0.4)^6(0.6)^{10-6}}{6!(10-6)!}=\frac{120\times 0.00462963\times 0.694444}{720\times 24}=0.1115$$

$$p(7)=\frac{10!(0.4)^7(0.6)^{10-7}}{7!(10-7)!}=\frac{120\times 0.00462963\times 0.694444}{5040\times 6}=0.0425$$

$$p(8)=\frac{10!(0.4)^8(0.6)^{10-8}}{8!(10-8)!}=\frac{120\times 0.00462963\times 0.694444}{40320\times 2}=0.00005$$

$$p(9)=\frac{10!(0.4)^9(0.6)^{10-9}}{10!(10-9)!}=\frac{120\times 0.00462963\times 0.694444}{363880\times 1}=0.00003$$

$$p(10)=\frac{10!(0.4)^{10}(0.6)^{10-10}}{10!(10-10)!}=\frac{120\times 0.00462963\times 0.694444}{3628800\times 1}=0.0001$$

Therefore, p (at least six of them have banned liquids) = 0.2007 + 0.1115 + 0.0425 + 0.00005 + 0.00003 + 0.0001 = 0.3549.

Poisson Distribution

An ordered or ranked series of spatial outcomes that are truly the result of random processes can be expressed using Poisson probabilities. Generally, a Poisson distribution is used under the following circumstances: (1) when there is a specified interval for an event (equally segregated spatial areas or temporal sequences), and it is possible to count how many events have occurred; (2) when the events occur independently of each other both in space and time; (3) when each of the two outcomes in an event is virtually zero; (4) when there are low-occurrence events, rare events, isolated events, or a low-density pattern; and (5) when the average rate is known for a specified number of occurrences for an event. In spatial analysis, we apply Poisson probability distribution to study the degree of randomness in point spatial patterns.

The Poisson probability distribution can be expressed mathematically as follows:

$$P(X)=\frac{e^{-\lambda}\lambda^X}{X!}$$

TASK 3.1 USING THE POISSON DISTRIBUTION FUNCTION

To illustrate the use of the Poisson distribution function, let us consider several examples of electrical storms in four major cities, a lightning strike occurrence in Kampala, and lightning deaths in the United States.

1. On average, electrical storms occur on about 31 days per year in New York. Suppose we observe 28 days a year. What will be the probability that we observe electrical storms?

$$X = 28: p(28) = \frac{2.71828^{-31} 31^{28}}{28!} = 0.0647$$

   The probability will be 6.45%.

2. On average, electrical storms occur on about 21 days per year in both Paris and Rome. Suppose we observe 14 days a year. What will be the probability that we observe electrical storms?

$$X = 14: p(14) = \frac{2.71828^{-21} 21^{14}}{14!} = 0.0282$$

   The probability will be 2.82%.

3. On average, electrical storms occur on about 16 days per year in London. Suppose we observe 14 days a year. What will be the probability that we observe electrical storms?

$$X = 14: p(14) = \frac{2.71828^{-16} 16^{14}}{14!} = 0.093016$$

   The probability will be 9.3%.

4. On average, the residents of Kampala, Uganda, hear thunderstorms 240 days per year, one of the highest rates in the world. Suppose we observe 200, 220, or 250 thunderstorm occurrences per year, what will be their probabilities? The probabilities will be 0.008 (0.08%), 0.0114 (1.1%), and 0.0205 (2.05%), respectively.
5. On average, lightning kills about 100 Americans and inflicts another 500 injuries per year. Suppose we observe 84 and 95 death occurrences, and/or 486 and 490 injury occurrences. What will be the probabilities for these occurrences? The probabilities for the death occurrences will be 0.0112 (1.1%) and 0.0360 (3.6%), respectively; likewise, for injuries the probabilities will be 0.0148 (1.5%) and 0.0163 (1.6%), respectively.

6. Another completed example is provided in Table 3.7.

TABLE 3.7

Worktable for Poisson Probabilities of Observed Road Fatalities per Every 100,000 Inhabitants per Year Occurrence for Selected Countries

Country	Number of Road Fatalities per Year	Observed Frequency in a Year	Observed Probability of Occurrence
China	5.1	3	0.1347
Eritrea	48.4	27	0.0003
Ghana	29.6	27	0.0680
Ireland	4.06	3	0.1924
Kenya	34.4	27	0.0324
Mauritius	11.1	18	0.0154
Nigeria	32.3	27	0.0483
South Africa	33.2	27	0.0412
South Korea	12.7	18	0.0352
Sweden	2.9	3	0.2237
Uganda	24.7	27	0.0689
United States	12.3	18	0.0295

Source: Data from World Health Organization (WHO) *Global Status Report on Road Safety*, 2009. Geneva, Switzerland, WHO.

where e represents the exponential constant value (2.71828), λ is the mean frequency, X is the number of occurrences, and $X!$ is the factorial. In MS Excel, the $p(X)$ formula would look like this: = (((Exp (–λ) × (λ∧X)/(FACT (X))))). One could use the Poisson distribution function.

Normal Distribution

Many times, we use binomial and Poisson distributions to describe discrete random variables, but to adequately describe continuous probability of variables, we use the normal distribution. We can describe the probability of a normal distribution using the mean and standard deviation. The distribution of a random variable can be visually represented using a histogram plot. In a normal distribution, the values in a histogram should form a normal curve. However, it should be known that the distribution of a random variable displayed in a histogram can spread out in numerous ways that depict the three measures of center, mean, median, and mode. These numerous ways include spreading out toward the center, left, and right. The distribution may also depict multiple modes in the random variable, sometimes there is only one mode, or two different modes. When the distribution of a random variable is toward the center without any bias toward the left or right, it is typically described as normally distributed. Its distributional shape reflects a bell curve, implying that

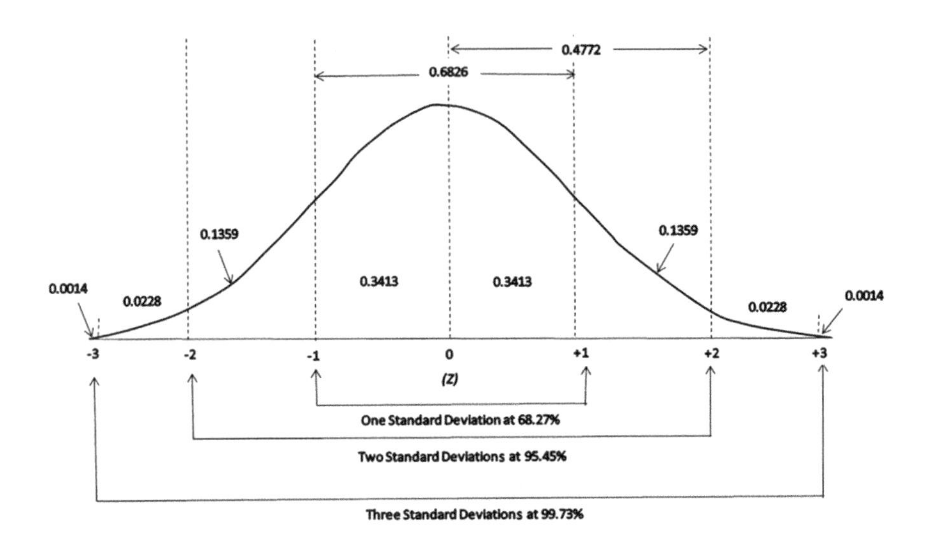

FIGURE 3.5
Different segments of the Z-value under the normal curve.

its mean is equal to the median and mode, and it is symmetrical from the center. Thus, as indicated in Figure 3.5, a bell-shaped curve has 50% of the values that are less than the mean (left segment of the normal curve), and the other 50% of the values are greater than the mean (right segment of the normal curve). In statistical analysis, if a normal distribution is evident in any set of observations, then it is possible to derive several useful conclusions from it.

Mathematically, a normal distribution of a random variable, x, can be defined as follows:

$$p(x) = \frac{1}{\sqrt{2\pi\sigma}} e^{-(x-\mu)^2/2\sigma^2}$$

where both π and e are constants, which are equal to 3.14159 and 2.71828, respectively; μ is the mean; and σ is the standard deviation of x. In MS Excel, the $p(x)$ formula would look like this: = (1/(SQRT (3.14159 × 2)) × ((EXP (– (((x–μ)/(σ)∧2)/2))))). One could use the normal distribution function.

We can describe the distribution of data values (e.g., **R** variable) using the standard deviation. We derive values that are within one standard deviation ($\bar{x} - \sigma \leq \mathbf{R} \leq \bar{x} + \sigma$), two standard deviations ($\bar{x} - 2\sigma \leq \mathbf{R} \leq \bar{x} + 2\sigma$), and three standard deviations ($\bar{x} - 3\sigma \leq \mathbf{R} \leq \bar{x} + 3\sigma$). Also, to be successful in testing hypotheses or comparing different observations, we can derive a set of statistics called the *Z-score*. The Z-score is a standard normal transformation that offers a better metric for comparing such observations. The Z-score is derived as follows: $\frac{x_i - \bar{x}}{\sigma}$ using the mean and standard deviation of the random variable. Once the Z-score is calculated, we can look for the probability $p(Z < z)$ in the standard normal distribution Z-score.

TASK 3.2 USING THE NORMAL DISTRIBUTION

1. Suppose that the tree height in samples from Chicago neighborhoods has a mean of 38.3 m and a standard deviation of 17.2 m. What is the probability that trees in a randomly selected tree sample will be (a) less than 51 m, (b) more than 51 m, and (c) between 26 and 66 m?

   a. Normal distribution probability of less than 51 m

      We can solve this problem in MS Excel using this normal distribution probability formula = NORMDIST (x, mean, standard deviation, TRUE).

      This returns $p(x < 51) = 0.771618$, that is, 77.2%.

   b. Normal distribution probability of more than 51 m

      We simply subtract 1 from the probability reported earlier: $p(x > 51) = 1 - p(z < 0.771618) = 1 - 0.771618 = 0.228382$, that is, 22.8%.

   c. Normal distribution probability between 26 and 66 m

      We solve this by obtaining the probabilities for 26 and 66 m just as we did earlier, then subtract the larger probability from the smaller probability to obtain the answer 26, which returns a probability of 0.239068, and 66 m returns a probability of 0.946983.

$$p(26 < x < 66) = 0.946983 - 0.239068$$
$$= 0.707915, \text{ that is, } 70.8\%.$$

2. Suppose that the blood lead levels among children from New York City have a mean of 8.3 μg of lead per deciliter of blood (μg/dL) and a standard deviation of 4.6 μg/dL. What is the probability that blood lead levels among children in a randomly selected blood testing sample will be (a) less than 10 μg/dL, (b) more than 10 μg/dL, and (c) between 5 and 20 μg/dL?

   a. Normal distribution probability of less than 10 μg/dL

      We can solve this problem in MS Excel using this normal distribution probability formula = NORMDIST (x, mean, standard deviation, TRUE).

      This returns $p(x < 10) = 0.644147$, that is, 66.4%.

   b. Normal distribution probability of more than 10 μg/dL

      We simply subtract 1 from the probability reported earlier:

$$p(x > 10) = 1 - p\ (z < 0.644147) = 1 - 0.644147$$
$$= 0.355853, \text{ that is, } 35.6\%.$$

c. Normal distribution probability between 5 and 20 μg/dL

We solve this by obtaining the probabilities for 5 and 20 μg/dL just as we did earlier, then subtracting the larger probability from the smaller probability to obtain the answer 5, which returns a probability of 0.236566, and 20 μg/dL, which returns a probability of 0.994512.

$$p(5 < x < 20) = 0.994512 - 0.236566$$

$$= 0.757946, \text{ that is, } 75.8\%.$$

TASK 3.3 USING *Z*-SCORE TO ASSESS THE RELATIVE POSITION IN DATA DISTRIBUTIONS

Z-scores can be used to describe how the distributions of observations fit within the standard normal distribution or to compare different normal distributions with similar standard deviations. These scores can also be used during a data screening process to detect univariate outliers. These outliers are *Z*-scores with values that are three standard deviations below or above the mean in a data distribution. On detection of outliers, one may wish to speculate on the underlying reasons for their occurrence, such as (1) they may simply be the result of errors in data entry, or failure to specify missing value codes in the data; (2) the specific cases may have come from a different population but were inadvertently included as a member of the current sample; and (3) they could be legitimate cases with extreme values that far exceed the norm in the rest of the sample data. The role of the data scientist is to rule out the first two situations and then proceed to further examine the distributional patterns and causes of the cases that are rightfully classified as outliers. To illustrate the applications, let us explore the normal probability distribution for the Illinois corn and soybean production data using Chapter3_Data_folder (Data files: *Illinois_cnty_agricultural_statistics* or *agricul_ILL_stats3*). This data was introduced in Chapter 2.

1. Derive and compare the *Z*-scores for corn and soybean production ($n = 102$). (**TIP**: add a *Z*-score field in the attribute table; convert raw score to *Z*-score using this formula (*X*-mean)/standard deviation). The results would look like what is presented in Table 3.8. The normal distribution curve for corn and soybean production is presented in Figure 3.6.

TABLE 3.8

Z-Scores for Corn and Soybean Production

CNTY_FIPS	Z-s=Score (corn)	Z-Score (Soybean)
189	−0.457798016	1.352524176
027	−0.503243101	0.312944136
157	−0.803096471	0.230400614
191	−0.594164224	0.661305116
119	−0.353950951	0.682067837
...	...	...
069	−1.272850448	−1.608114855
043	−1.238009835	−1.612017622

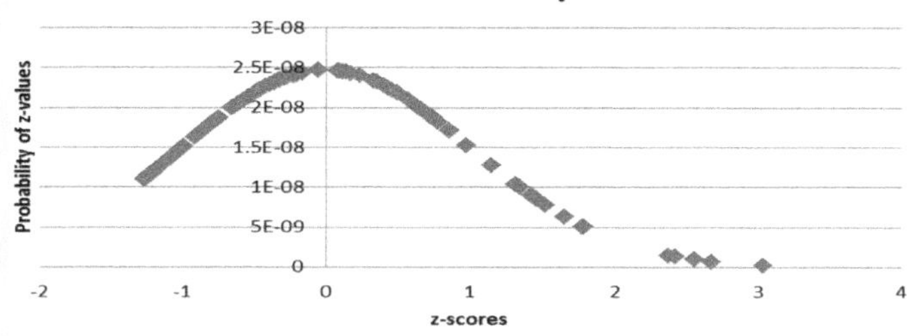

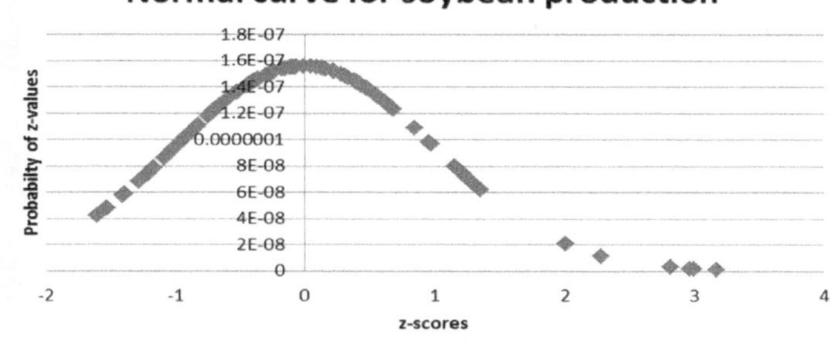

FIGURE 3.6
Upper panel represents the standard normal curve for corn production, while the lower panel represents the standard normal curve for soybean production in Illinois state.

2. Now map the Z-scores for both using the standard deviation and natural breaks classification methods. Set standard deviation at 1 SD interval size. The maps for standard deviation and natural classification methods would look like those in Figure 3.7.

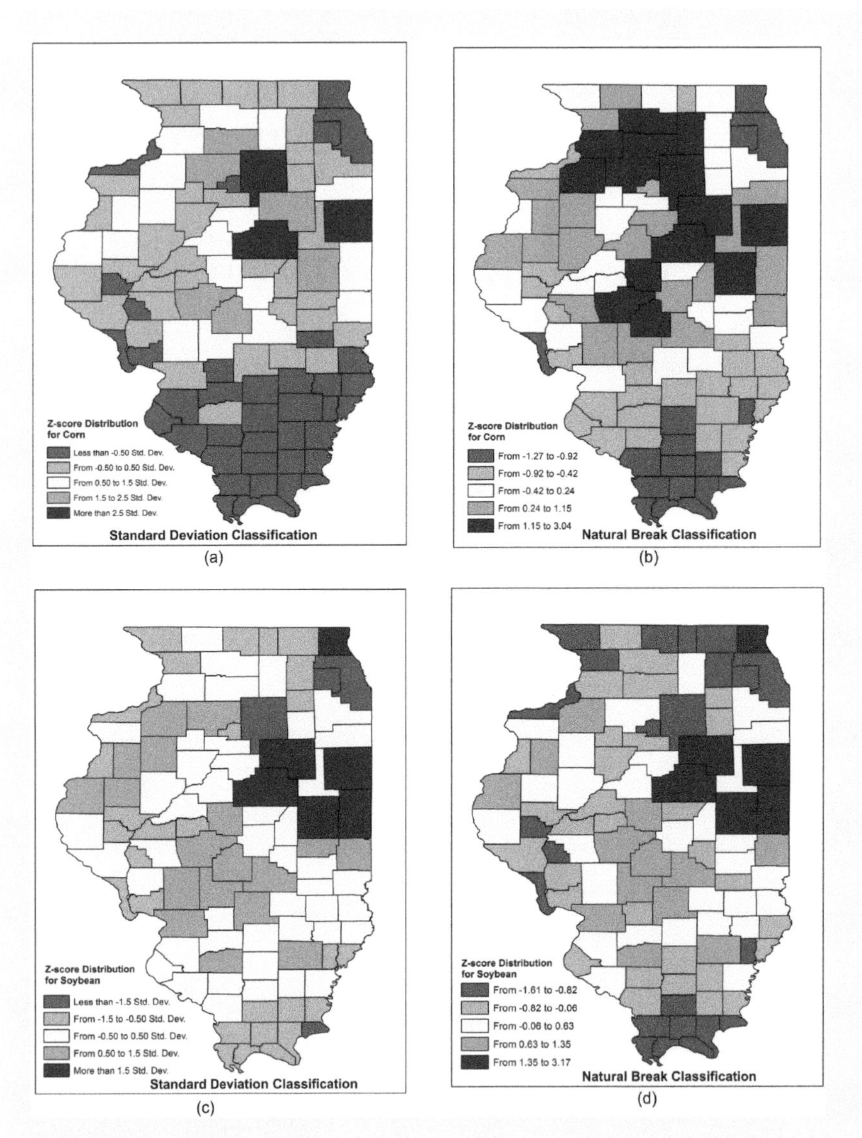

FIGURE 3.7
Spatial distribution of corn and soybean production in the Illinois.

Comment on the spatial patterns of the Z-scores in relation to corn and soybean production in Illinois state. Do outliers exist for corn and soybean production? If yes, describe the spatial distribution of these outliers. Speculate on why these outliers exist. There are wide variations in crop acreage at the county level in Illinois state, as it is evident in the standard deviation.

TASK 3.4 GENERATE AND INTERPRET TRADITIONAL DESCRIPTIVE STATISTICS

1. The data for completing this Challenge Assignment are located in Chapter3_Data_folder. Navigate to the *agricul_ILL_stats3. shp* file. Open the database file using MS Excel, or run this in R. Explore this dataset and generate some descriptive statistics for the following fields/columns: *NO_FARMS07, AVG_SIZE07, CROP_ACR07*, and *AVG_SALE07*. Fill in the correct statistics for each of these fields in Table 3.9.
2. In Table 3.10, compile these statistics for corn, soybean, and wheat grain production for each of their four fields.
3. Use the results from the three sampling designs from Chapter 2, Task 2.4 to generate three tables on additional descriptive statistics for corn, soybean, and wheat production. We know that one important factor in choosing an appropriate sampling method is the standard error, that is, using the sample as a method of estimating the population. Another factor is the standard deviation, indicating how much variation exists from the mean. So, based on these two factors, what is the most appropriate sampling design? Explain.
4. Comment on the distribution of summary statistics in questions 1 and 2.

TABLE 3.9
Descriptive Statistics for the Agricultural Variables

	Minimum	Maximum	Sum	Mean	Standard Deviation
NO_FARMS07					
AVG_SIZE07					
CROP_ACR07					
AVG_SALE07					

TABLE 3.10
Statistics for Corn, Soybean, and Wheat Grain

	Minimum	Maximum	Sum	Mean	Standard Deviation
Corn					
Soybean					
Wheat					

TASK 3.5 GENERATE AND INTERPRET DESCRIPTIVE SPATIAL STATISTICS

Noise-level data from Permanent Noise Monitor Locations were obtained from the City of Chicago Department of Aviation website. The stations are located in the surrounding communities of the airport. Noise-level data are reported in the Aircraft Noise Reports and are normally averaged on a monthly basis. The Noise Report summarizes measurements from each of the 34 permanent noise monitors. Currently, the dataset covers a 7-year study period (2004–2010). We can assess the environmental impacts of aircraft noise disturbance on the surrounding neighborhoods using the data. Most studies suggest that noise levels considered bearable for most human inhabitants range from 60 to 70 dB on average day/night sound levels decibel (dB measures the ratio of a physical quantity, in this context the signal-to-noise ratio).

1. Explore Noise_Project and Study_Area_Outline feature classes from Noise_OHare_Geodatabase.gbd located in Chapter3_Data_folder.
2. Generate the spatial mean (mean center), median center, standard distance, and directional distribution using the Noise_Project point feature class. Comment on the spatial distribution of the noise level events surrounding O'Hare International Airport.
3. State a working hypothesis regarding the spatial distribution of noise levels in the study region. Also, suggest a few factors that might influence the spatial distribution of noise levels in the study region.

To read a File Geodatabase file in R, ensure that "sf" or "rgdal" packages are installed as follows:

```
install.packages("sf")
install.packages("rgdal")
#Type in this command each command on its own line and
wait for a response, review what is on screen before
typing in the next one.
library(rgdal)
list.files(path.expand("Noise_OHare_Geodatabase.gdb"))
[1] Study_Area_Census_BLGP.lyr" "Noise_OHare_Geodatabase.
gdb"

# To read layers in the file geodatabase, use this
command
fc_list = ogrListLayers("Noise_OHare_Geodatabase.gdb")
print(fc_list)
```

```
# There are five layers in this Noise_OHare_Geodatabase.
gdb.

# Read a specific layer in the Noise_OHare_Geodatabase.
gdb, use this command
library(sf)
luca <- st_read("Noise_OHare_Geodatabase.gdb", layer =
"Study_Area_Census_BLGP")
read_sf("Noise_OHare_Geodatabase.gdb", layer =
"Study_Area_Census_BLGP")
library(sf)
luca <- st_read("Noise_OHare_Geodatabase.gdb", layer =
"Study_Area_Census_BLGP")
## The sample R codes for data processing are located in
Chapter3-R-Code, and the dataset that is used to implement
the code is stored in Chapter 3_Data_folder. The data
folder has two files. This will set you up and get you
ready to work with R. Use earlier commands for Spatial
Measures of Dispersion Section.
```

Conclusion

In this chapter, we learned the important concepts that underlie traditional statistical analysis and the essential role they play in spatial statistics. The descriptive measures that summarize the center and spread of distributions were presented for all data types, including those that are specifically tailored for spatial data. These were followed by a presentation of the fundamentals in probability theory and the primary forms of theoretical distributions that characterize both discrete and continuous variables. Following are some challenge exercises to help underscore these key concepts.

Worked Examples in R and Stay One Step Ahead with Challenge Assignments

Exploring *Z*-Score to Assess the Relative Position in Data Distributions Using R

```
#Type each command in sequence and on its own line and wait
for a response before typing in the next one.
library(rgdal)
library(dplyr)
```

```
x <- readOGR('agricul_ILL_stats3.shp', layer='agricul_ILL_stats3')
n <- x@data$CornProduc #corn production
n

# Upon extraction, you should get results that look like this.

[1] 15126400 12151200 17399100 15752000 64128000 59202500
44160000 23218000
[9] 27878400 8131500 36727500 22451800 34760000 22300400
60027000 18208500
[17] 34713900 62125000 26232500 22833600 33536800 69920000
33160000 28911400
[25] 31559000 32939400 39474000 24770200 49720000 42112000
26591200 20083800
[33] 13229600 22212900 17285400 18330000 8116800 33539200
47564000 30139200
[41] 22808400 28880000 23777000 45451500 17748000 8881500
20077000 32703300
[49] 31240600 32106500 17601200 16480800 3236100 11050000
11556700 15184000
[57] 14586300 12136000 14128000 8347600 15174500 7049000
8184400 8580000
[65] 8066000 12762900 12167500 11294200 4498200 7276500
6884700 13497000
[73] 4755300 10320000 6816000 7919200 6555400 4870600 3877600
10475400
[81] 6058800 1750000 331000 1026800 1955800 1086300 3685200
1152000
[89] 2622400 28315000 15640800 16274000 13401700 1228200
42596000 27380000
[97] 560000 17427600 43848000 893800 44435500 49595000

m <- x@data$SoyProduct #soybean production
m

# Upon extraction, you should get results that look like this.

[1] 1569500 2054400 4429800 2335900 268600 5635000 7516600
3539100
[9] 4540800 1045000 5428500 3264800 5808800 3192600 11773500
2514300
[17] 5952800 11397500 5296500 3572400 5432400 12316500 5179200
4926600
[25] 6360000 5648400 10020500 3180100 11850000 5955000 5415000
4037600
[33] 2281600 4169700 2865600 2430400 1187700 5217300 5951700
7160000
[41] 5594700 5231200 3526000 7303400 3432300 1804800 5305000
7374500
```

```
[49] 6683600 5759200 3807000 4600000 576000 3058000 2373500
6647000
[57] 4627600 4312000 5425200 4005000 5940800 5067600 3751500
4780600
[65] 4752000 4995000 5280800 5887600 1717800 2692800 3220000
7658700
[73] 3748000 5203800 3912000 4783500 3956000 3188000 3147600
2857700
[81] 2392000 1068600 72700 624200 896800 620100 1396000
1192000
[89] 1457400 3087000 1747200 1821300 1587600 9326400 3271500
2082500
[97] 259600 1922800 4540800 62700 3601800 3960600

#Derive their Z-scores using formula (X-mean)/(standard
deviation).
#corn production
n_mean <- mean(n);n_mean #mean
## [1] 20892069

n_sd <- sd(n);n_sd #standard deviation
## [1] 16153562

n_zscore <- ((n-n_mean)/n_sd);n_zscore #Z-score

## The sample R codes for understanding basic statistical
measures are located in Chapter3-R-Code and the dataset that
is used to implement the code is stored in Chapter3_Data_
folder. Please run these codes and continue to practice on
your own. This will set you up and get you ready to work
with R.
```

Stay One Step Ahead with Challenge Assignments

Concept: Recall in this chapter that we learned about basic concepts of statistical measures that are often used in analyzing data distributions, including descriptive statistics, spatial descriptive statistics, and random variables and probability distribution. In addition to what is covered in this book, we conduct an in-depth examination of these concepts using specific data that was generated for this purpose.

Task: Apply concepts learned in this chapter to explore and understand summarized data that was collected from an urban food environment in low- and high-income settings of Chicago. We continue to explore an earlier hypothesis stated in Chapter 2.

The basic statistical and spatial analysis was conducted using data exclusively obtained from the Chicago GIS data portal, American Community Survey, U.S Census Bureau, and food store location and operational hours website. Table 3.11 gives a summary of basic statistics, while Table 3.12 gives the demographic and food access profile. Figure 3.8 provides the key spatial statistical measures.

1. Using information from Table 3.11, compare and contrast five metrics used in identifying low- and high-income settings.
2. Critique the metrics used for identifying the two urban settings.
3. Using information from Table 3.12, rank racial/ethnic groups for low- and high-income settings in descending order. Comment briefly about the distributions of food access among the seven racial/ethnic groups.
4. Comment on the temporal food access for low- and high-income settings.
5. About one-third and one-fifth of children live within a low- and high-income setting, respectively, what implications does this have on food access? What other insights can be made about the data presented in Table 3.12?
6. What does the width of confidence interval tell us about data distributions given in both in Tables 3.11 and 3.12?
7. Interpret results in Tables 3.11 and 3.12.
8. Using information presented in Figure 3.1, describe the spatial distribution of food access.
9. Interpret results in Figure 3.8.
10. Critique the statistical and spatial measures used for summarizing these data distributions. Suggest alternative methods and strategies that could be applied to analyze these distributions.

Approach/strategy for solving task: Conduct a spatial query using relevant data layers compiled for your study area. In Table 3.11, neighborhood-level socioeconomic characteristics were used to define low- and high-income settings, grocery stores, farmers' markets, census data at census-tract level, and urban farms/gardens. This was mostly achieved using intersection spatial query between two relevant data layers. Compile results in an MS Excel spreadsheet. Summarize your findings/observations into a meaningful short report.

Possible solutions: Answers will vary in regard to statistical descriptors, distributions, and associated critiques, but this exercise is reasonably simple and straightforward to complete.

TABLE 3.11

Summary Statistics for the Five Metrics Used in Identifying Low- and High-Income Settings in the City of Chicago

Class	Variables	Mean	Standard Deviation	Class	Variables	Mean	Standard Deviation
Low	% Households below poverty	27.49	10.42	**High**	% Households below poverty	12.69	5.32
	% Aged over 25 years without high school diploma	26.22	9.77		% Aged over 25 years without high school diploma	11.06	5.55
	% Aged over 16 years and unemployed	19.12	6.66		% Aged over 16 years and unemployed	9.64	4.01
	% Aged under 18 years or over 64 years[a]	39.15	4.16		% Aged under 18 years or over 64 years[a]	30.78	7.86
	Per capita income ($)	16,826	581.05		Per capita income	38,130	2996.35

Source: Data covering a period of 2013–2014 were obtained from the City of Chicago data portal and the American Community Survey, U.S. Census.

[a] Note the minor differences between Tables 3.11 and 3.12 in % Aged under 18 years or over 64 years. The American Community Survey is based on a sample and has margins of error, but the 2010 census data in Table 3.12 are compiled from census counts.

TABLE 3.12

Demographic and Food Access Profile in Low- and High-Income Settings in the City of Chicago, Illinois

Class	Variables	Proportion (%)	Confidence Interval (CI)	Class	Variables	Proportion (%)	Confidence Interval (CI)
Low				**High**			
	Number of neighborhoods	44.90% (44/98)	35.43 to 54.75		Number of neighborhoods	55.10% (54/98)	45.25 to 64.57
	Number of grocery stores	56.92% (288/506)	52.57 to 61.17		Number of grocery stores	43.08% (218/506)	38.83 to 47.43
	Total population	48.99% (1,182,613/2,413,890)	48.93 to 49.05		Total population	51.01% (1,231,277/2,413,890)	50.95 to 51.07
	Race/ethnicity				*Race/ethnicity*		
	White	21.07% (249,210/1,182,613)	21.00 to 21.14		White	46.67% (574,695/1,231,277)	46.58 to 46.76
	African American	41.13% (486,455/1,182,613)	41.04 to 41.22		African American	18.37% (226,132/1,231,277)	18.30 to 18.44
	American Indian	0.27% (3,182/1,182,613)	0.26 to 0.28		American Indian	0.27% (3,295/1,231,277)	0.26 to 0.28
	Asian	2.15% (25,456/1,182,613)	2.12 to 2.18		Asian	3.59% (44,174/1,231,277)	3.56 to 3.59
	Hawaiian or Pacific Islander	0.03% (382/1,182,613)	0.03 to 0.03		Hawaiian or Pacific Islander	0.05% (671/1,231,277)	0.05 to 0.05
	Other	11.56% (136,751/1,182,613)	11.50 to 11.62		Other	9.65% (118,774/1,231,277)	9.60 to 9.70
	Multi-race	2.01% (23,793/1,182,613)	1.98 to 2.04		Multi-race	2.49% (30,616/1,231,277)	2.46 to 2.52
	Hispanic	21.76% (257,386/1,182,613)	21.69 to 21.83		Hispanic	18.92% (232,920/1,231,277)	18.85 to 18.99

(Continued)

TABLE 3.12 (Continued)

Demographic and Food Access Profile in Low- and High-Income Settings in the City of Chicago, Illinois

Class	Variables	Proportion (%)	Confidence Interval (CI)	Class	Variables	Proportion (%)	Confidence Interval (CI)
	Gender				*Gender*		
	Male	47.57% (562,513/1,182,613)	47.48 to 47.66		Male	48.80% (600,809/1,231,277)	48.71 to 48.89
	Female	52.43% (620,100/1,182,613)	52.34 to 52.52		Female	51.20% (630,468/1,231,277)	51.11 to 51.29
	Age				*Age*		
	Children (younger than 18 years)	30.69% (362,938/1,182,613)	30.61 to 30.77		Children (younger than 18 years)	22.11% (272,226/1,231,277)	22.04 to 22.18
	Adults (18–64 years)	59.04% (698,263/1,182,613)	58.95 to 59.13		Adults (18–64 years)	66.63% (820,432/1,231,277)	66.55 to 66.71
	Seniors (older than 64 years)	10.27% (121,412/1,182,613)	10.22 to 10.32		Seniors (older than 64 years)	11.26% (138,619/1,231,277)	11.20 to 11.32
	Operating hours				*Operating hours*		
	Range	8–24 hours			Range	10–24 hours	
	Operating more than 14 hours	7.99% (23/288)	5.38 to 11.70		Operating more than 14 hours	16.97% (37/218)	12.57 to 22.51
	Number of farmers' markets	33.33% (8/24)	17.97 to 53.29		Number of farmers' markets	66.67% (16/24)	46.71 to 82.03
	Number of urban farms/gardens	72.73% (8/11)	43.44 to 90.26		Number of urban farms/gardens	27.27% (3/11)	9.74 to 56.56

Source: Data covering a period of 2013–2014 were obtained from the City of Chicago data portal and U.S Census Bureau. Operational hours for most of the food stores were compiled from http://www.hours-locations.com/ (last accessed on May 8, 2015).

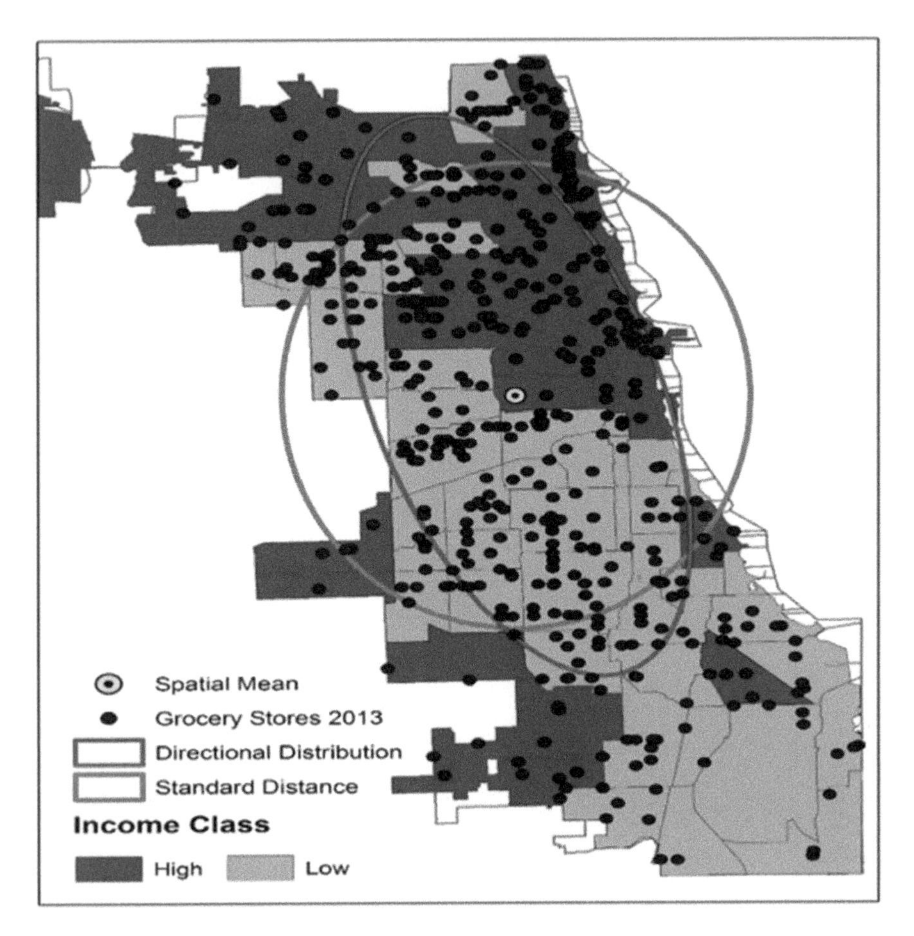

FIGURE 3.8
Spatial distribution of grocery stores in Chicago.

Review and Study Questions

1. What are descriptive statistics? Use concrete examples to illustrate their use in statistical analysis.
2. Using measures of center and spread, explain the distinction between traditional descriptive statistics and their counterparts in spatial statistics.
3. What are the benefits of exploring statistical and spatial distributions?
4. Distinguish between descriptive and inferential statistics.
5. Distinguish between exploratory and confirmatory data analysis.

Glossary of Key Terms

descriptive statistics: A useful starting point in statistical analysis. These consist of tabular, graphical, and statistical summaries that describe the general attributes of the data in a given study.

exploratory data analysis: An approach that enables a data scientist to thoroughly screen the data to uncover the underlying structure, identify outliers and anomalies, and test the key statistical assumptions prior to more advanced statistical analysis.

inferential statistics: These consist of statistical techniques that use information generated from sample data to draw conclusions about the general population. Analysis can be based on direct or point estimates of the population or could be based on indirect approaches that include confidence bands or hypothesis testing.

multivariate statistics: The detailed and simultaneous assessment of two or more variables in a database for a range of purposes including data explanation, prediction, classification, and data reduction.

univariate statistics: The detailed assessment of all cases within a single variable to describe the data distribution using measures of center, spread, relative position, and shape/normality.

***Z*-score:** This standardized score is obtained by subtracting the sample mean from the raw score and dividing the value by the sample standard deviation. It has several applications in statistical analysis including the detection of univariate outliers.

References

Besag, J. 1974. Spatial interaction and the statistical analysis of lattice systems. *Journal of the Royal Statistical Society, Series B (Methodological)* 36(2): 192–225.

Besag, J., Milne, R., and S. Zachary. 1982. Point process limits of lattice processes. *Journal of Applied Probability* 19: 210–216.

Cormack, R.M. 1977. The invariance of Cox and Lewis's statistic for the analysis of spatial patterns. *Biometrika* 64: 143–144.

David, H. A. 1998. Statistics in U.S. universities in 1933 and the establishment of the Statistical Laboratory at Iowa State. *Statistical Science* 13: 66.

Fisher, R.A. 1935. *The Design of Experiments*. Edinburgh, Scotland: Oliver and Boyd.

Matheron, G. 1963. Principles of geostatistics. *Economic Geology* 58: 1246–1266.

Mercer, W.B. and A.D. Hall. 1911. The experimental error of field trials. *Journal of Agricultural Science* 4: 107–132.

Varberg, D.E. 1963. The development of modern statistics. *The Mathematics Teacher* 56(4): 252–257.

4

Engaging in Exploratory Data Analysis, Visualization, and Hypothesis Testing

Learning Objectives

1. Explore trends in the development and use of data visualization methods.
2. Understand how to create and interpret graphical summaries.
3. Understand the uses and applications of hypothesis testing.
4. Learn how to compute and interpret tests of independent sample means.
5. Learn how to compute and interpret chi-square tests.

Introduction

In Chapters 2 and 3, we suggested that a good place to start with data analysis is to compute the descriptive measures that summarize the data distribution. In Chapter 3, we devoted coverage to statistical summaries that best describe the center, spread, shape, and relative position of the observations while also presenting the optional measures that apply to spatial data. In a similar fashion, we now explore the use of graphical summaries and data visualization methods as complementary tools in data exploration and analysis. As the familiar adage goes, "a picture is worth more than a thousand words," so becoming adept in the growing field of data visualization will significantly enrich our analytical skills. These tools will enable us to explore and visualize data in ways that would help us discern new information that would otherwise not be readily apparent when using conventional statistical tools. Data visualization methods are integral to what we might call "value-added" statistics in the sense that they enable us to go from large amounts of diverse forms of data to analyze, synthesize, and graphically display meaningful information with the expectations of possibly constructing and conveying new knowledge for use in decision-making. These methods are effective in exploring differences between phenomena, identifying expected as well as unexpected patterns, detecting clusters, revealing new relationships, and more. Drawing from several areas including spatial data mining, machine learning, geographic information systems (GISs), and cognitive science are many approaches in data visualization with applications in several domains. For example, we can use visualization tools and methods to simulate various real-world environments where users can test different scenarios; provide exploratory functions; practice/provide a real-world environment/experience; represent two-dimensional (2D) and three-dimensional (3D) environments; show spatial relationships; model different scenarios, for example, an urban environment; integrate real-time applications (wearable computers) with virtual environments, enable real-time applications, provide timely information/updates; support landscape viewing and drafting; engage human visual systems; and support the formulation of study hypotheses. The visualization community has also focused on developing visualization algorithms, tools, methods, and strategies, such as the social network analysis method, which is currently used for visualizing online social networks (Hoff et al. 2002; Heer and Boyd 2005; Perer and Shneiderman 2006; Luo et al. 2011; Luo and MacEachren 2014).

Given the cognitive and inherently subjective nature of synthesizing and interpreting the graphical displays, it is often best to validate the visual findings through hypothesis testing. There are also times when the results derived from hypothesis testing and statistical validation are best depicted through visual plots, charts, graphs, and maps to communicate the findings to the intended audience. As such, the processes of data exploration and visualization are closely aligned with hypothesis testing methods, a linkage that forms an integral part of spatial analysis and one that is clearly recognized and valued by geographers. Our plan in this chapter therefore is twofold. First, we explore the emerging field of data visualization and the contributory role of cartography and GISs in the development of these tools. This discussion is accompanied by examples of how standard plots are derived and the interpretation of the derived images. The second half of the chapter is devoted to the key steps in hypothesis testing. For hypothesis testing, our focus is on student's *t*-test and chi-square (χ^2) statistics, which are among the most commonly used significance tests. The examples presented in the chapter are foundational, with the primary goal of introducing the reader to the core concepts and tools in data visualization. Thereafter, in subsequent chapters of the book, we share examples that entail the use of more advanced visualization tools and statistical validation methods.

Exploratory Data Analysis, Geovisualization, and Data Visualization Methods

Data visualization, geovisualization, visual analytics, and exploratory data analysis (EDA) are all part of a growing domain of data-rich analytical, graphical, and interactive methods that are now available for screening, exploring, and synthesizing information. In the era of big data, these approaches are increasingly capable of converting diverse, dynamic, and complex forms of data into valuable information, and presenting this information in a comprehensible format that is beneficial to end users. A survey of the emerging literature on EDA, geovisualization, and data visualization may lead one to believe that these are three disparate fields with different end goals. However, scrutiny of the embedded tools and applications reveals many similarities in the core goals and objectives. These include the following: (1) data representation, (2) feature exploration and identification, (3) pattern recognition, (4) human–computer interaction, (5) knowledge construction and storage, and (6) effective communication and transmission of knowledge. These commonalities are elaborated on in the ensuing sections.

Data Visualization

The term *data visualization* is a relatively new and encompassing term for all visualization methods that are currently in use, even as more techniques are being developed. In an earlier article by Lengler and Eppler (2007), a visualization method was appropriately defined as "a systematic, rule-based, external, permanent and graphical representation that depicts information in a way that is conducive to acquiring insights or communicating experiences" (p. 1). This definition adequately captures the analytical goals noted in the preceding section, including the need for representation, knowledge acquisition, and effective communication. In the same article, the authors compiled a comprehensive listing of more than 100 visualization methods with the intention of pooling together the multiple streams of analytical procedures that are being developed in several areas. Calling this listing of methods a "periodic table of visualization methods," the authors readily acknowledged that data visualization draws from several disciplines, including statistics, human-computer interaction, cartography, graphic design, and architecture.

Geographic Visualization

Although the foundational role of cartography in data visualization was not explicitly recognized in the Lengler and Eppler (2007) study, several other studies have effectively outlined the valuable contributions of this field and geography as a whole in the development of these methods. For example, Nöllenburg (2007) explored the driving forces in visualization noting that geographic visualization has played an important role in human history well before the advent of the computer. Likewise, a seminal article written earlier by MacEachren and Kraak (2001) outlined the role of geovisualization techniques, the research prospects, and challenges that lay ahead. Drawing from their work on the International Cartographic Association's (ICA) Commission on Visualization and Virtual Environments, they defined geovisualization as an integration of "approaches from visualization in scientific computing (ViSC), cartography, image analysis, information visualization, exploratory data analysis (EDA), and geographic information systems (GISystems) to provide theory, methods and tools for visual exploration, analysis, synthesis and presentation of geospatial data" (p. 1.). Four themes and related challenges were cited in this article as relevant in the development of geovisualization tools: (1) *representation* of geospatial data, (2) *integration* of visuals with computational methods, (3) *interface design* for geovisualization environments, and (4) the *cognitive/usability* aspects (MacEachren and Kraak 2001).

Within the last decade, several studies have highlighted the foundational role of GIS and cartography in visual analytics, including the increasing role of interactive spatial mapping (Jacquez et al. 2005), spatiotemporal analysis (Andrienko et al. 2010), and visualization of spatial data (in R using ggmap by Kahle and Wickham 2013). Several advocates have appealed for ongoing research to expand the range of visual analytic GIS tools that are accessible to both amateur and professional data scientists with a core set of features and capabilities to handle large, complex spatial and temporal data (Guo 2007; Andrienko et al. 2010). Efforts are also underway to create multidisciplinary teams to address key challenges such as the following: (1) to achieve scalability of geovisual tools to handle the data size, variety, dimensionality, and synergistic linkages; (2) to promote interoperability and consistency in semantics, semiotics, and use interactions; (3) to advance visualization of complex spatial and temporal dimensions; (4) to be able to seamlessly link data exploration with validation; and (5) to provide ongoing support for knowledge capture and manipulation (Andrienko et al. 2007). These studies also emphasize the need for cognitive features that are required to ensure that the end users can decipher the implicit knowledge embedded in these visuals, while using the information to stimulate the generation of new ideas.

Two important visualization concepts that influence how graphical methods are applied to accomplish visualization tasks are expressiveness and effectiveness (Oyana et al. 2011). Expressiveness is defined as the graphical methods used to convey meaning without leaving out any facts or unintentionally adding or implying facts. Effectiveness measures how well the selected graphical method conveys meaning relative to other methods. Also, according to a popular MacEachren's 3D cartographic-visualization conceptual model, three major components guide the geovisualization process: private visual thinking, levels of interaction, and public human communication (MacEachren and Kraak 1997). Private visual thinking normally refers to situations where visualization scientists explore their own data. And when the results are effectively communicated to the public using well-designed maps or charts, then we can describe this as a public visual communication. Both private and public visual thinking processes have different levels of interaction. In private visual thinking, for example, the level of interaction is normally high because of the nature of data exploration and knowledge discovery process, while in public visual communication the level of interaction is low.

In summary, the process of geovisualization entails aspects of human cognition, communication, and formalism linked by interactive visualization. Data exploration tasks involve making sense of the unknown through visualization techniques. During the data exploration phase, there are high levels of interaction and engagement, and when the known is determined and effectively communicated to a wide audience, we can determine whether the knowledge or information was successfully conveyed. The geovisualization process activities involve effective encoding and decoding of data or information through a scientific process that requires a solid understanding of human cognition.

New Stunning Visualization Tools and Infographics

The use of dashboards and infographics to communicate data and information have been popularized in this unprecedented period of big data analytics, data science, and visual analytics tools and algorithms. Incredible visuals representing huge amounts of data and information are now commonplace. For example, authors of infographics use it to deliver valuable information; to improve learners' ability to discern data and knowledge about society, behavioral sciences, and any other disciplines; to study patterns and trends; or to effectively present their narratives/stories.

Exploratory Approaches for Visualizing Spatial Datasets

Even as new and high-level visualization techniques are rolled out, it is important to have a foundational knowledge of the traditional approaches that are used to graphically summarize data. These belong to a suite of applications that are classified as EDA. The philosophy in employing EDA methods is to maximize insights into a dataset, search for fundamental clues about a dataset, and uncover the hidden structure underlying a dataset. These techniques provide the analytical means by which useful aspects of a dataset can be presented in an understandable format. Pioneering work in EDA was completed by Tukey (1977), and since then, there have been significant contributions and improvements in the exploration and presentation of a dataset. EDA methods also provide the means through which we can learn about potential relationships and/or differences among groups of observations in the data, and then formulate a study hypothesis (Tukey 1977; Chambers et al. 1983; Tufte 1983).

In visualizing a spatial dataset, the statistician has to determine the appropriate mark (select visual variables or decide on the best combination of visual elements that effectively depict the dataset) that will encode the data, size and scale, and the dimensions to be explored. It should be noted that simpler graphs offer a higher ability than complex graphs to effectively communicate information to a wide audience. Complex graphs pose a number of challenging problems. For example, a large number of visual elements may be used, which create clutter. Also, they could simply compromise computing performance once the limits of the viewing platform are reached.

A variety of EDA techniques are available for exploring data, ranging from plotting raw data to presenting them in a format that maximizes the natural pattern recognition that matches the viewer's abilities. Commonly used graphs include histograms, pie charts, bar charts, line graphs, boxplots, scatterplots, and maps. Throughout the text, we showcase examples of these different visualization methods, focusing on the more advanced approaches. For now, however, we begin with some of the commonly used forms of visualization in data analysis: histograms, boxplots, scatterplots, and matrices. We explore the use of parallel coordinate plots (PCPs) as a high multidimensional visualization tool. We also use several other graphics to explore and present a number of spatiotemporal datasets. This will help us to learn and further deepen our knowledge on visualization concepts and methods.

Histogram: One of the simplest means of generating a graphical summary is by plotting the frequency distribution of a single variable (univariate) that is measured on an interval scale. It reveals the center of the data (mean, median, and mode), the spread of the data (dispersion or unevenness), the shape and distribution (skewness) of the data, and evidence of potential outliers in the data. A probability or a goodness-of-fit test curve can also be drawn on a histogram to verify the distributional model. Figure 4.1 depicts the distributional patterns of obesity rates observed within the counties in the states of New York and Mississippi. On the left panel is a histogram showing the distribution for New York and the normal probability curve. There appears to be a slight negative skew, a leptokurtic pattern with a few outliers to the left of the distribution. On the right panel is the histogram for the counties in Mississippi. The distribution is mesokurtic with observations trending toward a more normal distribution. Overall, on average, obesity rates are almost 10% higher in Mississippi than in New York.

Boxplot: Another visual approach that is useful for summarizing a set of observations measured on an interval scale is the boxplot. The boxplot shows the shape of the distribution, the center of the data, and its dispersion. It is

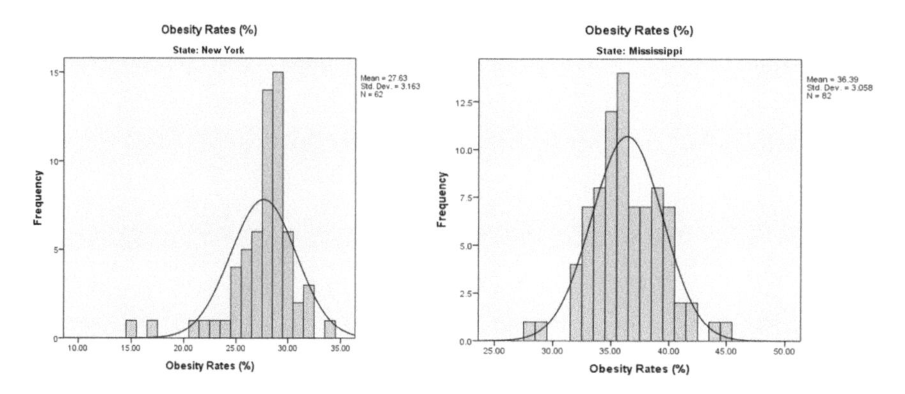

FIGURE 4.1
The distribution of obesity rates in New York and Mississippi.

sometimes called a five-number graphical summary, because the diagram specifically captures five statistical measures: the minimum and maximum values (range), lower and upper interquartiles, and the median. This plot can be used to indicate whether the distribution is skewed or not, and the presence of outliers. Plotting the distribution for two or more groups allows for a comparative assessment of the observed patterns. For example, Figure 4.2 shows the distribution of obesity rates (percent) across counties within the two states of New York and Mississippi. There is slightly more variability in New York, and the results confirm the presence of several extreme scores (denoted as circles) and outliers (stars) in New York State. Specifically, two counties (New York and West Chester) are outliers with significantly lower levels of obesity (below three standard deviations of the mean). Three other counties have lower levels (Tompkins, Queens, and Nassau), but they are not considered to be outliers in the distribution. It may be worth comparing these results to the preceding histograms to see which of the two plots is most effective in communicating the findings. The observations can be confirmed by reviewing the descriptive measures shown in Table 4.1.

Scatterplot: This is a visual representation that shows the direction and strength of a relationship between the two variables (the dependent Y against independent X). Specifically, the scatterplot explores whether the values of Y vary systematically with the corresponding values of X. The plot can be based on raw scores obtained for the two variables, or it can be based on residuals obtained after fitting a regression model. The patterns reveal statistical relationships or associations between two variables that manifest themselves by any nonrandom structure in the plot. Y is plotted on the *vertical axis* of the graph and represents the dependent/response variable, whereas X is plotted on the *horizontal axis* of the graph and represents the independent/predictor variable. The plot can also serve as a useful diagnostic tool for assessing causal associations between variables. If

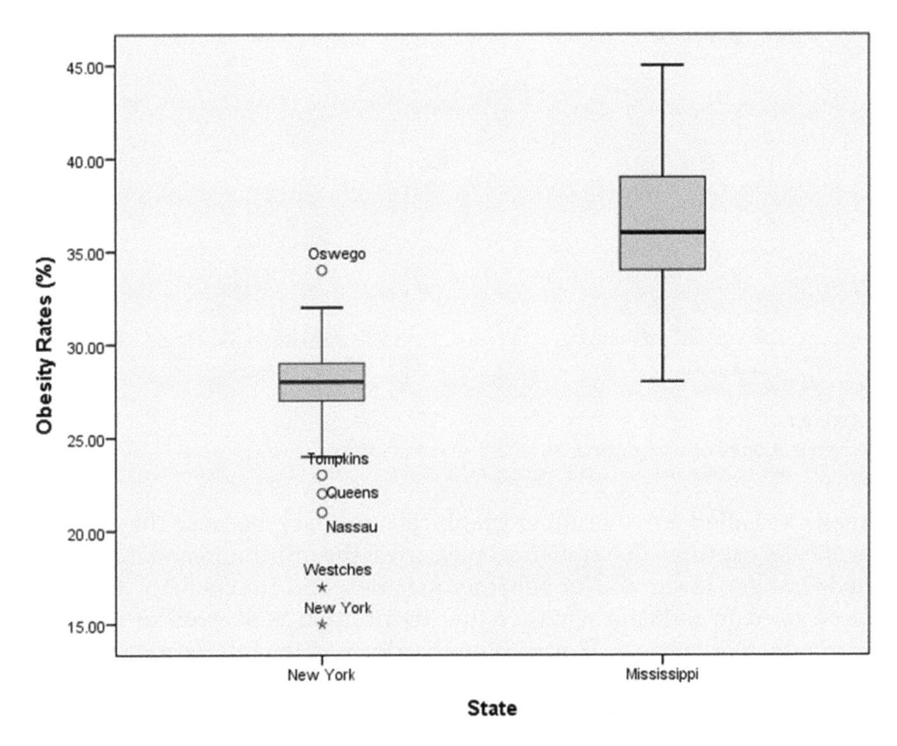

FIGURE 4.2
Obesity rates within the two states of New York and Mississippi.

a strong association exists in the data, then it suggests an underlying cause-and-effect mechanism. However, because this plot does not necessarily confirm the presence of a cause-and-effect relationship, it is still incumbent on the statistician to draw knowledge from the underlying science to determine whether there is causality or not.

To visually explore the relationships among several pairs of variables, the best approach to use is a scatterplot matrix. In a single display, the scatterplot matrix can depict the relationships among all possible pairs of variables selected for analysis. As with a typical correlation matrix, the scatterplot matrix is symmetrical with the same number of rows and columns as there are variables. The on-diagonal elements are blank and not reported because a variable's relationship with itself is one. Scanning across the lower half (or upper half) of the matrix, however, one is able to assess the direction (whether it is positive or negative), and the potential strength of the observed relationship among the variables. It is also possible to identify potential outliers (bivariate) in the scatterplot matrix. Data points that are farthest away from other points in the distribution could well be extreme cases or outliers and are, therefore, worthy of further examination. Finally, the significance of the observed relationships can be validated using correlation or regression analysis.

TABLE 4.1

Descriptive Statistics for Health Risk Behaviors in New York and Mississippi

	Descriptive Statistics						
	N	Mean	Standard Deviation	Skewness		Kurtosis	
New York Counties	**Statistic**	**Statistic**	**Statistic**	**Statistic**	**Standard Error**	**Statistic**	**Standard Error**
Pct_Obese	62	27.6290	3.16349	−1.724	0.304	5.017	0.599
Pct_Smoker	61	20.4426	6.92225	−1.013	0.306	1.919	0.604
Pct_Inactive	62	25.3871	3.02099	−0.366	0.304	0.665	0.599
Pct_ExcessiveDrinking	62	16.8387	5.21408	−1.303	0.304	3.555	0.599
Pct_Unemployed	62	8.5484	1.12785	0.801	0.304	2.195	0.599
Pct_LimitAccessHlthFood	62	6.8548	7.27188	0.811	0.304	−0.386	0.599

	Descriptive Statistics						
	N	Mean	Standard Deviation	Skewness		Kurtosis	
Mississippi Counties	**Statistic**	**Statistic**	**Statistic**	**Statistic**	**Standard Error**	**Statistic**	**Standard Error**
Pct_Obese	82	36.3902	3.05810	0.169	0.266	0.533	0.526
Pct_Smoker	82	22.5488	6.40053	−2.023	0.266	5.576	0.526
Pct_Inactive	82	34.0854	3.17864	−0.342	0.266	−0.156	0.526
Pct_ExcessiveDrinking	82	10.1829	3.15887	−0.815	0.266	1.357	0.526
Pct_Unemployed	82	12.0512	2.62294	0.702	0.266	0.928	0.526
Pct_LimitAccessHlthFood	82	10.5976	11.62358	1.463	0.266	1.331	0.526

TASK 4.1 INTERPRETING SCATTERPLOT MATRICES

Figure 4.3 depicts the scatterplot matrix derived by exploring healthy behaviors and obesity rates within counties in New York and Mississippi. Six variables are included in the analysis: percent obese, percent residents with limited access to healthy foods, percent residents who are inactive, percent smokers, percent residents who drink excessively, and percent unemployed. Examine the plots depicted in this matrix, and explain the observed strength and direction between obesity and the other variables. What relationship appears to be the strongest and the weakest?

Using the latter half of the matrix, obesity rates appear to be most strongly associated with the percent of inactive residents in these counties. The relationship appears to be positive, meaning that counties that have a higher proportion of inactive residents are likely to have

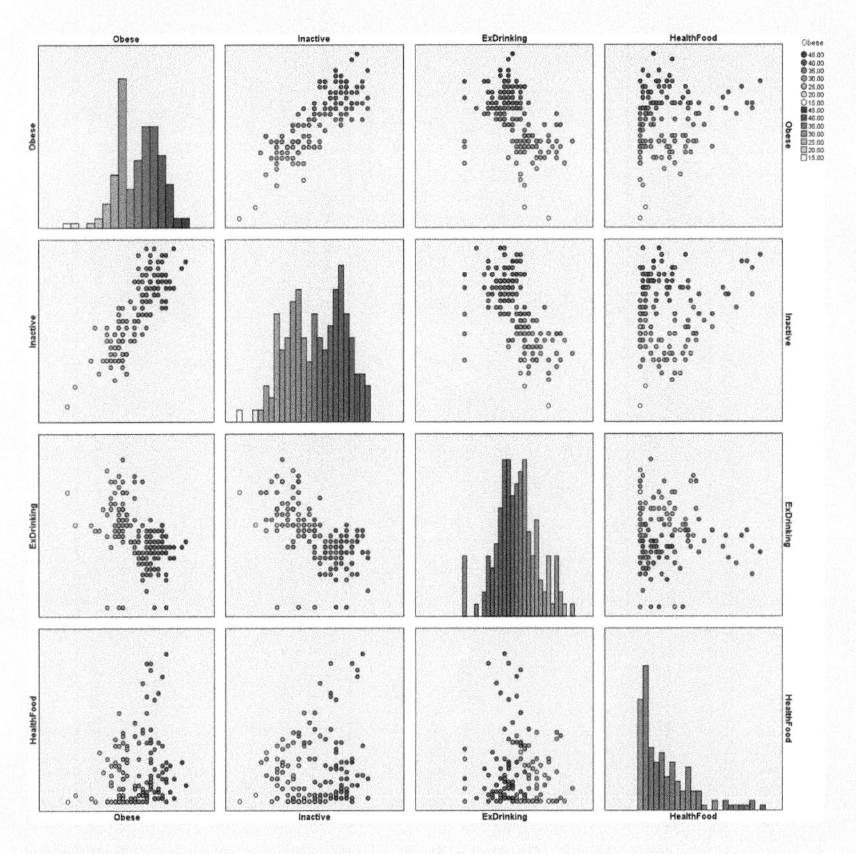

FIGURE 4.3
A scatterplot matrix depicting obesity rates and health risk behaviors in New York and Mississippi.

higher obesity rates. Similarly, there appears to be a strong positive association between obesity rates and unemployment rates in the two states. The relationship between smoking rates and the other variables appears to be the weakest, and the direction of these relationships is unclear in the plots. To validate these visual patterns observed in the matrix, one would need to formulate statistical hypotheses and test these using Pearson's correlation test.

Parallel Coordinate Plots: Another means of visually exploring data is by using a PCP. The PCP was made popular in data-mining research and exploration by Inselberg (1985) and has since become a commonly used application in several domains including remote sensing, hazard assessment, climate change modeling, and spatial epidemiology (Inselberg and Dimsdale 1990; Edsall 2003; Huh and Park 2008; Inselberg 2009; Ge et al. 2009). PCP is most effective when examining the multidimensional attributes and relationships within large continuous datasets, though it can also be applied to categorical data. Among the several touted benefits of using this visualization technique are its ability to represent complex spatial and spatiotemporal data (Edsall 2003), its interactive nature and uniform treatment of multiple dimensions (Siirtola and Raiha 2006), its conceptual simplicity and compact appearance (Huh and Park 2008), and its ability to visualize uncertainty and potential outliers in a large dataset (Ge et al. 2009).

TASK 4.2 CONSTRUCTING AND INTERPRETING RESULTS FROM PARALLEL COORDINATE PLOTS

Table 4.2 presents housing data drawn from the U.S. Census. The decadal data spans from 1900 to 2010 for all states including Washington, DC. When reviewing the boxplots for various years, one finds that there are subtle differences observed between the various times with some trending toward more normal distributions, whereas others depict relatively skewed distributions. Given the temporal sequencing of this data, another useful and graphically compact way of depicting the trends is by using the PCP.

There are 12 dimensions (variables) that need to be displayed, with each dimension representing a snapshot of home ownership patterns observed during the decennial census. The PCP is most valuable when the dimensions are continuous scaled variables (interval/ratio) as in the case of the home ownership data. Also, having the same units of measurement for all variables (such as percentages in the housing

TABLE 4.2

U.S. Housing Data from 1900 to 2010, Census Bureau

FIPS	State	1900	1910	1920	1930	1940	1950	1960	1970	1980	1990	2000	2010
1	Alabama	34.4	35.1	35	34.2	33.6	49.4	59.7	66.7	70.1	70.5	72.5	74.1
4	Arizona	57.5	49.2	42.8	44.8	47.9	56.4	63.9	65.3	68.3	64.2	68	68.9
5	Alaska	47.7	46.6	45.1	40.1	39.7	54.5	61.4	66.7	70.5	69.6	69.4	68.5
6	California	46.3	49.5	43.7	46.1	43.4	54.3	58.4	54.9	55.9	55.6	56.9	57
8	Colorado	46.6	51.5	51.6	50.7	46.3	58.1	63.8	63.4	64.5	62.2	67.3	68.4
9	Connecticut	39	37.3	37.6	44.5	40.5	51.1	61.9	62.5	63.9	65.6	66.8	70.5
10	Delaware	36.3	40.7	44.7	52.1	47.1	58.9	66.9	68	69.1	70.2	72.3	76.5
11	Washington, DC	46.8	44.2	42.5	42	43.6	57.6	67.5	68.6	68.3	67.2	70.1	70.9
12	Florida	30.6	30.5	30.9	30.6	30.8	46.5	56.2	61.1	65	64.9	67.5	67.4
13	Georgia	0	0	0	0	0	33	41.1	46.9	51.7	53.9	56.5	59.5
16	Idaho	71.6	68.1	60.9	57	57.9	65.5	70.5	70.1	72	70.1	72.4	75.5
17	Illinois	45	44.1	43.8	46.5	40.3	50.1	57.8	59.4	62.6	64.2	67.3	69.1
18	Indiana	56.1	54.8	54.8	57.3	53.1	65.5	71.1	71.7	71.7	70.2	71.4	72
19	Iowa	60.5	58.4	58.1	54.7	51.5	63.4	69.1	71.7	71.8	70	72.3	72.4
20	Kansas	59.1	59.1	56.9	56	51	63.9	68.9	69.1	70.2	67.9	69.2	67.4
21	Kentucky	51.5	51.6	51.6	51.3	48	58.7	64.3	66.9	70	69.6	70.8	71.2
22	Louisiana	31.4	32.2	33.7	35	36.9	50.3	59	63.1	65.5	65.9	67.9	71.9
23	Maine	64.8	62.5	59.6	61.7	57.3	62.8	66.5	70.1	70.9	70.5	71.6	74
24	Maryland	40	44	49.9	55.2	47.4	56.3	64.5	58.8	62	65	67.7	69.6
25	Massachusetts	35	33.1	34.8	43.5	38.1	47.9	55.9	57.5	57.5	59.3	61.7	65.1
26	Michigan	62.3	61.7	58.9	59	55.4	67.5	74.4	74.4	72.7	71	73.8	74.5
27	Minnesota	63.5	61.9	60.7	58.9	55.2	66.4	72.1	71.5	71.7	71.8	74.6	72.9
28	Mississippi	34.5	34	34	32.5	33.3	47.8	57.7	66.3	71	71.5	72.3	75.5

(Continued)

TABLE 4.2 (Continued)

U.S. Housing Data from 1900 to 2010, Census Bureau

FIPS	State	1900	1910	1920	1930	1940	1950	1960	1970	1980	1990	2000	2010
29	Missouri	50.9	51.1	49.5	49.9	44.3	57.7	64.3	67.2	69.6	68.8	70.3	72
30	Montana	56.6	60	60.5	54.5	52	60.3	64	65.7	68.6	67.3	69.1	70.7
31	Nebraska	56.8	59.1	57.4	54.3	47.1	60.6	64.8	66.4	68.4	66.5	67.4	70.2
32	Nevada	66.2	53.4	47.6	47.1	46.1	48.7	56.3	58.5	59.6	54.8	60.9	62.4
33	New Hampshire	53.9	51.2	49.8	55	51.7	58.1	65.1	68.2	67.6	68.2	69.7	76
34	New Jersey	34.3	35	38.3	48.4	39.4	53.1	61.3	60.9	62	64.9	65.6	65.9
35	New Mexico	68.5	70.6	59.4	57.4	57.3	58.8	65.3	66.4	68.1	67.4	70	69.1
36	New York	33.2	31	30.7	37.1	30.3	37.9	44.8	47.3	48.6	52.2	53	54.4
37	North Carolina	46.6	47.3	47.4	44.5	42.4	53.3	60.1	65.4	68.4	68	69.4	70.1
38	North Dakota	80	75.7	65.3	58.6	49.8	66.2	68.4	68.4	68.7	65.6	66.6	65.7
39	Ohio	52.5	51.3	51.6	54.4	50	61.1	67.4	67.7	68.4	67.5	69.1	69.7
40	Oklahoma	54.2	45.4	45.5	41.3	42.8	60	67	69.2	70.7	68.1	68.4	69.6
41	Oregon	58.7	60.1	54.8	59.1	55.4	65.3	69.3	66.1	65.1	63.1	64.3	68.2
42	Pennsylvania	41.2	41.6	45.2	54.4	45.9	59.7	68.3	68.8	69.9	70.6	71.3	72.2
44	Rhode Island	28.6	28.3	31.1	41.2	37.4	45.3	54.5	57.9	58.8	59.5	60	62.9
45	South Carolina	30.6	30.8	32.2	30.9	30.6	45.1	57.3	66.1	70.2	69.8	72.2	74.4
46	South Dakota	71.2	68.2	61.5	53.1	45	62.2	67.2	69.6	69.3	66.1	68.2	69.6
47	Tennessee	46.3	47	47.7	46.2	44.1	56.5	63.7	66.7	68.6	68	69.9	71.1
48	Texas	46.5	45.1	42.8	41.7	42.8	56.7	64.8	64.7	64.3	60.9	63.8	65.4
49	Utah	67.8	64.8	60	60.9	61.1	65.3	71.7	69.3	70.7	68.1	71.5	74.1
50	Vermont	60.4	58.5	57.5	59.8	55.9	61.3	66	69.1	68.7	69	70.6	74.3
51	Virginia	48.8	51.5	51.1	52.4	48.9	55.1	61.3	62	65.6	66.3	68.1	69.7
53	Washington	54.5	57.3	54.7	59.4	57	65	68.5	66.8	65.6	62.6	64.6	65.5
54	West Virginia	54.6	49.5	46.8	45.9	43.7	55	64.3	68.9	73.6	74.1	75.2	78.7
55	Wisconsin	66.4	64.6	63.6	63.2	54.4	63.5	68.6	69.1	68.2	66.7	68.4	70.4
56	Wyoming	55.2	54.5	51.9	48.3	48.6	54	62.2	66.4	69.2	67.8	70	73.8

data) makes it easier to display these dimensions, though this is not a requirement to run the procedure.

To create the PCP, each dimension will be portrayed on a vertical axis. For the housing data, as there are 12 dimensions (variables), there will be 12 vertical axes that are parallel to one another (see Figure 4.4). The spacing between these vertical lines has to be consistent, though many studies have devised ways to enhance this spacing to improve the interpretability of the plot. Next, the data points (or coordinates) on each axis should be plotted to represent the measurements taken for each unit of analysis in the dataset. So for this dataset, one would plot the home ownership pattern for each state as observed in 1900, 1910, 1920, through 2010. Finally, horizontal lines are used to connect these data points across the vertical axis to produce the PCP. Figure 4.4 depicts the line segments for each state plotted across the 12 dimensions (years). Each data item (or unit of analysis) has been graphed across multiple dimensions by using a single line of connecting segments that is called a *polyline*. In Figure 4.4, each polyline has been color coded to improve interpretability. Table 4.3 shows the statistical distribution of housing data, and Figure 4.5 shows a single-dimension PCP plot of housing data.

The plot shows greater similarity in home ownership patterns in the early years, with more variability between the states in the latter years (Figure 4.5). It is safe to assume, therefore, that there is a higher correlation in home ownership patterns during the initial years of data

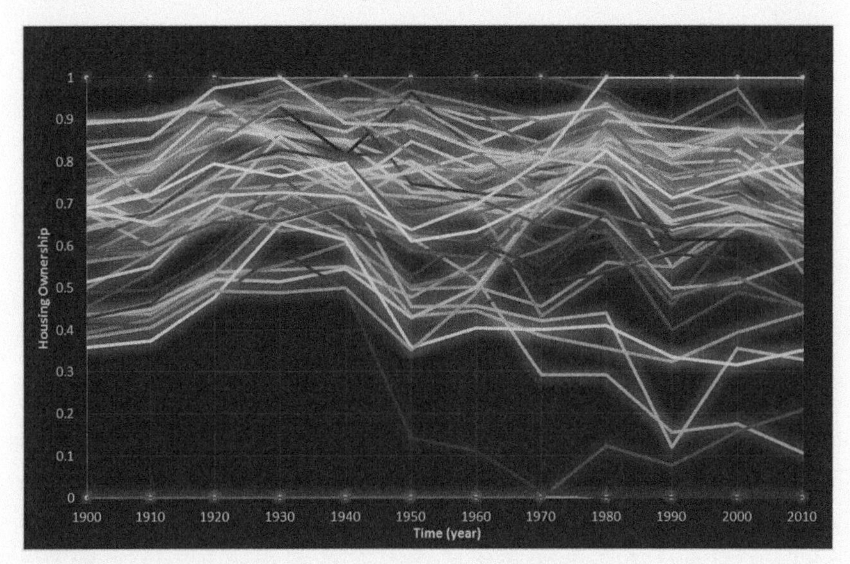

FIGURE 4.4
Parallel coordinate plot showing multidimensional distribution of housing data 1900 to 2010.

TABLE 4.3

Statistical Distribution of Housing Ownership between 1900 and 2010

Year	Observations (n)	Minimum (%)	Maximum (%)	Mean (%)	Standard Deviation (%)
1900	49	0	80.00	49.90	14.52
1910	49	0	75.70	49.04	13.75
1920	49	0	65.30	47.66	11.93
1930	49	0	63.20	48.42	11.13
1940	49	0	61.10	45.35	10.16
1950	49	33.00	67.50	56.57	7.52
1960	49	41.10	74.40	63.45	6.37
1970	49	46.90	74.40	65.26	5.60
1980	49	48.60	73.60	66.80	5.29
1990	49	52.20	74.10	66.18	4.80
2000	49	53.00	75.20	68.12	4.62
2010	49	54.40	78.70	69.77	4.84

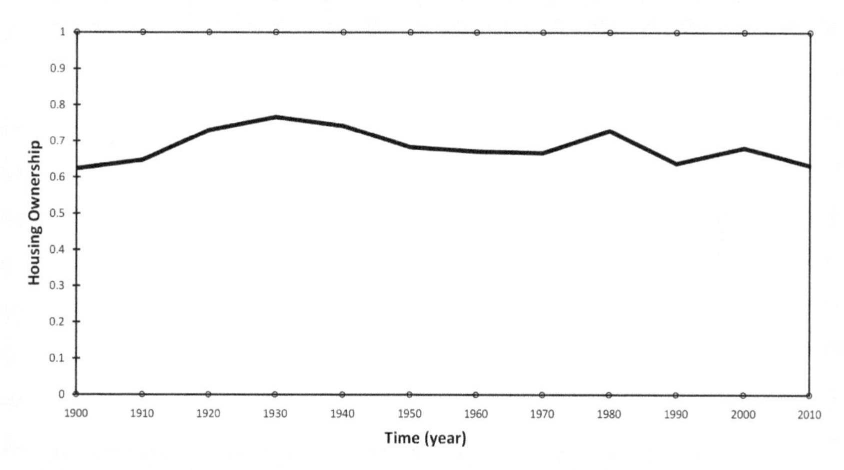

FIGURE 4.5

Parallel coordinate plot showing a single dimension of housing ownership data.

compilation (1900s through 1940s), and the strength of this association gets progressively weaker in the latter years (1980s through 2010). There are ongoing efforts to enhance the PCP, including the use of approaches such as data brushing, strumming, color customization, and classification methods (Edsall 2003), changing the spacing to detect clusters among variables, and using smoother curves rather than straight-line segments to connect the axes (Huh and Park 2008). Other important features to consider using PCPs and other data visualization tools include algorithms that enhance the human-computer interaction. Interactive tools that give the user direct control and ability to query, store, update, analyze, and present the data are often the most effective. Siirtola and Raiha (2006) rightfully contend that this active interaction and manipulation is what facilitates discovery and the construction of new knowledge. Drawing from the previous work of Shneiderman (1996, 1997), these scholars present seven critical features to have in PCPs and other data visualization tools:

1. *Overview*: The ability to gain an overview of a large dataset.
2. *Zoom*: The ability to zoom in on key areas of interest.
3. *Delete/Filter/Mask*: The ability to delete, filter, or mask uninteresting items from the collection of data points.
4. *Data brushing/getting details on demand*: Querying the data and highlighting an item or group of observations for further scrutiny, or for comparisons with other observations in the distribution.
5. *Relate*: The ability to view and understand relationships among items or multiple dimensions in the data.
6. *History*: The ability to save the history of actions taken during the analysis and to conduct progressive refinements by undoing, replaying, and modifying those actions.
7. *Extract*: The ability to extract a subgroup of information for more detailed analysis.

Visualizing Multidimensional Datasets: An Illustration Based on U.S. Educational Achievements Rates, 1970–2012

A series of visualization procedures including those described previously were applied to the U.S. education dataset compiled from the U.S. Census Bureau (1970, 1980, 1990, and 2000) and the 2006–2010, 2007–2011, and 2008–2012 data derived from the American Community Surveys. The multidimensional and multitemporal datasets were processed using different techniques and algorithms, and the results are summarized in Figures 4.6 through 4.10. Figure 4.6 shows the regionalized distribution of educational levels (9th grade and higher) using a regionalization algorithm. Table 4.4 shows a statistical distribution of education levels.

Figure 4.7 shows the regionalization of educational levels based on a ranking process of the percent of individuals who have attained college education or higher between the period covering 1970 and 2010. Figures 4.8 and 4.9 show

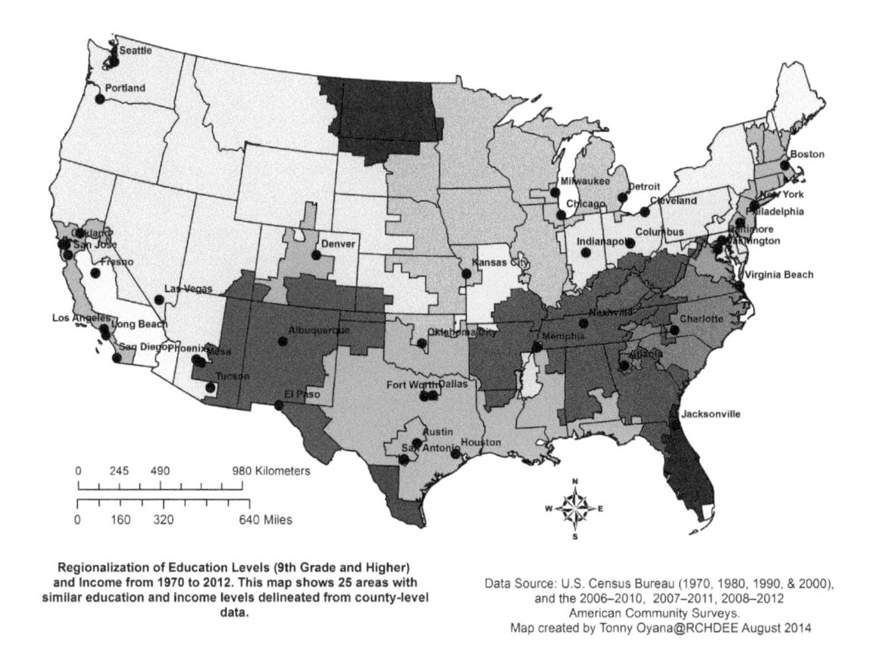

FIGURE 4.6
Regionalization of educational levels.

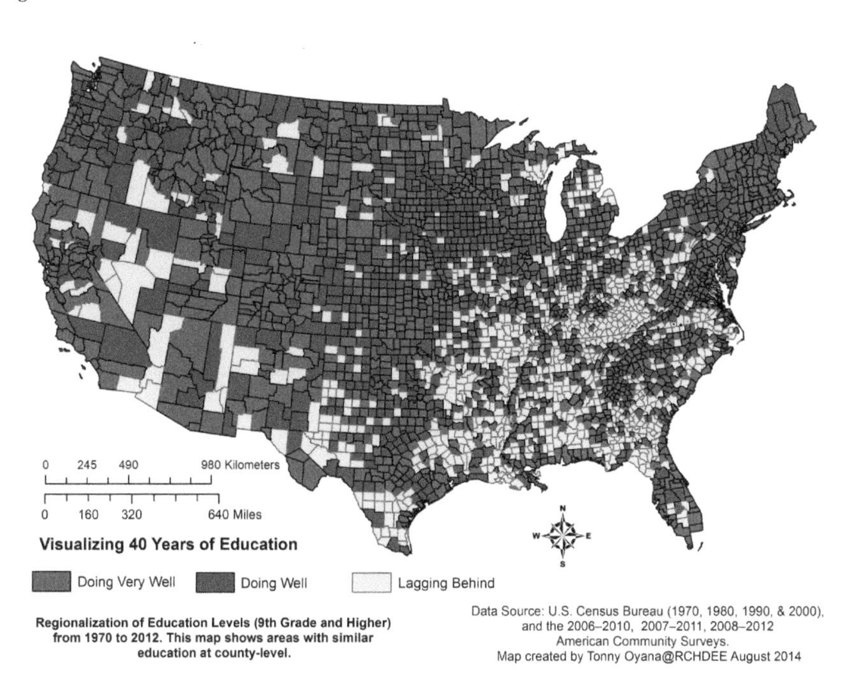

FIGURE 4.7
Regionalization of educational levels using a ranking statistical procedure.

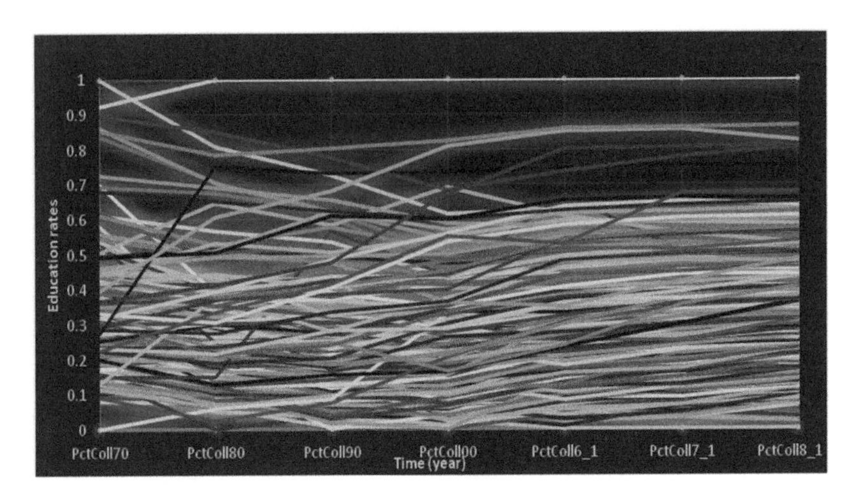

FIGURE 4.8
Parallel coordinate plot showing multidimensional distribution of educational achievement rates for 3108 counties.

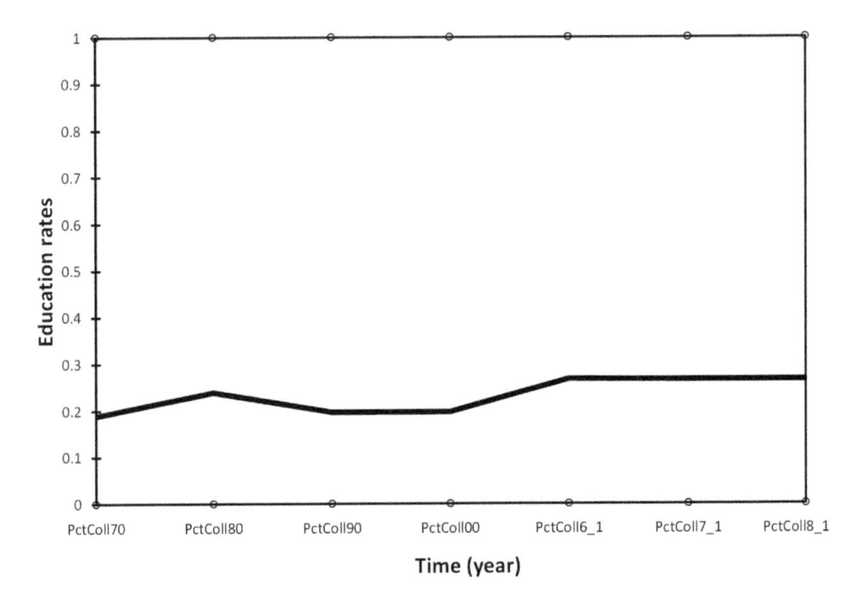

FIGURE 4.9
Parallel coordinate plot showing a single dimension of educational achievement rates.

a representation of educational achievement rates using single and multiple PCP dimensions at the county level. Figure 4.10 shows a visual exploration of local relationships between poverty and educational achievement variables using a local entropy algorithm. The different charts and figures illustrate how one can use the visualization process to gain fundamental insights on educational achievement rates during the lengthy study period.

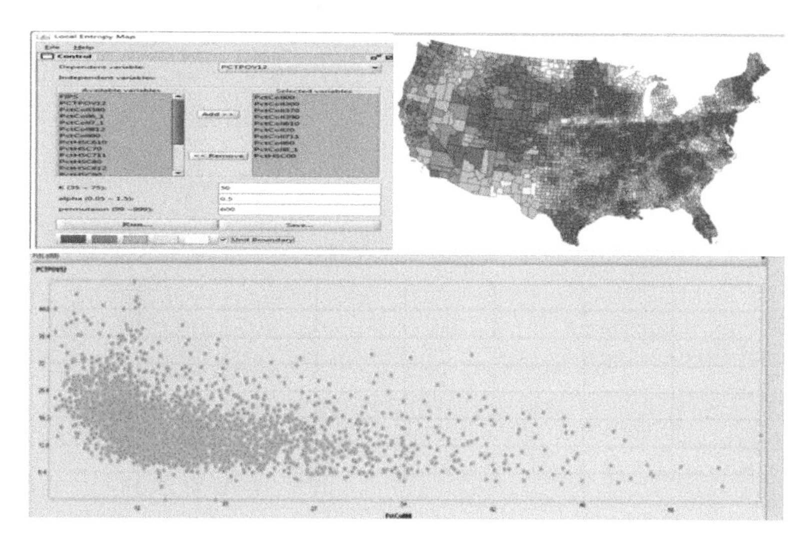

FIGURE 4.10
A screenshot shows a visual exploration of local relationships between poverty and educational achievement variables using local entropy algorithm.

TABLE 4.4

Statistical Distribution of Educational Achievement Rates between 1970 and 2012

Year	Observations (n)	Minimum (%)	Maximum (%)	Mean (%)	Standard Deviation (%)
PctColl1970	3108	0.00	38.60	7.29	3.95
PctColl1980	3108	0.00	47.80	11.43	5.44
PctColl1990	3108	3.70	53.40	13.48	6.57
PctColl2000	3108	4.90	63.70	16.50	7.80
PctColl06_10	3108	0.00	70.96	18.99	8.67
PctColl07_10	3108	0.00	72.00	19.21	8.72
PctColl08_12	3108	0.00	72.79	19.43	8.76

Finally, as noted in earlier sections of this chapter, the conclusions drawn from all of these graphical displays (histograms, boxplots, scatterplots, PCPs, or otherwise) must be validated using statistical significance tests. We discuss these statistical validation processes in the following sections.

We can explore the Annual USA Flu Trend Dataset from Google using this sample R-Code as follows:

Step I: Data Exploration and Visualization Using R

```
#Type each command in sequence and on its own line. Cut the
entire code and type Enter button and wait for a response
```

```
before typing in the next one. Ensure that you have reviewed
what is on screen before typing in the next one.
setwd("../ChapterData/Chapter4_Data_Folder")
USAFludata=read.csv('USAFludata.csv',header=TRUE, sep=",")
str(USAFludata)
plot(USAFludata)
```

Step II: Visualizing Multidimensional Data using Parallel coordinates plot (PCP) to gain further insights.

```
# Required Libraries
 library(hrbrthemes)

library(GGally)
library(viridis)

# Importing the dataset into R

USAFludata=read.csv("USAFludata.csv") # read csv file
USAFludata

# Plot the data using PCP.
ggparcoord(data=USAFludata,
 columns=1:13, groupColumn=14, order="anyClass",
 showPoints=TRUE,
 title="Parallel Coordinate Plot for the USA Flu Data",
 alphaLines=0.3
 ) +
scale_color_viridis(discreteTRUE) +
theme_ipsum()+
theme(
 plot.title=element_text(size=10)
)

# We can scale, present and compare the same dataset in
multiple ways, thus extracting fundamental insights. In this
example, we present four commonly used options. Original gally
code is adopted and modified from R Graph Gallery.

# Example 1. No scaling (globalminmax) is done to the dataset
ggparcoord(data=USAFludata,
 columns=1:13, groupColumn=14, order="anyClass",
 scale="globalminmax",
 showPoints=TRUE,
 title="No Scaling of USA Flu Trend Data",
 alphaLines=0.5
 ) +
scale_color_viridis(discreteTRUE) +
theme_ipsum()+
theme(
```

```
 legend.position="none",
 plot.title=element_text(size=13)
) +
xlab("")

# Example 2. Scaling (uniminmax) is standardized using a
minimum of 0 and maximum 1 to the dataset
ggparcoord(data=USAFludata,
 columns=1:13, groupColumn=14, order="anyClass",
 scale="uniminmax",
 showPoints=TRUE,
 title="Standardize USA Flu Trend Data to Min=0 and Max=1",
 alphaLines=0.5
 ) +
scale_color_viridis(discreteTRUE) +
theme_ipsum()+
theme(
 legend.position="none",
 plot.title=element_text(size=13)
) +
xlab("")

# Example 3. Scaling (std) is normalized univariately (i.e.,
subtract mean & divide by sd) to the dataset
ggparcoord(data=USAFludata,
 columns=1:13, groupColumn=14, order="anyClass",
 scale="std",
 showPoints=TRUE,
 title="Normalize univariately the USA Flu Trend Data (i.e.,
subtract mean & divide by sd)",
 alphaLines=0.5
 ) +
scale_color_viridis(discrete=TRUE) +
theme_ipsum()+
theme(
 legend.position="none",
 plot.title=element_text(size=13)
) +
xlab("")

# Example 4. Scaling (centered) is standardized and centered
(i.e., subtract mean & divide by sd) to the dataset
ggparcoord(data=USAFludata,
 columns=1:13, groupColumn=14, order="anyClass",
 scale="center",
 showPoints=TRUE,
 title="Standardize and center variables of USA Flu Trend
Data",
 alphaLines=0.5
 ) +
```

```
scale_color_viridis(discreteTRUE) +
theme_ipsum()+
theme(
 legend.position="none",
 plot.title=element_text(size=13)
) +
xlab("")
```

Hypothesis Testing, Confidence Intervals, and *p*-Values

Hypothesis testing is the process of carrying out statistical assessments of a sample and using the results to make inferences about population parameters. The true values of population parameters are often unknown and may be assessed either directly using point estimation techniques, or indirectly through hypothesis testing. The latter is the most common approach in inferential statistics, and it embodies the classic tradition of deductive reasoning or the *a priori* approach described in Chapter 2. Hypothesis testing begins by formulating a hypothetical statement or proposition about the true value of the population parameter. The proposed statement could be based on prior information about the population parameter generated from previous studies, observations from data exploration using visualization tools, results from a pilot project, or purely based on theoretical grounds. The analysis then proceeds with the statistical evaluation of the sample data for use in validating or denying the proposed statement. Three things are important when performing hypothesis testing: (1) the formulation of a hypothesis set consisting of both null and alternative hypotheses, (2) the decision regarding the test criteria and the level of statistical significance, and (3) the choice of the appropriate statistical test to evaluate the formulated hypothesis.

The hypothesis set consists of two competing claims that are made about the true value of the population parameter. The first claim, the null hypothesis designated as H_0, describes the hypothetical state of affairs. This null is the statement under statistical investigation; as the name implies, it is a negation and is often contrary to the research hypothesis or the opposite of what a data scientist believes to be true. The alternative hypothesis is a statement of the research hypothesis, or a conjecture of what a data scientist hopes to establish as true based on the empirical observations drawn from the sample. This alternative hypothesis is designated as H_A, and following the statistical analysis, it will be accepted as the true statement when the null is rejected. Both hypotheses must be formulated in such a way that they are mutually exclusive of each other but collectively exhaustive of all of the possible values of the true estimate of the population parameter.

TASK 4.3 HYPOTHESIS TESTING USING STUDENT'S *T*-STATISTICS

The student's t-statistic can be used to test one sample means, or test the difference in means obtained from two samples. Examples of research projects that require the test of two sample means include (1) comparison of physical or cultural characteristics of two regions, or two spatial units; (2) evaluation of the effectiveness of a new drug among the treatment (experimental study group) versus a control (placebo group); (3) before and after studies such as examining the effectiveness of a weight loss prevention program; (4) population health disparities between minority and nonminority groups; and (5) health impacts of anthropogenic versus natural hazards.

The test of two sample means may be based on independent samples or paired samples. Independent samples t allow for the comparison of means drawn from two samples in which the selection of the observations from the first sample has no bearing on the observations selected in the second sample. The samples are completely independent and unrelated to each other. For example, in Chicago, one could choose to compare the prevalence of lead poisoning among minority children and nonminority children. For a paired samples t-test, the two samples may be related, say from sets of twins, married couples, or having measurements taken repeatedly (but under different scenarios) from the same observations to generate the data. For example, one could decide to examine water quality in the Susquehanna River before and after Hurricane Sandy. The same monitoring stations or sample points will be used to generate the data at different times.

In the current task, let us explore the application of the independent samples t-test. We will use the obesity data generated for two states: New York and Mississippi. We will rely on the descriptive statistics reported earlier in Table 4.1.

Hypothesis testing also requires a data scientist to predetermine the level of statistical significance at which to evaluate the null hypothesis. To do so, it is vital to have some knowledge of the probability distribution that measures the likelihood of obtaining a certain value out of all possible outcomes. The significance level (denoted as α) represents a fixed probability of wrongly rejecting the null hypothesis when it is true. The most commonly selected probabilities are 0.01 or 0.05, respectively, signifying a 1% or 5% chance of making the inferential (or type I) error. The other kind of error (type II) occurs when we do not reject a null hypothesis that is false (denoted by 1-beta). Another relevant piece of information required to evaluate the hypothesis is deciding on the tails of the probability distribution, and whether one is working with a one-tailed or two-tailed test. Invariably, this depends on the overall objectives of the research and the formulation of the hypothesis sets. If a data scientist has a sense of the specific direction in which the true value of the parameter is likely to fall, a directed or pointed hypothesis set will be formulated. Such a hypothesis set will call for a one-tailed significance test that uses either the upper or lower tails of the probability distribution. A nondirectional hypothesis set in which the population estimate is likely to fall within the lower and upper tails of the probability distribution will call for a two-tailed significance test.

The third and perhaps the most critical decision to make in hypothesis testing is choosing the appropriate test to analyze the data. Several factors come into play here, including the nature of the research question, the sample size, the measurement scale of the variables, and whether or not the data conform to the key assumptions of the statistical test. There are several statistical techniques for testing all population parameters. Let us work through, but most examples are drawn from tests of sample means, proportions, and tests of associations. Examples include the use of student's *t*-tests and χ^2 test statistics. Let us work through a few examples to illustrate the application.

Step 1: Formulating the hypothesis set

H_O: There are no statistical differences in mean obesity rates observed between New York and Mississippi. The observed means of the two states are not significantly different:

$$Ho : \mu_{NY} = \mu_{MS}$$

H_A: There are statistically significant differences in mean obesity rates observed between New York and Mississippi. The observed means of the two states are significantly different:

$$Ha : \mu_{NY} \neq \mu_{MS}$$

Step 2: Establishing the level of significance

The hypothesis set formulated earlier is nondirectional and, therefore, calls for a two-sided significance test that will enable us to work with both tails of the probability distribution. We will conduct the test based on a fixed probability of 0.05.

Step 3: Applying the appropriate test

The test of an independent sample *t* is based on four assumptions: (1) The criterion variable should be measured on an interval/ratio scale. In the previous example, the criterion variable is percent obesity, measured on the ratio scale. (2) Data values drawn from the two groups are independent from each other. In the previous example, the data from New York are statistically independent from Mississippi. (3) The samples (location 1 and 2) must be drawn from normally distributed populations. This is the normality assumption and can be validated during the data screening procedures. (4) The two sampled populations must have similar/equal variances. This is the homogeneity of variance test and can also be validated during data screening using Levene's test of equal variance.

Assuming that the samples are approximately normal with equal variance, let us use the following equation to compute the t-test:

$$t = \frac{\bar{x}_1 - \bar{x}_2}{\sqrt{\left((n_1-1)s_1^2 + (n_2-1)s_2^2 / n_1 + n_2 - 2\right)\left((1/n_1) + (1/n_2)\right)}}$$

where

$\bar{x}_1$ is the sample mean for location 1, and $\bar{x}_2$ is the sample mean for location 2

s_1^2 is the sample variance for location 1, and s_2^2 is the standard deviation for location 2

The degrees of freedom (df) for this is $n_1 + n_2 - 2$. For the previous example, the df is 142.

As this is a two-tailed significance test, the critical value would be determined by dividing the alpha value (α) of 0.05 by two. Therefore, the critical t value obtained from a t distribution table at 0.025 with 142 degrees of freedom is 1.977. So, we will reject the null hypothesis if the observed t is less than the critical t of –1.977, or greater than the critical t of +1.9766.

Computation

Using the information from Table 4.1 presented earlier, the t-statistic is computed as follows:

$$t = \frac{27.6 - 36.3}{\sqrt{\left((62-1)3.2^2 + (82-1)3.1^2 / 62 + 82 - 2\right)\left((1/62) + (1/82)\right)}}$$

Statistical Conclusion

As the observed t of –16.73 is less than the critical value of –1.977, the null hypothesis must be rejected. Therefore, one can conclude that there is a statistically significant difference in obesity rates observed between the two states. The average obesity rate of 27.6% is significantly different from the average of 36.3% observed in Mississippi.

Step 1: Formulating the hypothesis set

H_0: There is no statistically significant relationship between the four independent categories of corn and soybean production.

H_A: There is a statistically significant relationship between the four categories of corn and soybean production.

TASK 4.4 HYPOTHESIS TESTING USING χ^2 STATISTICS

Similar to the student *t*-test, the χ^2 is a versatile significance test that is used to evaluate statistical hypotheses. Although the *t*-test focuses on means generated for criterion variables that are continuous, the χ^2 test is best for analyzing the categorical variables. One of the applications of χ^2 is contingency analysis. That is, when data from two or more variables are organized into categories, we might be interested in knowing whether the distribution of the categories observed in one variable is *contingent* on the other variable. The goal is to investigate whether the distributions of categories are likely to occur together or whether they are statistically independent.

To illustrate this, we use the Illinois agricultural data generated in Chapter 2. We divide two variables for corn (randomly sample 82 records, so $n = 82$) and soybean production ($n=102$) into four categories of 80th percentile and above (tier 1), between 50th and 80th percentiles (tier 2), between 20th and 50th percentiles (tier 3), and below 25th percentile (tier 4).

Step 2: To derive the χ^2 test of independence, we use the following equation:

$$\chi^2 = \sum_{i=1}^{n}\sum_{j=1}^{k}\frac{\left(O_{ij} - E_{ij}\right)^2}{E_{ij}}$$

where n, k are the number of categories for respective variables; O_{ij} is the observed number of cases; and E_{ij} is the expected number of cases.

Step 3: Now fill in the frequency for each tier in Table 4.5, *and use it to calculate your χ^2 in* Table 4.6.

Step 4: Use this information to derive the frequency of observed data and expected values

For cell a, the expected value would be $(a + b + c + d)(a + e)/N$

For cell b, the expected value would be $(a + b + c + d)(b + f)/N$

For cell c, the expected value would be $(a + b + c + d)(c + g)/N$

For cell d, the expected value would be $(a + b + c + d)(d + h)/N$

For cell e, the expected value would be $(e + f + g + h)(a + e)/N$

For cell f, the expected value would be $(e + f + g + h)(b + f)/N$

For cell g, the expected value would be $(e + f + g + h)(c + g)/N$

For cell h, the expected value would be $(e + f + g + h)(d + h)/N$

Degrees of freedom = (number of columns – 1)(number of rows – 1), degrees of freedom = 3, critical values at $p < .05 = 7.81$.

TABLE 4.5

Use This Information to Derive the Frequency of Observed Data and Expected Values

	Tier 1	Tier 2	Tier 3	Tier 4	Rows Total
Corn production	A	B	C	D	a+b+c+d
Soybean production	E	F	G	H	e+f+g+h
Columns total	a+e	b+f	c+g	d+h	N=a+b+c+d+e+f+g+h

TABLE 4.6

Fill in the Correct Values to Complete/Derive Chi-Square Using the Last Column

Observed (O)	Expected (E)	|O–E|	(O–E)²	(O–E)²/E
a				
b				
c				
d				
e				
f				
g				
h				

Chi-square = sum of last column.

If the observed χ^2 is less than the critical value, then we accept the null hypothesis. If not, then we accept the alternative hypothesis. Look up your critical values at $p < .05$.

Step 5: Make a decision regarding the null or alternative hypothesis

Conclusion

In Chapter 3, we made the distinction between descriptive and inferential statistics and proceeded to explore the statistical measures that are used to describe data points in a given distribution. In this chapter, our goal was to examine the data visualization tools that typically accompany these preliminary stages of data analysis and exploration. Along with discussing the trends in the development of these techniques, we also learned how to apply standard plots in summarizing data. This was followed by learning how to formulate and test hypotheses as part of the process of statistical validation. Following are some challenge exercises to help underscore these key concepts and procedures.

Worked Examples in R and Stay One Step Ahead with Challenge Assignments

Generating Graphical Data Summaries

Here we illustrate how to generate graphical data summaries using U.S. housing data from 1900 to 2010, Census Bureau. As with any project, the starting point is to load the data we will be using and the required libraries.

```
#Set working directory
setwd("../ChapterData/Chapter4_Data_Folder")

#load in the excel file
library(readxl)
p <- read_excel("us_housing_data.xlsx")

#Extract the housing data for the respective years from the
excel data
Years <- cbind(p$`1900`,p$`1910`,p$`1920`,p$`1930`,p$`1940`,p
$`1950`,p$`1960`,p$`1970`,p$`1980`,p$`1990`,p$`2000`,p$`2010`)
colnames(Years) <- c(1900,1910,1920,1930,1940,1950,1960,1970,
1980,1990,2000,2010)

#Graphically summarize the data using a boxplot and a
histogram;
boxplot(Years, col="orange")
```

TASK 4.5 COMPUTING AND INTERPRETING TESTS OF INDEPENDENT SAMPLE MEANS

1. Using the descriptive statistics reported in Table 4.1, choose any three of the following variables and formulate a statistical hypothesis set to evaluate the differences between the states of New York and Mississippi:
   a. Percent smoker
   b. Percent inactive
   c. Percent involved in excessive drinking
   d. Percent unemployed
   e. Percent with limited access to health foods

What conclusions can you draw based on the independent samples *t*-test? Which of these variables should best be analyzed using a one-tailed test? Explain.

```
#Let's now generate a histogram for one period (In this case,
2010)
hist(p$`2010`, main="Histogram for Housing Ownership in
2010", xlab="Housing Ownership", col="grey")

## The sample R codes for data exploration and visualization
are located in Chapter4-R-Code, and the dataset that is used
to implement the code is stored in Chapter3_Data_folder. Run
these codes and continue to practice on your own. This will
set you up and get you ready to work with R.
```

TASK 4.6 VISUALIZING LAND USE CHANGES IN A RAPIDLY CHANGE URBAN AREA IN MBALE, UGANDA

For this task, we use Figure 4.11 to visualize urban land use change.

1. Name the land use categories in 1973, 2000, and 2005.
2. Describe the urban changes that occurred between 1973 and 2005.
3. What will this urban area look like in 2020?
4. Suggest two possible hypotheses that may inform the study of urban land use change in this study region.

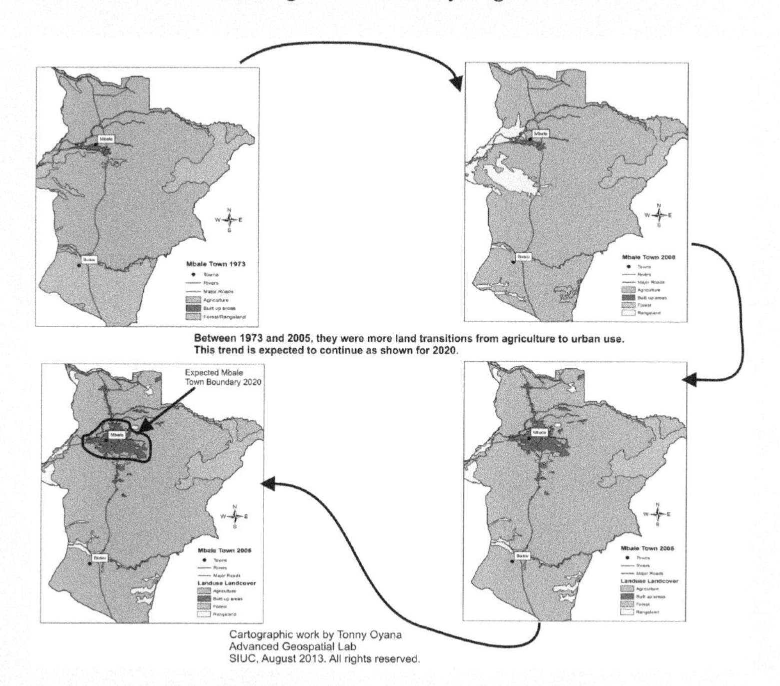

FIGURE 4.11
A land use map showing urban changes in Mbale Town, Uganda.

TASK 4.7 VISUALIZING CRIME TRAJECTORIES USING A 3D SPACE-TIME DIMENSION

For this task, we use Figure 4.12 to visualize crime trajectories for the city of Spokane, Washington.

1. Name the top seven places with consistently high crime rates throughout the study period.
2. Name the seven places with consistently low crime rates throughout the study period.
3. Name the place and year where we had observed the highest crime rate. Provide a possible explanation for this spike.

Suggest two possible hypotheses that may inform the study of crime trajectories in this study region.

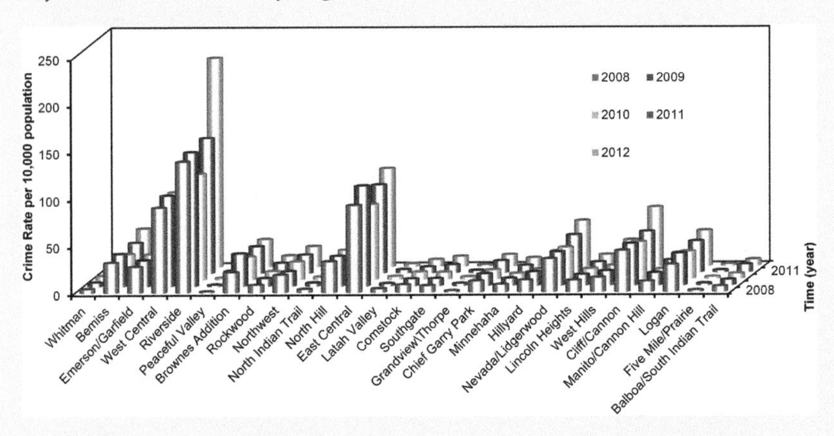

FIGURE 4.12
Three-dimensional representation of crime rate per 10,000 population between 2008 and 2012 for the city of Spokane, Washington.

Stay One Step Ahead with Challenge Assignments

Concept: Recall that in this chapter we learned about visualization concepts and principles that are used in exploratory data analysis, hypothesis generation, and data science reporting applications/communication. Human cognitive elements that are essential in encoding and decoding visuals were also discussed. In this online exercise, we conduct an in-depth examination of derived images to further advance our knowledge and understanding of these concepts.

Task: A number of derived images are presented in relation to a dataset that was collected for an urban food environment in low- and high-income settings of Chicago. Just like we did in the online exercises for Chapters 2 and 3, we continue to explore this dataset.

Question 1: Indicate whether the following statements are True or False using Figure 4.13.

a. There is a north-south directional bias in the spatial distribution of grocery stores. True or False

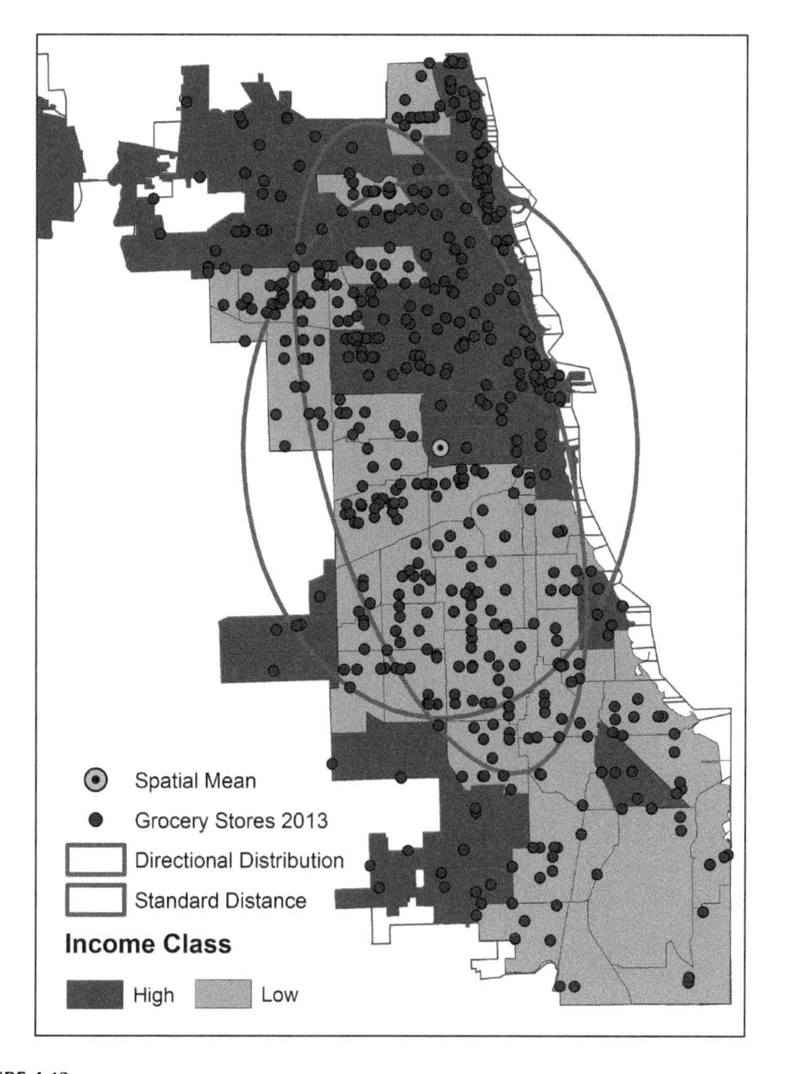

FIGURE 4.13
Three spatial measures for grocery stores in Chicago neighborhoods overlaid with income class. Note that all spatial measures were weighted using operation hours.

b. The spatial mean is located in the low-income setting. True or False
c. Standard distance captures different income settings in the north and south sides. True or False

Question 2: Indicate whether the following statements are True or False using Figure 4.14.

a. Low- and high-income settings are predominately evident in the south and north sides, respectively. True or False
b. Fourteen percent of the census tracts in Chicago have more than two food stores. True or False
c. Fifty-six percent of the census tracts in Chicago have at least a food store. True or False
d. Forty-four percent of the census tracts in Chicago have more than one food store. True or False

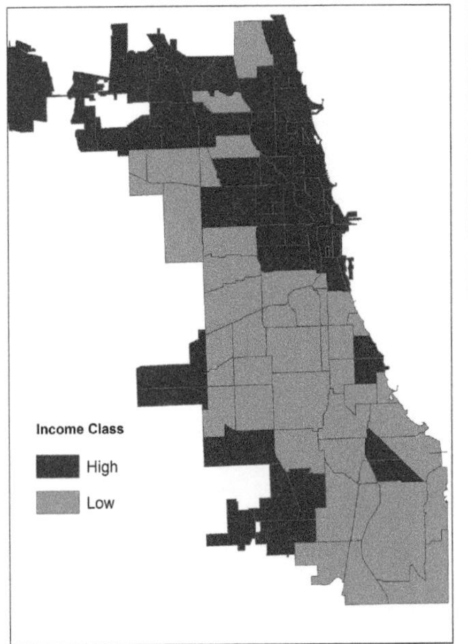

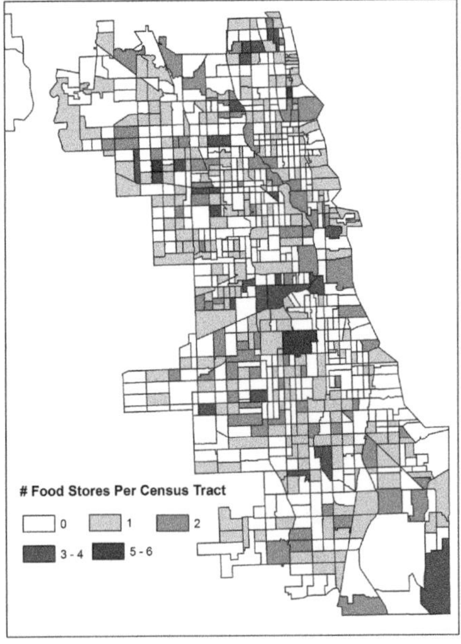

FIGURE 4.14
Spatial distribution of income settings at neighborhood level and the number of food stores at census-tract level. The left panel indicates the distribution of income settings, while the right panel shows the distribution of food stores. Thirty percent of the census tracts have one food store, 10% two food stores, and 4% more than three food stores. An in-depth analysis of the food access attributes at census-tract level shows that most low-income settings have more than three chain discounts, while many of the large supermarkets and chain full food services are in the high-income settings.

Question 3: Indicate whether the following statements are True or False using Figure 4.15.

a. Food deserts are evidently located in the southern part of Chicago. True or False
b. There is an abundance of food stores in the western and eastern parts of Chicago. True or False
c. Operational hours are not evenly distributed in the north and far north of Chicago. True or False

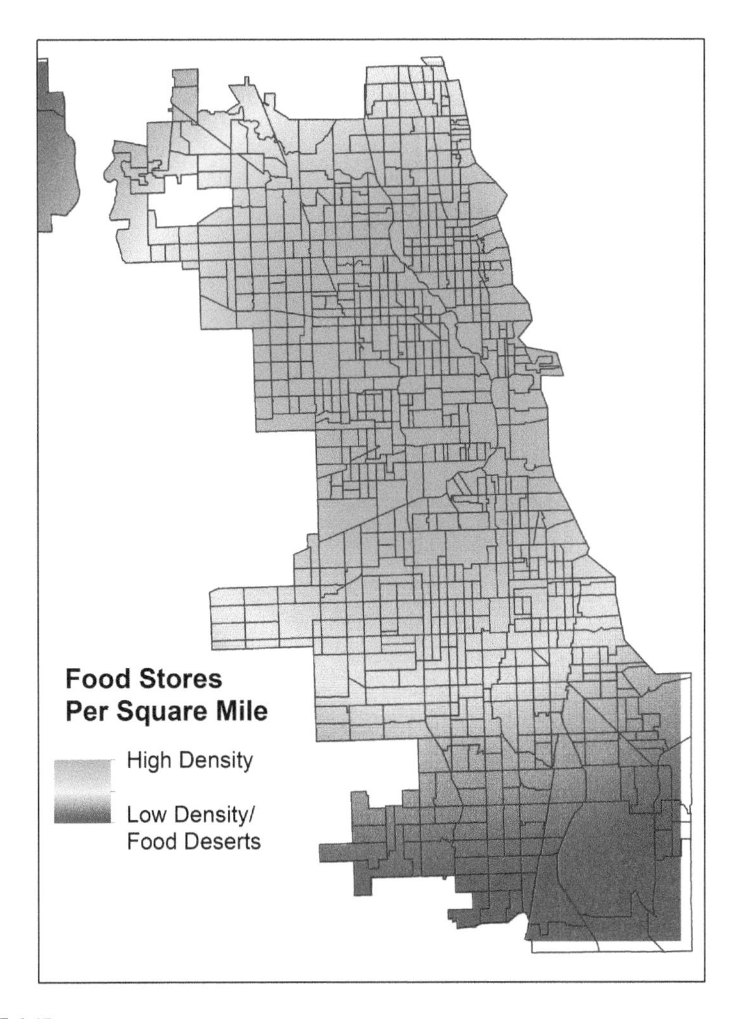

FIGURE 4.15
The surface map shows the spatial distribution of food stores per square mile. There is a low density of food stores in the south area, while there is a high density of food stores in the central and north areas.

Question 4: Indicate whether the following statements are True or False using Figures 4.16 and 4.17.

a. The farthest point in the south has a food store. True or False
b. High-income settings have more food stores that operate for 14 hours or more than the low-income settings. True or False

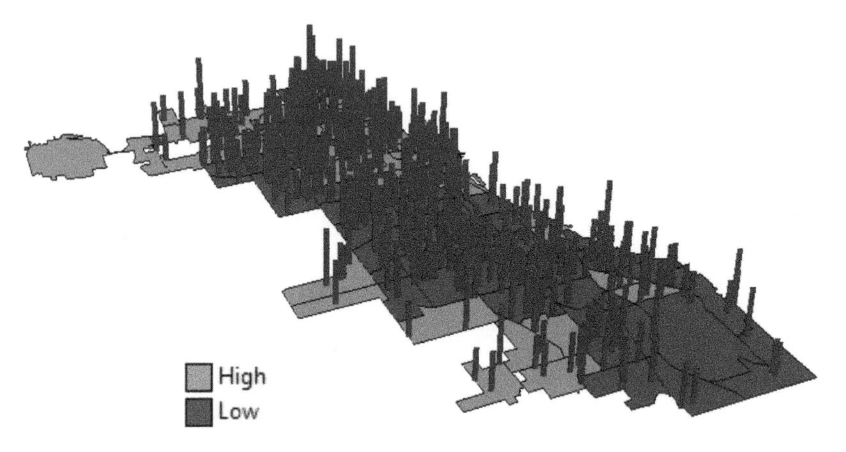

FIGURE 4.16
Three-dimensional visualization of the temporal dimension based on operational hours of grocery stores.

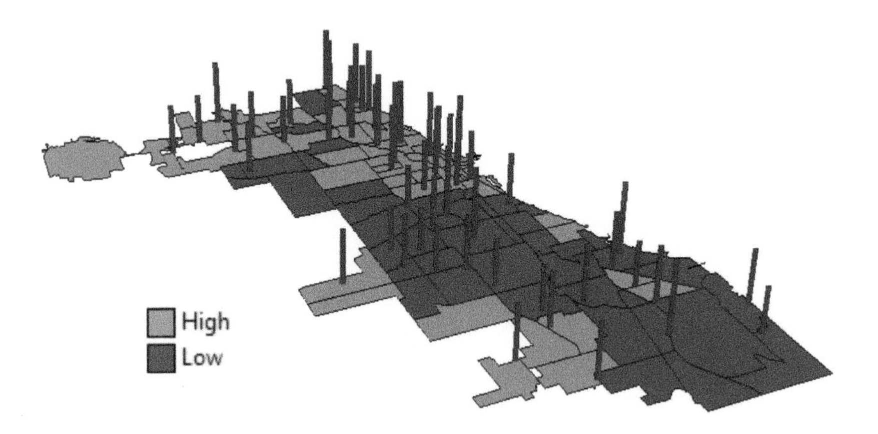

FIGURE 4.17
Three-dimensional visualization of the temporal dimension based on grocery stores with 14 hours of operation or more.

Question 5: Using insights gained through questions 1 through 5, state two testable hypotheses regarding the spatial distribution of food access in the city of Chicago.

Approach/strategy: Exploratory visual analysis was conducted using census tracts, neighborhoods, and community areas. Results are presented in Figures 4.13 through 4.17. Use your newly acquired visualization knowledge and skills to explore spatial and temporal patterns of the derived images. Synthesize the information, insights, or knowledge into testable hypotheses about food access in different income settings of Chicago.

Possible solutions: Answers to these questions should be fairly simple and straightforward. However, a bigger challenge would be to independently conduct a visual data exploratory analysis for a study of your choice.

Review and Study Questions

1. What are the major objectives of data visualization tools? Using one of the graphical approaches described in this chapter, explain how these objectives are attained.
2. What are the similarities and differences between scatterplot matrices and parallel coordinate plots?
3. It has been noted that a crucial feature in data visualization approaches is interaction. Using one of the graphical tools discussed in this chapter, explain the ways in which one can ensure human-computer interaction to effectively analyze a given dataset.
4. Using your research area of interest, specify two research questions that call for
   a. Exploring differences between two groups of observations
   b. Exploring relationships between two categorical variables
5. For question 4a, what is the null hypothesis (H_0) and the alternative hypothesis (H_A)?
6. For question 4b, what is the null hypothesis (H_0) and the alternative hypothesis (H_A)?
7. What statistical test would be ideal for evaluating the hypothesis set in 4a? Will this be a one-tailed or two-tailed significance test? Similarly, what statistical test would be ideal for evaluating the hypothesis set in question 4b? Will this be a one-tailed or two-tailed significance test?

Glossary of Key Terms

boxplots: A five-number graphical summary consisting of the minimum, maximum, median, lower and upper quartile values. As an EDA tool, it effectively summarizes a large dataset and depicts the central tendency and variability of the distribution. It also helps to uncover extreme scores and outliers in the distribution.

chi-square test: A goodness-of-fit test that compares the observed to the expected values in a data distribution. It can also be used as a contingency analysis test to explore the association between two or more categorical variables.

data/information visualization: This process entails effective selection of a set of marks in a graphic: exploration, synthesis, presentation, and communication.

exploratory data analysis: An approach that enables a data scientist to thoroughly screen the data to uncover the underlying structure, identify outliers and anomalies, and test the key statistical assumptions prior to more advanced statistical analysis.

hypothesis testing: The systematic process of evaluating a claim or statement about the true value of a population parameter using data drawn from a sample.

inferential statistics: Statistical techniques that use sample data to draw conclusions about the general population. Analysis can be based on direct or point estimates of the population, or they can be based on indirect approaches that include confidence bands and hypothesis testing.

parallel coordinate plots: A graphical tool for presenting and exploring large datasets with multiple dimensions that are measured using continuous and categorical data. Plots include parallel vertical axes representing the individual dimensions and polylines representing the observations.

scatterplots: A graphical device for depicting the strength and direction of the associations between two or more variables.

student's *t*-test: A commonly used significance test with several applications including the test of one sample mean, and the test of two sample means.

References

Andrienko, G., N. Andrienko, P. Jankowski, D. Keim, M.-J. Kraak, A. MacEachren, and S. Wrobel. 2007. Geovisual analytics for spatial decision support: Setting the research agenda. *International Journal of Geographical Information Science* 21(8): 839–857.

Andrienko, G., N. Andrienko, U. Demsar, D. Dransch, J. Dykes, and S. Fabrikant. 2010. Space, time and visual analytics. *International Journal of Geographical Information Science* 24(10): 1577–1600.
Chambers, J., W. Cleveland, B. Kleiner, and P. Tukey. 1983. *Graphical Methods for Data Analysis*. Belmont, CA: Wadsworth.
Edsall, R.M. 2003. An enhanced geographic information system for exploration of multivariate health statistics. *The Professional Geographer* 55(2): 146–160.
Ge, Y., S. Li, V.C. Lakhan, and A. Lucieer. 2009. Exploring uncertainty in remotely sensed data with parallel coordinate plots. *International Journal of Applied Earth Observation and Geoinformation* 11: 413–422.
Guo, D. 2007. Visual analytics of spatial interaction patterns for pandemic decision support. *International Journal of Geographical Information Science* 21(8): 859–877.
Heer, J. and D. Boyd. 2005. Vizster: Visualizing online social networks. *IEEE Symposium on Information Visualization, INFOVIS*, Minneapolis, MN, October 23–25.
Hoff, P.D., A.E. Raftery, and M.S. Handcock. 2002. Latent space approaches to social network analysis. *Journal of the American Statistical Association* 97(460): 1090–1098.
Huh, M.-H. and D. Park. 2008. Enhancing parallel coordinate plots. *Journal of Korean Statistical Society* 37: 129–133.
Inselberg, A. 1985. The plane with parallel coordinates. *Computational Geometry of the Visual Computer* 1: 69–97.
Inselberg, A. 2009. *Parallel Coordinates: Visual Multidimensional Geometry and Its Applications*. New York, NY: Springer.
Inselberg, A. and B. Dimsdale. 1990. Parallel coordinates: A tool for visualizing multidimensional geometry. *Proceedings of Visualization '90 IEEE Computer Society*, Los Alamitos, CA, pp. 361–378.
Jacquez, G.M., P. Goovaerts, and P.A. Rogerson. 2005. Space-time intelligence systems: Technology, applications and methods. *Journal of Geographical Systems* 7(1): 1–5.
Kahle, D. and H. Wickham. 2013. ggmap: Spatial visualization with ggplot2. *R Journal* 5(1): 144–161.
Lengler, R. and M. Eppler. 2007. Towards a periodic table of visualization methods for management. *GVE '07: Proceedings of the IASTED International Conference on Graphics and Visualization in Engineering*, pp. 83–88. Last accessed via https://dl.acm.org/doi/10.5555/1712936.1712954 on January 20, 2020.
Luo, W. and A.M. MacEachren. 2014. Geo-social visual analytics. *Journal of Spatial Information Science* 8: 27–66.
Luo, W., A.M. MacEachren, P. Yinm, and F. Hardisty. 2011. Spatial-social network visualization for exploratory data analysis. In *LBSN '11: Proceedings of the 3rd ACM SIGSPATIAL International Workshop on Location-Based Social Networks*. Chicago, IL, ACM, pp. 65–68. https://doi.org/10.1145/2063212.2063216
MacEachren, A.M. and M.-J. Kraak. 1997. Exploratory cartographic visualization: Advancing the agenda. *Computers and Geosciences* 23(4): 335–343.
MacEachren A.M. and M. Kraak. 2001. Research challenges in geovisualization. *Cartography and Geographic Information Science* 28(1): 3–12.
Nöllenburg, M. 2007. Geographic visualization. In Kerren, A., A. Ebert, and J. Meyer (eds), *Human-Centered Visualization Environments*. Germany: Berlin-Heidelberg, pp. 257–294.

Oyana, T.J., R.I. Rushomesa, and L.M. Bhatt. 2011. Using diffusion-based cartograms for visual representation and exploratory analysis of plausible study hypotheses: The small and big belly effect. *Journal of Spatial Science* 56(1): 103–120.

Perer, A. and B. Shneiderman. 2006. Balancing systematic and flexible exploration of social networks. *IEEE Transactions on Visualization and Computer Graphics* 12(5): 600–700.

Shneiderman, B. 1996. The eyes have it: A task by data type taxonomy for information visualizations. *Proceedings of the 1996 IEEE Symposium on Visual Languages (VL'96). IEEE Computer Society*, p. 336.

Shneiderman, B. 1997. *Designing the User Interface: Strategies for Effective Human-Computer Interaction*. Boston, MA: Addison-Wesley Longman.

Siirtola, H. and K.J. Raiha. 2006. Interacting with parallel coordinates. *Interacting with Computers* 18: 1278–1309.

Tufte, E. 1983. *The Visual Display of Quantitative Information*. Cheshire, CT: Graphics Press.

Tukey, J. 1977. *Exploratory Data Analysis*. Reading, MA: Addison-Wesley.

5

Understanding Spatial Statistical Relationships

Learning Objectives

1. Generate and interpret correlation statistics.
2. Conduct exploratory spatial analysis among variables.
3. Define and run a spatial regression model.
4. Generate and analyze regression diagnostic measures.

Engaging in Correlation Analysis

Correlation analysis is used to evaluate whether two measured variables are contemporaneous, covary, or coexist in space and/or time. The two commonly used measures to quantify such relationships are the Pearson product-moment correlation (for interval/ratio scaled variables) and Spearman's rank correlation (for ordinal scaled variables). Both statistics are expressed by an r-value (correlation coefficient) that denotes the strength (0 to 1) of the relationship and the direction of the relationship (positive or negative). The measures are also accompanied by a significance value, or p-value, for use in testing the research hypothesis. There are several ways to compute the Pearson's correlation coefficient, r. One approach entails the use of an efficient statistical formula that bypasses the computation of standard deviations of the two variables X and Y, or the deviations from their respective means:

$$r = \frac{\sum XY - \frac{\sum X \sum Y}{N}}{\sqrt{\left(\sum X^2 - \frac{(\sum X)^2}{N}\right)\left(\sum Y^2 - \frac{\sum Y^2}{N}\right)}}$$

TASK 5.1 COMPUTING PEARSON'S CORRELATION COEFFICIENT

Figure 5.1a and b depicts the scatterplots of tar and carbon monoxide yields and nicotine and carbon monoxide yields in selected cigarette brands. The data were drawn from the 1999 Federal Trade Commission Report. Using these data, let us compute the relevant coefficients and assess whether there are any significant statistical relationships between the two variables (tar and nicotine) and carbon monoxide yields in selected cigarette brands.

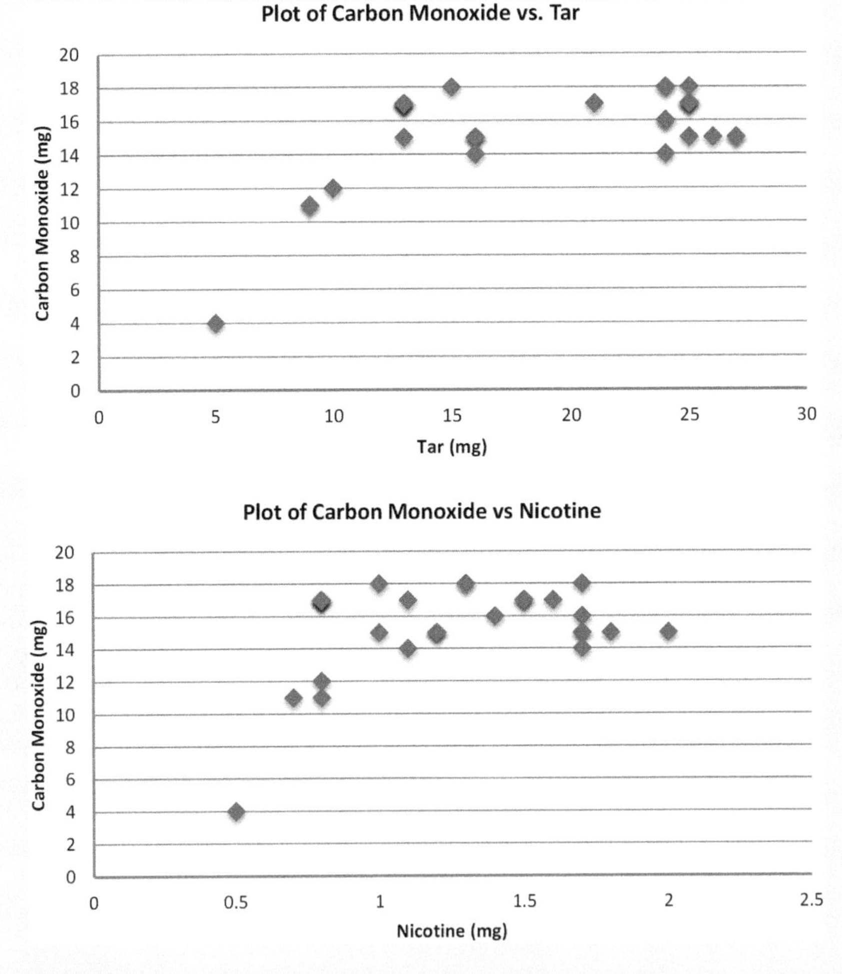

FIGURE 5.1
Relationships between (a) carbon monoxide and tar and between (b) carbon monoxide and nicotine.

Alternately, if one wishes to derive the mean and standard deviations of the variables, the ideal formula is as follows:

$$r = \frac{\frac{\sum_{i=1}^{n} X_i Y_i}{N} - \overline{xy}}{\sigma_x \sigma_y}$$

Tables 5.1 and 5.2 illustrate the computation of coefficients using the raw score formula introduced in the previous section. The results show that cigarette brands containing both tar and nicotine are likely to be positively related to the carbon monoxide yields, with a far stronger relationship observed among brands with higher nicotine levels than those with tar levels.

Having computed the correlation coefficients, let us now test whether the observed relationships are statistically significant. To conduct this test, we can use a student's *t*-distribution to determine whether the correlation coefficient is significant in each of the samples.

The *t*-statistic (t_{observed}) is given by $\sqrt{\frac{n-2}{1-r^2}}$, where the degree of freedom is $(n - 2)$, and we will conduct a two-tailed test at $\alpha = 0.05$. Using the *t*-distribution table, the critical *t*-value for 28 degrees of freedom is 2.048. For both samples, the population parameter under investigation in the correlation hypothesis is rho (designated as ρ).

For sample I, the null hypothesis is that there is no significant correlation between tar and carbon monoxide, meaning that $\rho_{\text{Tar.CO}}$ is not statistically different from zero.

$$\text{Sample I} = 0.5461354\sqrt{\frac{30-2}{1-0.5461354^2}}$$

We reject the null hypothesis, implying that there is a statistically significant correlation between tar and carbon monoxide. As noted earlier, the observed coefficient of $r = 0.546$ shows that there is a positive, relatively strong relationship between tar content and carbon monoxide levels in the cigarette brands.

Applying the same approach for sample II, the null hypothesis is that there is no significant correlation between nicotine and carbon monoxide. The population parameter, $\rho_{\text{Nicotine.CO}}$, is not statistically different from zero.

$$\text{Sample II} = 0.7584992\sqrt{\frac{15-2}{1-0.7584992^2}},$$

It is noted that $t_{\text{observed}} = 4.1965890$, and $t_{\text{critical}} = 2.160$; therefore, $t_{\text{observed}} > t_{\text{critical}}$, and we reject the null hypothesis, implying that there is a

TABLE 5.1

Worktable for Deriving Correlation Coefficients for Cigarette Brands, Sample I

Brand	Tar (X)	Carbon Monoxide (Y)	X^2	Y^2	XY
Basic	27	15	729	225	405
Commander	27	15	729	225	405
Bristol	26	15	676	225	390
English Ovals	25	15	625	225	375
Old Gold	25	18	625	324	450
Lucky Strike	25	17	625	289	425
Class A Dlx	25	17	625	289	425
Gen/Private Label	25	17	625	289	425
Tareyton Herbert	25	17	625	289	425
Camel Reg	24	16	576	256	384
Players Reg	24	14	576	196	336
Pall Mall	24	16	576	256	384
Chesterfield King	24	18	576	324	432
Pyramid King	24	18	576	324	432
Alpine	16	14	256	196	224
Alpine King	16	15	256	225	240
Alpine Lights	9	11	81	121	99
American Filters	16	15	256	225	240
Austin	13	17	169	289	221
Benson & Hedges	16	15	256	225	240
Best Choice	13	17	169	289	221
Bonus Value	13	17	169	289	221
Brandon	13	17	169	289	221
Brentwood	13	17	169	289	221
Bucks King	10	12	100	144	120
Cambridge	15	18	225	324	270
Camel	9	11	81	121	99
Canadian Players	13	15	169	225	195
Capri Menthol	5	4	25	16	20
Cardinal King	21	17	441	289	357
$\Sigma =$	**561**	**460**	**11755**	**7292**	**8902**

$$r = \frac{8902 - ((561 * 460)/30)}{(SQRT(11755 - ((561 \wedge 2)/30))) * (SQRT(7292 - ((460 \wedge 2)/30)))}$$

$$r = \frac{300}{549.314361} = \mathbf{0.5461354}$$

TABLE 5.2

Worktable for Deriving Correlation Coefficients for Cigarette Brands, Sample II

Brand	Tar (mg)	Carbon Monoxide (mg)	X^2	Y^2	XY
Carlton	2	2	4	4	4
Carolina Gold	16	15	256	225	240
Cavalier	9	12	81	144	108
Century	8	12	64	144	96
Charter	4	6	16	36	24
Chesterfield	19	13	361	169	247
Cimarron	21	17	441	289	357
Citation	9	12	81	144	108
City	11	13	121	169	143
Commander	23	13	529	169	299
Cost Cutter	9	12	81	144	108
Courier	14	16	196	256	224
Covington	10	14	100	196	140
Director's Choice	21	17	441	289	357
Doral	11	14	121	196	154
$\Sigma =$	187	188	2893	2574	2609

$$r = \frac{2609 - ((187 * 188)/15)}{\left(SQRT\left(2893 - ((187 \wedge 2)/15)\right)\right) * \left(SQRT\left(2574 - ((188 \wedge 2)/15)\right)\right)}$$

$$r = \frac{265.2667}{349.7257} = \mathbf{0.7584992}$$

significant difference between nicotine and carbon monoxide. We can infer from the sample data that the population parameter is statistically different from zero. The observed correlation of 0.758 suggests a very strong, positive, and significant relationship between the two variables.

As illustrated earlier, the application of correlation analysis is fairly straightforward and yields information that can be used to validate the strength, direction, and significance of relationships observed in a scatterplot. It is important to also point out that the Pearson's correlation analysis is a parametric test. This implies that the validity of the test results rests on meeting some key underlying assumptions such as linearity, and normality and measurement scale of variables. The two variables X and Y must be measured on an interval or ratio scale. In instances where one or both variables are ordinal or nominal in nature, other techniques such as Spearman's correlation, or *chi square* (χ^2) tests of association, must be used accordingly. Another assumption is that the observations must be randomly selected from a normally distributed population. Pearson's correlation is a robust technique, and the test of significance might be valid under modest departures from normality; however, it is best to confirm the assumption of normality prior to the analysis. A more notable requirement to watch for is the need to have a linear relationship between the two variables. This can be discerned in the scatterplots generated prior to computing the test statistic. If there are violations, it is best to consider other coefficients such as eta to test for the association between the variables.

Overall, there are several variants of the Pearson's correlation coefficient, all of which are formulated to address the different scales of variable measurement, varying properties of statistical data, and analytical objectives of a given study. There are also spatial variants such as the Moran's *I* statistic. This is a weighted correlation coefficient that is uniquely designed to incorporate the element of spatial dependency into the analysis. We discuss more on Moran's *I* and related coefficients later.

Ordinary Least Squares and Geographically Weighted Regression Methods

Whereas correlation analysis primarily focuses on the association between two or more variables, regression analysis can be used to explain the causal nature of the relationship, if any, and for predictive purposes. Regression analysis generates coefficients that represent the slope and intercept of a line that best fits the observed data points. Using standard analytical methods such as ordinary least squares (OLS), these two essential components can be generated to formally express the nature, strength, and direction of a statistical relationship (Demšar et al. 2008a,b; Burt et al. 2009). The relationship is confirmed if two things happen: (1) when there is a tendency for the *dependent* (or *response*) variable, *Y*, to vary with an *independent* (or *predictor*) variable, *X*, in a systematic fashion and (2) when there is a well-defined scattering of data points around the curve that depicts some type of model direction. The equation derived from a linear regression analysis can also be used to predict the values of a variable or estimate unknown values of one variable when given the values of the other. We normally predict variable estimates after successfully fitting the regression model.

Among the prerequisites for developing a causal model for regression, one of the most important considerations is the establishment of a functional linear relationship. This can be achieved through correlation, which is a necessary but insufficient condition for causality. One must first examine whether the response variable Y is significantly and strongly correlated with one or more predictor variables X_n. Establishing this relationship helps in delineating the "line of best fit" through observed values that most accurately model or predict the relationship between the response and predictor variables. Two other things to take into consideration when formulating a causal model are time precedence and nonspuriousness (Kenny 1979). For time precedence, one must ensure that if the variables X_n cause the response variable Y, then X_n must precede Y in time. Data scientists are advised to take this into consideration, especially during the data collection phase to avoid a temporal mismatch between the response and predictor variables. A third consideration for establishing a causal regression model is nonspuriousness. This necessitates a close review of the hypothesized associations to exclude potentially spurious variables, Z_n. Thus, if one claims that X causes Y, one must ensure that there is no spurious variable Z_i that is related to both X and Y such that if you control for Z_i, the relationship between X and Y disappears. Techniques such as scatterplots, Pearson's or Spearman's rank correlation, and partial correlation analysis are all useful strategies to test for functional linear relationships and detect any hidden, intervening, or spurious relationships prior to the formal specification of a regression model.

In modeling spatial relationships using regression, it is best to take a two-tiered approach that involves the use of both OLS and geographically weighted regression (GWR). The first-tier modeling helps in identifying the most important predictors that may explain the spatial processes in an area (Fotheringham et al. 1996; Nakaya et al. 2005; Demšar et al. 2008a,b; Harris et al. 2011a,b; Nakaya 2011). It provides a global model of the response variable or process using OLS. This is followed by testing whether the errors (residuals) in the global model are randomly distributed. To explore the pattern of spatial dependency, the most common means is by computing the Moran's I statistic, a measure of spatial autocorrelation that determines whether or not the errors/residuals in the model are independent. If they are not, this could be a problem of model misspecification. The second-tier modeling in regression provides local models of the response variable by fitting regression equations with variable regression coefficients that account for spatial variability. After fitting the model, the errors/residuals must also be evaluated for spatial autocorrelation just as noted in the first-tier model using OLS.

The spatial regression models are normally used to model spatial covariance structures. We use the OLS model to effectively identify the strongest predictors in any given model while taking into consideration the residuals. In general, we can write a simple regression/bivariate model as follows:

$$Y = \beta_0 + \beta_1 X + \varepsilon$$

Suppose we have four independent variables (X_i); then, a specific OLS model for modeling the relationship is as follows:

$$Y = \beta_0 + \beta_1 X_1 + \beta_2 X_2 + \beta_3 X_3 + \beta_4 X_4 + \varepsilon_i$$

where Y is the observed response variable, β_0 is the intercept, β_i variables correspond to the regression coefficients associated with each of the predictor variables X_i, and the error term is represented as ε_i.

If the OLS model is properly specified, and there is evidence of spatial autocorrelation in the dependent variable, we can proceed with fitting a GWR model. In this example, the GWR model for four independent variables is given as follows:

$$Y = \beta_0(u_i, v_i) + \beta_1(u_i, v_i)X_1 + \beta_2(u_i, v_i)X_2 + \beta_3(u_i, v_i)X_3 + \beta_4(u_i, v_i)X_4 + \varepsilon_i$$

where Y is the observed response variable, and the regression coefficients, β_i, are to be estimated at a location for which the spatial coordinates are provided by the variables u and v. The model enables the computation of the raw and standardized regression coefficients (βweights) and the standardized residuals for use in differentiating local spatial variations.

The primary assumptions of a traditional regression model are as follows:

1. The dependent variable is a *linear* function of a *specific* set of independent variables, plus the error term; this underscores the notion of linearity, and the need for a correct specification of the model. The equation for a bivariate model is $Y = \beta_0 + \beta_1 X + \varepsilon$.
2. The errors (or residuals) must have a zero mean and *constant variance*. (The expectation of homoscedasticity is implied here.)
3. The errors (residuals) must be independent, which means that the value of one error is not affected by the value of another error. (The expectation of nonautocorrelation, spatially and/or temporally, is implied here.)
4. For each value of X, the errors have a normal distribution about the regression line. (This is called the expectation of normality.)
5. No strong or perfect linear relationships must exist between the independent variables. (This expectation of *non-multi-collinearity* requires that the independent variables must not be highly correlated with each other.)

Although all of the assumptions are important for building regression models, some are more robust than others to model violations. Also, these assumptions are amendable when dealing with variables that have a spatial dimension. In the following sections, we review the procedures for fitting spatial regression models and the diagnostic measures used to ensure that the models are statistically valid.

Procedures in Developing a Spatial Regression Model

Depending on how you decide to start your regression analysis, there are seven major necessary steps to fit a model:

1. Using the data screening and visualization methods introduced in Chapters 3 and 4, examine the response, *Y*, and predictor variables, *X*, and investigate the nature of the potential association using scatterplots.
2. Check and determine whether the predictor variables are collinear, and identify the measures that show evidence of multicollinearity. Usually, a correlation analysis is a great starting point for demonstrating this.
3. Formulate a regression model based on hypothesized relationships.
4. Run the model, and determine the direction and strength of the hypothesized relationships by analyzing the test statistics.
5. Select the best regression model.
6. Test for lack of fit using a residual/scatterplot or histogram by ordering the residuals.
7. Review the fitness statistics by looking at the spread of the plot, evaluating observed values around the regression line, and examining how accurate the independent variables are in predicting *Y*.

Let us apply these procedures in Task 5.2.

TASK 5.2 USING SPATIAL REGRESSION TO ASSESS THE DETERMINANTS OF WELL-BEING SIGNIFICANCE IN THE CITY OF CHICAGO

The main goal of this spatial regression analysis is to test for and explore spatial variations in well-being significance. The indicators of well-being significance are often complex and hard to define. In this example, we rely on several factors that have been used in the past as proxy measures and determinants of well-being. Specifically, Table 5.3 contains a list of 10 factors or conditions that burden individuals or communities and prevent them from achieving good quality of life, overall well-being, and socioeconomic success. The factors have been compiled from a variety of data sources, including the American Community Survey, U.S. Census Bureau website, and the City of Chicago's geographic information system (GIS) data repository. In this study, hardship index (HI), a proxy measure of well-being significance, will serve as the dependent/response variable.

The decision to go with a traditional or spatial regression model can be made by first exploring the presence or absence of spatial autocorrelation in the dependent variable. If it is determined that there is spatial dependency in the variable, then sufficient reason exists to proceed with a spatial regression model. Otherwise, if there is no spatial autocorrelation, an OLS model should be considered rather than fitting the data with a spatial regression model. In the current study, the test of spatial autocorrelation was based on Moran's *I*, a coefficient that measures the intensity of spatial clustering among observations. The dependent variable (HI) has a Moran's *I* index of 0.547, a *Z*-score of 7.73, and a p-value < 0.00000. From this test, we find that the HI is positively autocorrelated, with a moderately high spatial clustering pattern. This enables us to proceed with the spatial regression model. The model will identify influential predictors that

TABLE 5.3

Potential Factors that May Explain the Spatial Differences in Socioeconomic Conditions

Context Description	Variables
Well-being significance	*Dependent/Response*
	Hardship index (HI)
	Independent/Predictor/Explanatory
Crowded housing	Percent of occupied housing units with more than one person per room (HS)
Poverty	Percent of households living below the federal poverty level (POV)
Unemployment	Percent of persons aged 16 years or older in the labor force that are unemployed (UEM)
Education	Percent of persons aged 25 years or older without a high school diploma (EDU)
Economically inactive population	Percent of the population under 18 or over 64 years of age (AEA)
Average income	Per capita income (INC), U.S. dollars
Race/ethnicity	White (W), Black (B), Hispanic (H), Asian (A)
Safety	Proximity to police stations (PL), distance in feet
Health care	Proximity to hospitals (HP), distance in feet

best explain the different socioeconomic conditions in the study region. Table 5.3 shows the factors that potentially account for the spatial differences in well-being conditions. From this list, we have 12 predictor variables that may help explain the dependent variable (Table 5.4).

Examining Relationships between Regression Variables

The relationships between regression variables are examined using Spearman's rank correlation, variance inflator factor (VIF), and scatterplots (Tukey 1977; Chambers et al. 1983; Tufte 1983). A Spearman's rank correlation matrix is shown in Table 5.4 for the correlation coefficients for all paired variables. Sixteen paired variables, shown in bold typeface, have been identified to exhibit some level of collinearity. Any pair of variables with a correlation of 0.70 or higher has been placed in the collinearity category for further scrutiny. In general, the correlation matrix suggests that most of the predictors are either moderately or marginally correlated. Percent of occupied housing (HS) and percent of persons aged 25 years or older without a high school diploma (EDU) are the most strongly correlated. The least correlated are among pairwise correlations for Asian (A), proximity to police station (PL), and proximity to hospitals (HP), with the exception of PL and HP.

Examining the Strength of Association and Direction of All Paired Variables Using a Scatterplot Matrix

The overall patterns among most of the variables suggest possible linear relationships (increasing/decreasing trends in both x and y variables—some are positively correlated, whereas others are negatively correlated); exceptions include the pairings that involve the Asian group or proximity to police stations (Figure 5.2). These appear to show neither clear (weak correlation) association nor direction.

Fitting the Ordinary Least Squares Regression Model

We need to fit the best OLS regression model to ensure that we have a properly specified model before moving ahead with the GWR model.

Primary Model

$$Y_{HI} = \beta_0 + \beta_1 HS + \beta_2 POV + \beta_3 UEM + \beta_4 EDU + \beta_5 AEA + \beta_6 INC + \beta_7 W$$
$$+ \beta_8 B + \beta_9 H + \beta_{10} H + \beta_{11} H + \beta_{12} H + \varepsilon$$

We need to determine if the primary model is statistically significant at $\alpha = 0.05$. We do this by investigating whether

$$H_O : \beta_0 = \beta_1 = \beta_2 = \beta_3 \ldots \beta_{12} = 0$$

$$H_A\text{: At least one } \beta \neq 0.$$

TABLE 5.4

A Correlation Matrix for all Paired Variables

	HS	POV	UEM	EDU	AEA	INC	HI	W	B	H	A	PL	HP
HS	1												
POV	0.32	1											
UEM	0.14	**0.76**	1										
EDU	**0.91**	0.42	0.32	1									
AEA	0.24	0.40	0.60	0.42	1								
INC	–0.55	–0.53	–0.61	**–0.71**	**–0.76**	1							
HI	0.69	**0.77**	**0.74**	**0.83**	0.68	**–0.84**	1						
W	–0.42	**–0.72**	**–0.75**	–0.57	–0.61	**0.72**	**–0.83**	1					
B	–0.17	0.65	**0.79**	–0.06	0.49	–0.36	0.45	**–0.73**	1				
H	**0.72**	–0.19	–0.27	**0.71**	0.02	–0.29	0.27	–0.06	–0.60	1			
A	–0.05	–0.07	–0.30	–0.04	–0.28	0.19	–0.18	0.23	–0.34	–0.10	1		
PL	–0.10	–0.23	–0.17	–0.09	0.11	–0.04	–0.13	0.23	–0.16	0.08	–0.19	1	
HP	0.02	–0.26	–0.06	0.04	0.31	–0.17	–0.01	0.16	–0.21	0.21	–0.15	0.52	1

Note: The 16 paired variables shown in bold typeface have been identified to exhibit some level of collinearity.

Abbreviations: A, Asian; AEA, percent of the population under 18 or over 64 years of age; B, Black; EDU, percent of persons aged 25 years or older without a high school diploma; H, Hispanic; HP, proximity to hospitals; HS, percent of occupied housing units with more than one person per room; INC, per capita income; PL, proximity to police station; POV, percent of households living below the federal poverty level; UEM, percent of persons aged 16 years or older in the labor force who are unemployed; W, White.

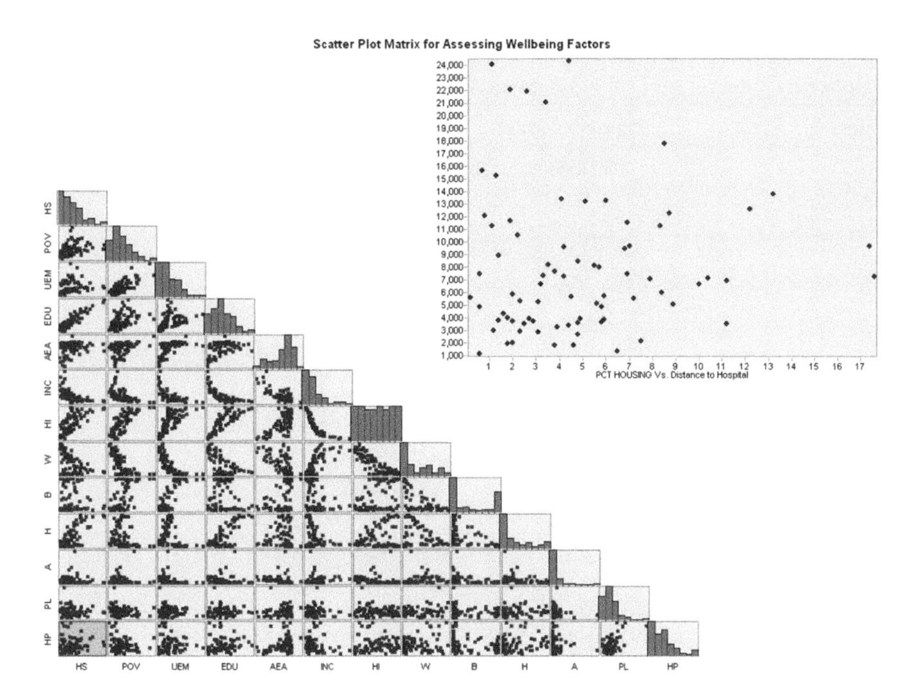

FIGURE 5.2
A scatterplot matrix and histogram showing all paired variables.

Given that the observed joint F-statistic is 258.03, and it is greater than the critical value at (12, 64) degrees of freedom, we can reject the null hypothesis and conclude that at least one regression coefficient is not equal to zero.

Examining Variance Inflation Factor Results

The VIF is another formal measure of detecting the presence of collinearity. It is used to eliminate—by adding or deleting a predictor variable—any potential redundancy among independent variables, X_i. VIF indicates how much the variance of the coefficient estimate is being inflated by multicollinearity. Simply put, the existence of this problem in a regression model suggests a large amount of standard error in the coefficient estimates. Most standard statistical textbooks suggest a VIF cutoff point greater than five to indicate a concern for collinearity. This is because the expected sum of squared errors in standardized regression coefficients is nearly five times as large as it would be if the predictor variables were uncorrelated. However, Neter et al. (1996) have suggested the examination of VIF values that greatly exceed 10. In ESRI's ArcGIS, the cutoff is placed at larger than 7.5 when examining an OLS model for the collinearity problem. This book recommends anything above the rule of thumb, that is, VIF values that exceed five should be critically reviewed when deriving the best model.

TABLE 5.5

Variance Inflator Factor (VIF) Values for the Three Ordinary Least Squares Models

Factors	Primary Model	Reduced Model	Best Model
HS	8.188543[a]	3.242602	1.063577
POV	4.261475		
UEM	6.238082[a]	3.758031	1.577382
EDU	23.531012[a]		
AEA	3.766619	2.419355	1.642695
INC	5.055292[a]		
W	>1000.0[a]		
B	>1000.0[a]	7.444437[a]	
H	>1000.0[a]	5.985656[a]	
A	283.243408[a]	1.720664	
PL	1.662634	1.504726	
HP	2.069709	1.858672	

[a] VIF values that exceed 5, a consecutive threshold is being applied to critically evaluate the presence of collinearity.

Abbreviations: A, Asian; AEA, percent of the population under 18 or over 64 years of age; B, Black; EDU, percent of persons aged 25 years or older without a high school diploma; H, Hispanic; HP, proximity to hospitals; HS, percent of occupied housing units with more than one person per room; INC, per capita income; PL, proximity to police station; POV, percent of households living below the federal poverty level; UEM, percent of persons aged 16 years or older in the labor force who are unemployed; W, White.

Table 5.5 summarizes the VIF values for the three OLS models that were generated using ESRI's ArcGIS.

The primary model has eight variables with VIF values that are larger than five (HS, UEM, EDU, INC, W, B, H, and A), reduced model has two variables with VIF values that exceed the threshold (Black and Hispanic), and best model shows a remarkable improvement of VIF values with the highest VIF value only being observed in AEA (1.643). This is far below the required threshold.

Reduced Model

After examining the well-being factors using a scatterplot and correlation analysis, the reduced model is as follows:

$$Y_{HI} = \beta_0 + \beta_1 HS + \beta_2 AEA + \beta_3 UEM + \beta_4 B + \beta_5 H + \beta_6 A + \beta_7 PL + \beta_8 HP + \varepsilon$$

We need to determine if the primary model is statistically significant at $\alpha = 0.05$. We do this by investigating whether

$$H_O : \beta_0 = \beta_1 = \beta_2 = \beta_3 \ldots \beta_8 = 0$$

TABLE 5.6

A Summary of the Ordinary Least Squares Results for the Three Models

Variables	Primary Model Coefficient Estimate (t-value)	Reduced Model Coefficient Estimate (t-value)	Best Model Coefficient Estimate (t-value)
HI	−88.1853 (−1.0549)	−39.206192 (−8.379899)[a]	−32.621131 (−6.612045)[a]
HS	1.0057 (2.518495)[a]	3.532561 (8.853233)[a]	4.374369 (16.390050)[a]
POV	0.623354 (6.802502)[a]		
UEM	0.647681 (3.573078)[a]	1.782008 (7.976269)[a]	2.105701 (12.456289)[a]
EDU	1.024476 (5.112745)[a]		
AEA	0.594223 (4.361404)[a]	0.837226 (4.828465)[a]	0.910480 (5.456300)[a]
INC	−0.000116 (−1.507690)		
W	67.405221 (0.781687)		
B	76.487877 (0.889839)	21.010896 (3.846550)[a]	
H	68.516060 (0.786059)	27.896733 (3.816535)	
A	73.666656 (0.818366)	50.236394 (4.509104)[a]	
PL	0.000018 (0.140818)	0.000302 (1.560332)	
HP	0.000016 (0.118153)	−0.000293 (−1.449490)	
AIC	468.93	533.37	550.25
r^2	0.976	0.9394	0.9173
Observations	77	77	77
Moran's I	−0.062 (−0.677)	−0.0204 (−0.2102)	0.0254 (1.1158)

[a] Statistically significant coefficient estimates.

Abbreviations: A, Asian; AEA, percent of the population under 18 or over 64 years of age; AIC, Akaike's information criterion (a measure of model performance with the smallest value preferred); B, Black; EDU, percent of persons aged 25 years or older without a high school diploma; H, Hispanic; HP, proximity to hospitals; HS, percent of occupied housing units with more than one person per room; INC, per capita income; PL, proximity to police station; POV, percent of households living below the federal poverty level; UEM, percent of persons aged 16 years or older in the labor force who are unemployed; W, White.

$$H_A\text{: At least one } \beta \neq 0.$$

Given that the observed joint F-statistic is 148.17, and it is greater than the critical value at (8, 68) degrees of freedom, we can reject the null hypothesis and conclude that at least one regression coefficient is not equal to zero.

All three regression models can explain more than 90% of the total variation in the well-being significance that is attributable to all the independent variables, X_i, as defined by model fit to the data (Table 5.6). Additionally, all three predictor variables identified in the best model have positive coefficients, implying that as these variables increase, the level of hardship in the community areas also increases. However, due to the severe concern of collinearity problems in the primary and reduced models, we must resolve this concern by finding a meaningful model.

Best Model

The best model after reviewing fitness statistics, lack of fit test, and analyzing other relevant collinearity diagnostics is as follows:

$$Y_{HI} = \beta_0 + \beta_1 HS + \beta_2 AEA + \beta_3 UEM + \varepsilon$$

We need to determine if the primary model is statistically significant at $\alpha = 0.05$. We do this by investigating whether

$$H_O : \beta_0 = \beta_1 = \beta_2 = \beta_3 = 0$$

$$H_A\text{: At least one } \beta \neq 0.$$

Given that the observed joint F-statistic is 281.95, and it is greater than the critical value at (3, 73) degrees of freedom, we can reject the null hypothesis and conclude that at least one regression coefficient is not equal to zero.

In selecting the best equation, we must also determine which of the independent variables, X_i, is statistically significant at $\alpha = 0.05$. We do this by investigating whether

$$H_O : \frac{\sigma^2 due\ to\ X_i}{\sigma^2 Res} = 1$$

$$H_A : \frac{\sigma^2 due\ to\ X_i}{\sigma^2\ Res} > 1$$

The Jarque-Bera statistics that measures whether model predictors are biased or not—a goodness-of-fit test that shows whether residuals are normally distributed at 2 degrees of freedom using a chi-square distribution—indicates the primary model was 1.701 (p-value < 0.427), reduced model was 2.219 (p-value < 0.329), and best model was 3.1615 (p-value < 0.164). We concluded that all the residuals in the three OLS models are normally distributed and unbiased.

Examining Residual Changes in Ordinary Least Squares Regression Models

Analyzing the residuals offers fundamental clues about the quality of the regression model (Figure 5.3). It is not only important to analyze these residuals after successfully fitting a model but also essential to check whether the residuals have a mean of zero and a standard deviation of 1.

The next step would be to identify and analyze observations that have standardized residuals greater than 2 (depict model underprediction) or

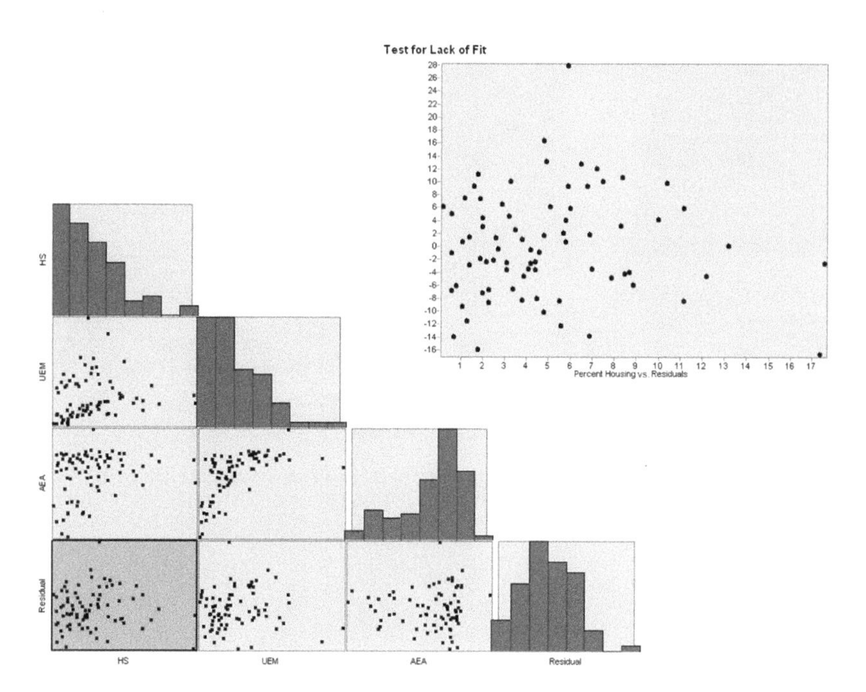

FIGURE 5.3
Residual plots and histograms showing identified HS, AEA, and UEM predictors.

negative standardized residuals less than –2 (depict model overprediction) (Figure 5.4):

1. *Residual analysis for the primary OLS model*: Community areas with standardized residuals greater than 2.0 are Fuller Park and Kenwood, and an area with standardized residuals less than –2 is Lake View. Fuller Park predominately consists of an over 90% Black population, has high unemployment, has a high percent of individuals who have no high school diploma, has a high percent of individuals who are economically inactive, and has a high crime rate. Kenwood is ethnically diverse with over 70% Black, 17% White, and 6% Asian people; it is a relatively economically vibrant community, which has a mixture of the elite, a rich population, and significant pockets of a poor population with high poverty and crime rates. Additionally, the area has a mixture of old and new housing developments. Lake View is a very rich neighborhood with 80% White, 8% Hispanic, 6% Asian, and 4% Black.
2. *Residual analysis for the reduced OLS model*: A community area with standardized residuals greater than 2.0 is Englewood, and an area with standardized residuals less than –2 is Calumet Heights. Englewood is a poor neighborhood with high poverty rates, high

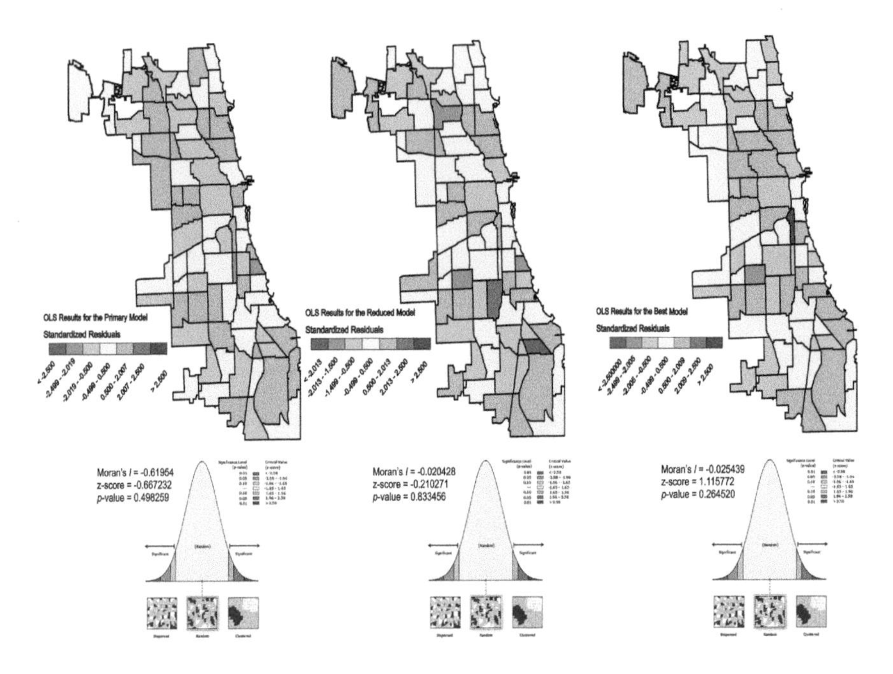

FIGURE 5.4
The spatial distribution of standardized residuals from the ordinary least squares models.

crime rates, and lack of medical care, and it has a population that is over 97% Black. Calumet Heights is a rich neighborhood; it is 90% Black and 4% Hispanic, has declining crime rates, and has a higher number of persons who are economically active.

3. *Residual analysis for the best OLS model*: A community area with standardized residuals greater than 2.0 is Armour Square, and an area with standardized residuals less than –2 is Gage Park. Armour Square is a relatively poor neighborhood with over 72% Asian (location of Chinatown), 11% Black, and 12% White. Gage Park is also a poor neighborhood with over 89% Hispanic, 5% Black, and 4% White; a high crime rate; a large number of persons without a high school diploma; and an economically inactive population.

In reviewing the geographic distribution of the globally verified predictor variables, we observed that all the standardized residuals greater than 2 are located in community areas that have an HI larger than or equal to 80 and a per capita income of less than $17,000 with the exception of Kenwood, which has an HI of 25 and a per capita income of about $38,000. Also, we observed that all the standardized residuals less than 2 are located in community areas that have an HI less than or equal to 57 and a per capita income of more than $28,000, with the exception of Gage Park which has an HI of 93 and a per capita income of about $12,000. We can therefore conclude that in the primary and reduced models, community areas with standardized residuals larger than 2 exhibit significant hardships, while those with less than 2 are characterized by minor hardships. However, the best model identifies two community areas that exhibit significant hardships. Given the evidence of over- and underpredictions within residuals in a few of the community areas, we can further conclude that the model is possibly missing explanatory variables to explain well-being significance in these areas.

The lower panel of Figure 5.4 shows test results for spatial autocorrelation. From the maps, we see different spatial patterns of standardized residuals in the study region. The red areas in the maps indicate that actual observed values are higher than the values the model predicted, whereas the blue areas show where the actual observed values are lower than the model predicted. All three models have no evidence of spatial autocorrelation in the regression residuals.

Fitting the Geographically Weighted Regression Model

Assuming spatial nonstationarity, a GWR model is conducted using three globally verified predictor variables from the OLS model. It is relevant for two primary reasons: (1) we detected that the dependent variable exhibited a significant amount of spatial dependency, so there is a need to construct local models to explain variations in well-being significance, and (2) we must analyze the geographic distribution of local regression coefficients at different local regression neighborhoods to understand the effects of predictor variables on local areas.

A GWR model for the three globally verified predictor variables is given as follows:

$$Y_{HI} = \alpha(u_i, v_i) + \beta_1(u_i, v_i)HS + \beta_2(u_i, v_i)UEM + \beta_3(u_i, v_i)AEA + \varepsilon_i$$

Examining Residual Change and Effects of Predictor Variables on Local Areas

The standardized residuals of the GWR model have a mixture of five community areas exhibiting major (Gage Park, Englewood, and Armour Square) and minor (Beverly and Calumet Heights) hardships (Figure 5.5). The negative standardized residuals show community areas with major hardship, whereas the positive values show the ones with minor hardship with the exception of Gage Park. Beverly is a diverse, rich neighborhood, which consists of 58% White, 34% Black, and 5% Hispanic populations. The other community areas are already described under the OLS regression model. What is notable in the GWR model is the identification of the Beverly community area that was not previously identified by the OLS regression model.

Residual analysis shows that community areas with standardized residuals greater than 2.0 are Beverly, Calumet Heights, and Gage Park, and the ones

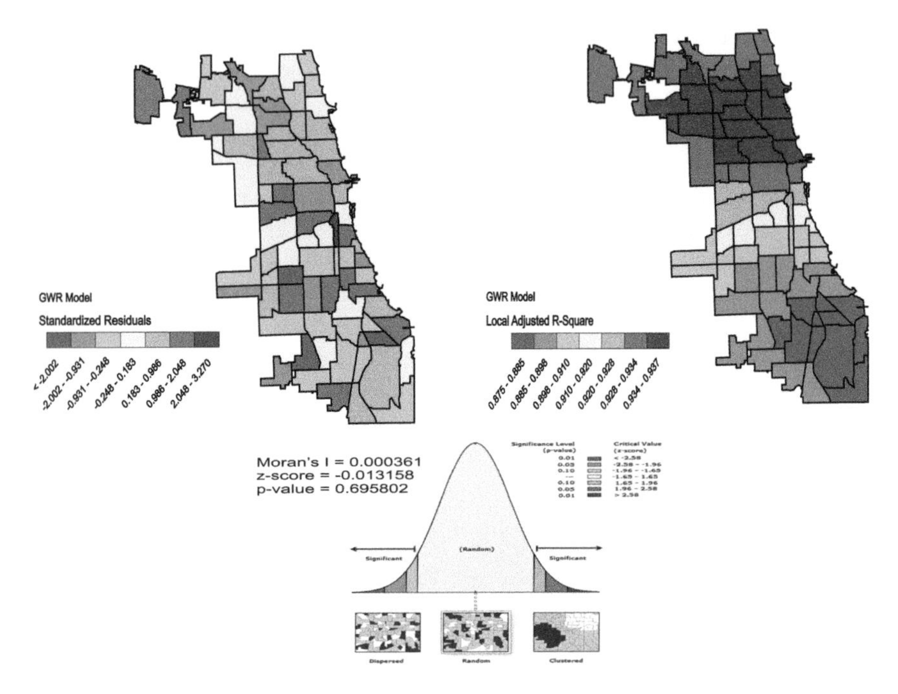

FIGURE 5.5
Mapping the standardized residuals and the local R^2 results from geographically weighted regression (GWR) results from the GWR model.

with standardized residuals less than –2 are Armour Square and Englewood. In Figure 5.5, the left panel shows the spatial distribution of standardized residuals in the GWR model. The middle panel shows test results of spatial autocorrelation, and the right panel shows local adjusted R^2. There is no spatial autocorrelation in the regression residuals.

With respect to the effects of predictor variables, there is an evident divide in variables that measure socioeconomic disparities between community areas located in the north and south. Negative regression relationships are more evident in the south than in the north. In reviewing the spatial patterns of the local coefficient estimates, we observed that negative estimates from the intercept were concentrated among community areas located in the central and downtown areas, whereas positive estimates were located in the south and north (Figure 5.6). The spatial patterns of the negative regression relationships observed among the local coefficient estimates of housing and economically inactive population variables are similar in spatial extent with the exception of unemployment. The spatial patterns for the negative regression relationships observed among the local coefficient estimates of housing and economically inactive population variables are predominately located in the south and the north, whereas the ones showing positive regression relationships are

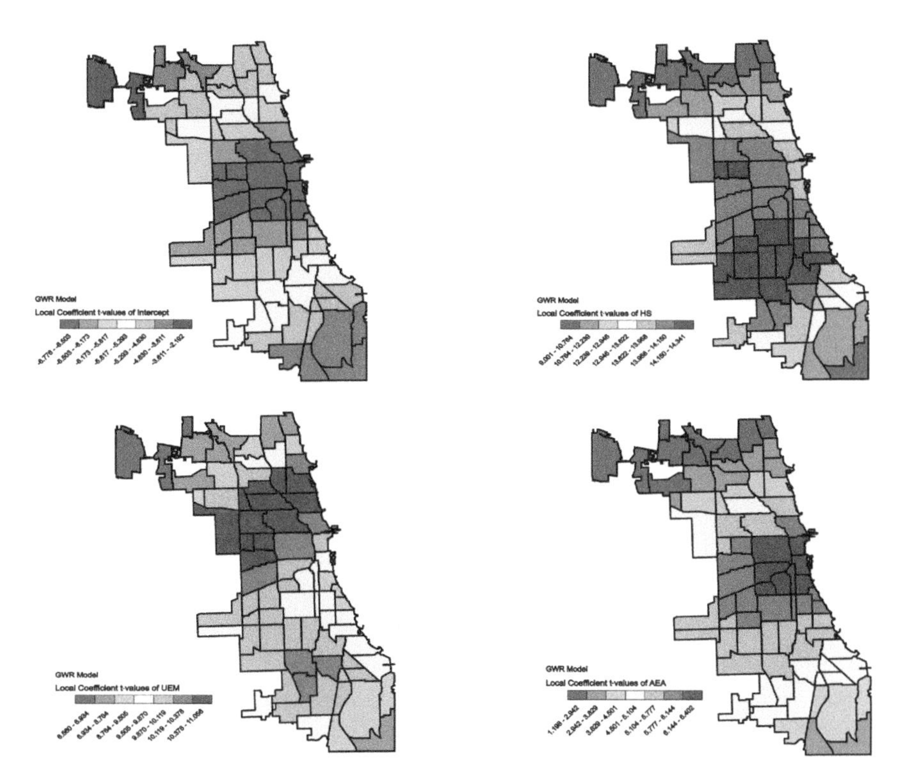

FIGURE 5.6
Spatial distribution of local coefficient *t*-values from the geographically weighted regression model.

located in the central and downtown areas. A critical examination of these spatial patterns suggests local variations in housing, unemployment, and economically inactive population among community areas located in the downtown, central, and lower southern regions. Overall, the GWR model is robust with an adjusted R^2 and Akaike information criterion (AIC) of 0.923 and 548.17, respectively.

Summary of Modeling Result

The fitted regression model explains the geographic variations of well-being significance for 77 community areas in Chicago. A meaningful model of well-being significance consists of three globally verified predictor variables (percent of occupied housing units with more than one person per room [HS], percent of persons aged 16 years or older in the labor force who are unemployed [UEM], and percent of the population under 18 or over 64 years of age [AEA]). The models pointed to significant global and local spatial variations in well-being significance. Three local models explained local variations in well-being significance. However, significant differences were evident in the globally verified model and the local models. For example, the analysis of local regression residuals identified the Beverly community area that was not previously identified by the global model. The adjusted R^2 for the primary model explained 97.6%, for the reduced model 93.9%, and for the best model 91.7% of the well-being significance. Although the AICs for the three models were small and robust, there was a slight increase in them. The AIC values ranged from 468 to 550. Residual plots were all normally distributed, suggesting that the models were unbiased. A few patterns were evident in the regression residuals, suggesting missing exploratory variables. Additional parameters should be considered to highlight the true influence of predictor variables on well-being in varying socioeconomic community areas.

- The GWR model explained about 92% of the local variations of well-being significance. The examination of local coefficient estimates indicated local variations in housing, unemployment, and economically inactive population among community areas located in the downtown, central, and lower southern regions.
- Overall, there were suggestive spatial relationships among three significant determinants of well-being significance, thus lending support to the existence of profound socioeconomic disparities between community areas located in the north and south. Negative regression relationships were even more apparent in the south than in the north.

Conclusion

In this chapter, we introduced multivariate analysis based on a full-scale exploration of the associations between a given set of response and predictor variables. Key steps that began with the use of correlation analysis as a precursor toward establishing causality were first presented. These were followed by systematically pooling together the appropriate statistical techniques and related diagnostics to analyze the data based on the key assumptions of regression and the underlying structure of the data. Completing the problem sets here will help you hone in on these essential skills, including knowledge of the analytical strategies that are used to overcome data challenges in regression.

Worked Examples in R and Stay One Step Ahead with Challenge Assignments

Sample R codes to Run OLS regression and GWR: Assuming we have now practiced R enough and can still recall earlier commands used in Chapters 1 through 4, we can now proceed with exploring Run OLS regression and GWR in R.

```
#Type each command in sequence and on its own line and wait
for a response before typing in the next one. We start by
loading the data we are going to use. This will be followed by
extracting the variables we need for the primary model
formulation.
```

Step 1: Exploring the strength and direction of relationships using a scatterplot and histogram

```
setwd("../K24901_Data_Folders/Chapter5_Data_Folder")
library(rgdal)
library(dplyr)
p <- readOGR('agricul_ILL_stats3.shp',
layer='agricul_ILL_stats3')

OGR data source with driver: ESRI Shapefile
Source: "agricul_ILL_stats3.shp", layer: "agricul_ILL_
stats3"with 102 features
It has 66 fields
Integer64 fields read as strings: NO_FARMS07 AVG_SIZE07
CROP_ACR07

plot(p, main="Spatial distribution of Agriculture Production
in the Illinois")
d <- data.frame(p)

a <- d[, c(29, 47, 57, 61, 65)]

head(a)
MED_AGE AVG_SIZE07 CornYield SoyYield WheatYield
0 41.6 277 163 43 0
1 37.8 255 183 48 59
2 33.3 252 177 46 69
3 34.1 394 176 47 0
4 38.1 397 192 34 63
5 39.6 402 199 49 74
aa <- data.matrix(a)
# Put histograms on the diagonal panel
panel.hist <- function(x, …)
+ {
+ usr <- par("usr"); on.exit(par(usr))
+ par(usr = c(usr[1:2], 0, 1.5))
```

```
+ h <- hist(x, plot = FALSE)
+ breaks <- h$breaks; nB <- length(breaks)
+ y <- h$counts; y <- y/max(y)
+ rect(breaks[-nB], 0, breaks[-1], y, col = "cyan", ...)
+ }
pairs(aa, pch = 19, upper.panel=NULL, diag.panel=panel.hist,
+ gap = 0, cex.labels=1.2, font.labels=1.5)

#Steps II and III, as well as all the GWR commands are
provided in Chapter5-R-Code. Datasets for this task stored in
Chapter5_Data_folder. Run these codes and continue to practice
on your own. This will increase your ability to easily work
with OLS and GWR models. You can also complete Tasks 5.3
through 5.5 using the same R commands.
```

Stay One Step Ahead with Challenge Assignments

Concept: In this chapter, recall that we learned about spatial statistical relationships, including correlation analysis, and OLS and GWR methods. In this task, we analyze semiparametric GWR model outputs.

Task: Analyze and interpret semiparametric GWR model outputs generated from fitted food access models. These models explain spatial variations in food access in different income settings of Chicago. Just like we did in the online exercises for Chapters 2, 3, and 4, we continue to understand the spatial statistical relationships using a food access dataset.

1. Describe the overall spatial patterns in Figures 5.7 through 5.9.
2. Compare and contrast the spatial pattern in the left and right panels of Figures 5.7 and 5.8.
3. Construct a profile for the under- and overpredicted community areas listed later.
4. Interpret the results of food access models presented in Figures 5.7 through 5.9.

Approach/strategy: Use maps and results generated from semiparametric GWR models. The models allow one to mix globally fixed terms and locally varying terms of explanatory variables concurrently. Food Access Models 1 and 2 were fitted using Gaussian GWR.

Food Access Model 1: $FoodStoreSize_i = \beta_0\,(X_i,Y_i) + \beta_1(X_i,Y_i)\mathrm{PCTPOV} + \beta_2(X_i,Y_i)\mathrm{PCTHSD} + \beta_3(X_i,Y_i)\mathrm{PCT18O64} + \beta_4(X_i,Y_i)\mathrm{PCTUEM} + \varepsilon_i$

Results: There were no geographically varying (local) coefficients, especially among the four independent variables (percent households below poverty, percent older than 25 years without high school diploma, percent aged under

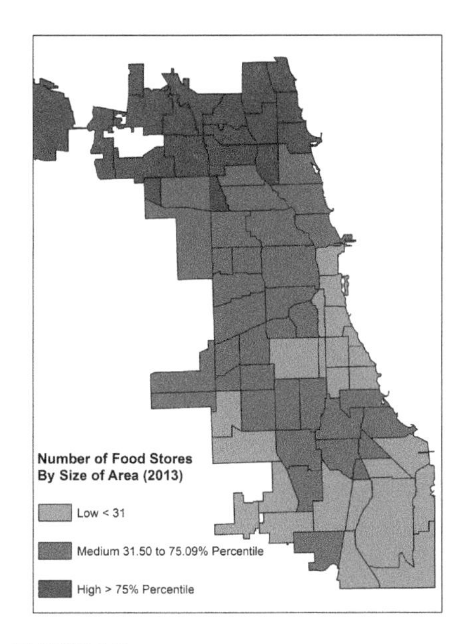

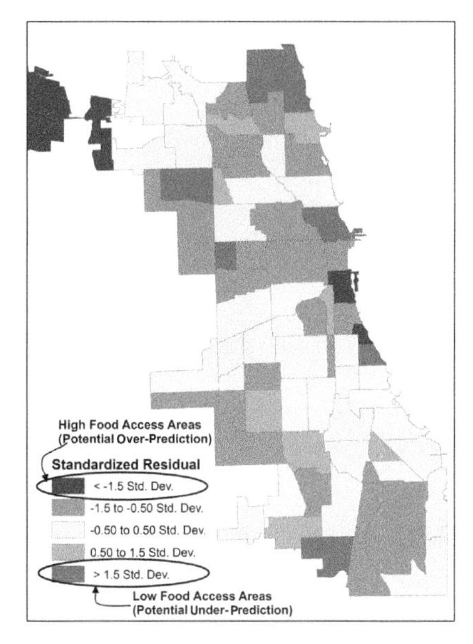

FIGURE 5.7
The left panel shows the spatial distribution of the food stores by area size, while the right panel shows the spatial pattern of standardized residuals. The map on the left panel was generated using ranked and synthesized *t*-observed values of the variables, and the map on the right panel was developed by mapping the standardized residual results of the geographically weighted regression model.

18 or over 64 years, and percent older than 16 years and unemployed) and their intercept. In the Food Access Model 1, the number of food stores per area size served as the dependent variable. The global model was sufficient to explain the spatial variations of food stores per area size in Chicago.

1. Community areas with standardized residuals more than 1.5 are Near North Side, West Pullman, Wet Garfield Park, Edgewater, Belmont Cragin, Rogers Park, Kenwood, and West Ridge.
2. Community areas with standardized residuals less than –1.5 are Near South Side, Oakland, and O'Hare.

Food Access Model 2: $FoodStorePple10k_i = \beta_0 (X_i,Y_i) + \beta_1(X_i,Y_i)\text{PCTPOV} + \beta_2(X_i,Y_i)$ $\text{PCTHSD} + \beta_3(X_i,Y_i)\text{PCT18O64} + \beta_4(X_i,Y_i)\text{PCTUEM} + \varepsilon_i$

Results: There were minor geographically varying (local) coefficients, especially among the four independent variables (percent households below poverty, percent older than 25 years without high school diploma, percent aged under 18 or over 64 years, and percent older than 16 years and unemployed) and their intercept. In the Food Access Model 2, the number of food stores per 10,000 served as the dependent variable. Both the global and local models were used to explain the spatial variations of food stores

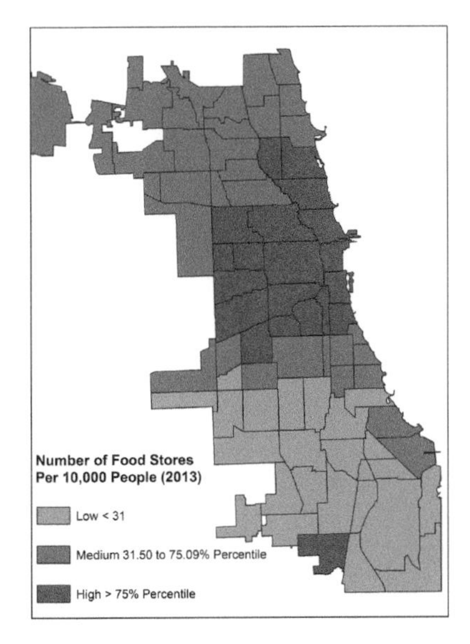

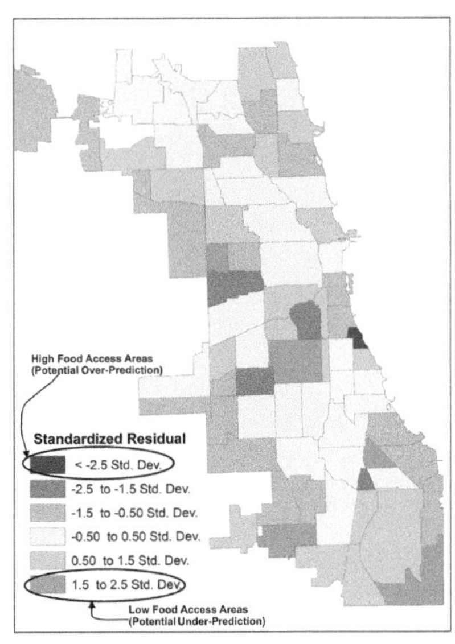

FIGURE 5.8
The left panel shows the spatial distribution of the food stores per 10,000 people, while the right panel shows the spatial pattern of standardized residuals. The map on the left panel was generated using ranked and synthesized *t*-observed values of the variables, and the map on the right panel was developed by mapping the standardized residuals results of the geographically weighted regression model.

per 10,000 people in Chicago. However, the summary statistics for varying (local) coefficients (defined in order) shows that percent older than 16 years and unemployed, percent aged under 18 or over 64 years, and percent older than 25 years without high school diploma were more influential.

1. Community areas with standardized residuals more than 1.5 are Hegewisch, West Garfield Park, Morgan Park, Avalon Park, and New City.
2. Community areas with standardized residuals less than –1.5 are North Lawndale, Gage Park, Bridgeport, Burnside, and Oakland.

Summary of modeling results: In both models, the south of Chicago lies within less than the 31% percentile. Based on models 1 and 2, three explanatory variables (percent older than 16 years and unemployed, percent aged under 18 or over 64 years, and percent older than 25 years without high school diploma) were determined to be more influential in explaining food access in Chicago. The best food access model had an AIC of –520.03 and R^2 of 0.30. In reviewing the food access results, it is essential to seek spatial patterns that are consistent, because different analyses tend to present different angles of the same data. The consistent patterns suggest an underlying spatial structure that must be explicitly interrogated.

TASK 5.3 HOW TO GENERATE AND INTERPRET CORRELATION STATISTICS

1. In this task, we investigate the relationships between 12 agricultural variables (*NO_FARMS07, AVG_SIZE07, AVG_SALE07, CornAcre, CornYield, CornProduction, SoyAcre, SoyYield, SoyProduction, WheatAcre, WheatYield,* and *WheatProduction*). The data for completing this challenge are located in Chapter5_Data_folder (data file: agricul_ILL_stats3.shp). You may use MS Excel or any statistical software that you are familiar with to conduct this correlation analysis. Generate a correlation matrix for these variables (n = 102). Review the correlation results to describe the relationships among these variables. Identify the four strongest and four weakest relationships.
2. Test whether there is a significant difference in correlation coefficients among the following variables: (1) corn acres and corn production, (2) average size and average sale, and (3) number of farms and average size.
3. What insights can we get from analyzing the associations in the agricultural variables?

Possible solutions: Most of the answers are embedded within the previous text. Other answers require some deep critical thinking and data interpretation. The outputs from the three spatial models should be systematically assessed, visually and statistically, to identify the best performing model. This was done for you, and it is contained in the summary report of modeling results, presented earlier. Although we attempted to uncover the underlying spatial relationships of food access in Chicago, it may be more complex than any of three spatial models. Therefore, further evaluation and caution are required in the interpretation of the under- and overpredicted community areas. A validation of the final food access model used to create Figure 5.3 should be done using field observations.

1. Generate diagnostic statistics and graphics (RMSE, R^2, and residual plots using histograms or scatterplots) for your best OLS and GWR models.
2. Analyze the model residuals in the OLS and GWR models, and perform a lack of fitness test (tip: present a histogram or scatterplot displaying the residuals).
3. Does the final model meet the underlying statistical assumptions for regression analysis?

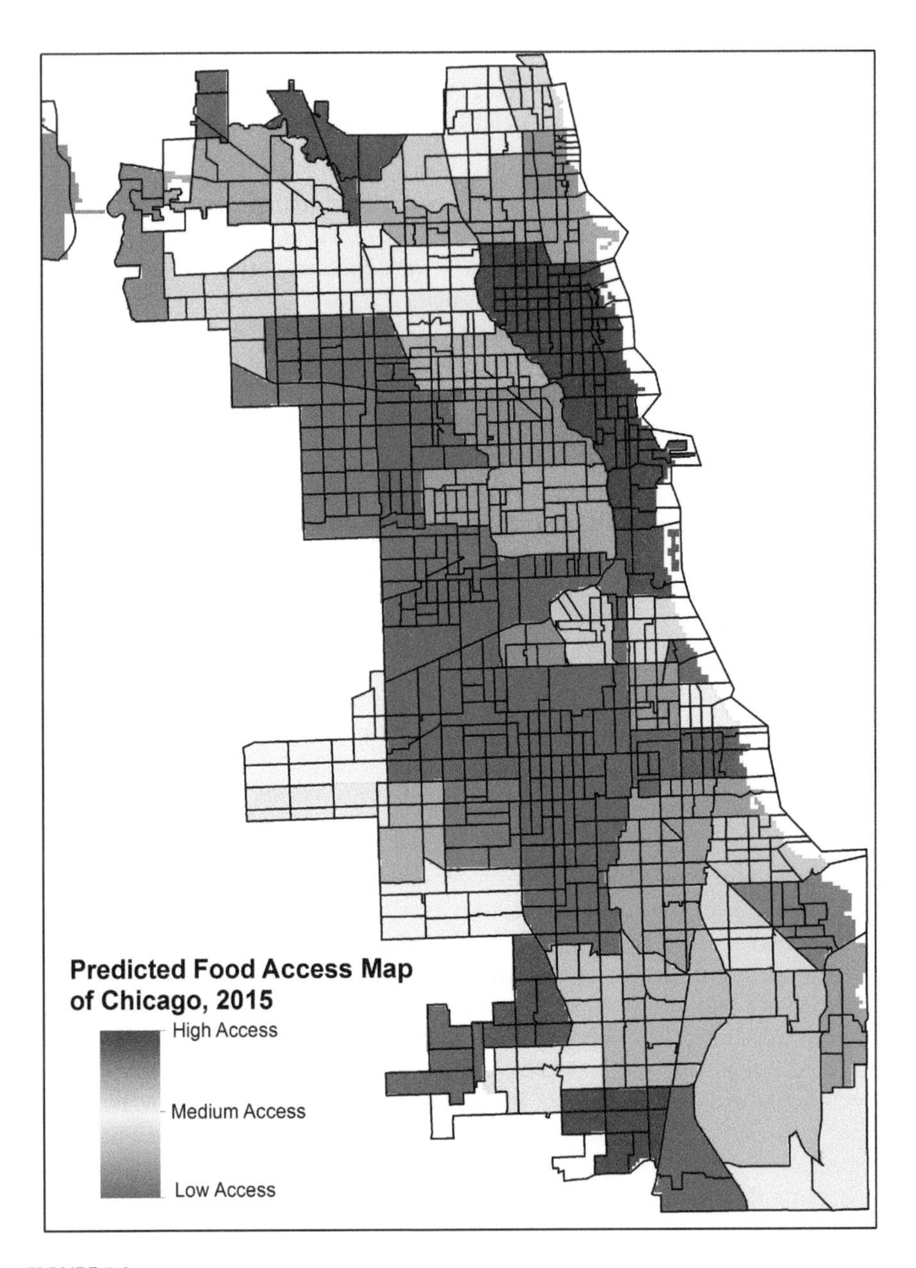

FIGURE 5.9
A final food access map for Chicago. This map was produced using Map Algebra, which combines surface maps of three predictors and their regression coefficients (i.e., percent older than 16 years and unemployed, percent aged under 18 or over 64 years, and percent older than 25 years without high school diploma).

TASK 5.4 HOW TO BUILD AND SUCCESSFULLY RUN OLS AND GWR MODELS

1. Use the following multivariate model to discover relationships between these variables: average farm size, median age, corn yield, soybean yield, and wheat yield (data file: Illinois_agriculture_model.shp). A GWR model (Geographically Weighted Regression (GWR) (2014), GWR Software 3.X and 4.x, Spatial Analysis Research Center, Arizona State University) is available for download at https://geodacenter.asu.edu/gwr_software.
   a. Dependent/response variable, *Y*: *AVG_SIZE07*
   b. Independent/predictor variables, X_n: *MED_AGE, CornYield, SoyYield,* and *WheatYield*
   c. Primary model of interest:

$$AVG\_SIZE07 = \beta_0 + \beta_1 MED\_AGE + \beta_2 CornYield + \beta_3 SoyYield + \beta_4 WheatYield + \varepsilon$$

2. Generate and compile histograms for the five variables (tip: use MS Excel).
3. Generate and compile scatterplots for the five variables. Modify the legend position, title, and axes. Submit the scatterplots in your final report (tip: use MS Excel).
4. Rewrite the multivariate model based on the strength and association observed in the scatterplots.
5. Run the OLS regression to find a properly specified model, and examine the output feature class residuals using the test for spatial autocorrelation.
6. Explain the OLS model and spatial autocorrelation results.
7. The GWR model, where X_1 is any one of the three independent/predictor variables that are statistically significant in the OLS model, and (u, v) represents the coordinates of each location: $AVG\_SIZE07(u, v) = \beta_0(u, v) + \beta_1(u, v)X_1 + \varepsilon(u, v)$.
   a. If you have a properly specified OLS model, and the test for spatial autocorrelation on residual variables shows that they are random, then run the GWR model.
   b. Examine the output feature class residuals using the test for spatial autocorrelation (tip: given the fact that there is significant evidence of global and local multicollinearity

among corn yield, soybean, and wheat yield variables, you can use only one of the explanatory variables from OLS models. This variable should be able to explain the variations in any of the three variables).

8. Explain the GWR model and spatial autocorrelation results.

TASK 5.5 HOW TO GENERATE AND ANALYZE DIAGNOSTIC STATISTICS

Diagnostic statistics include the root mean square error (RMSE) = SQRT(SSE/n – k), where SSE = $\sum(Y_{predicted} - Y_{observed})^2$, n is the number of observations, and k is the number of independent variables plus the intercept; **adjusted coefficient of determination (R^2)**; and **residual plots** using histograms or scatterplots (Figure 5.10).

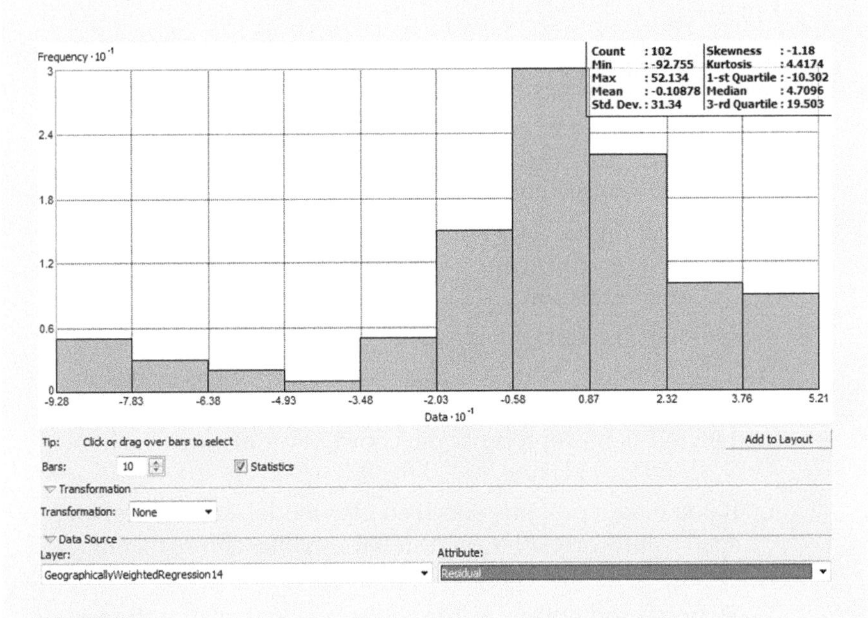

FIGURE 5.10
A histogram of ordered residuals of a geographically weighted regression model.

Review and Study Questions

1. What are the similarities and differences between correlation analysis and regression analysis?
2. Distinguish between a traditional OLS regression model and a spatial regression model. When is it appropriate to use a spatial regression model?
3. What are the key assumptions of a Pearson's correlation analysis?
4. Choose two assumptions of regression analysis, and explain how you would go about validating these assumptions using the regression diagnostic measures.
5. What are the best measures for evaluating the fit of a spatial regression model?

Glossary of Key Terms

Akaike information criterion: This is a statistical measure used to compare two or more competing regression models, and it enables one to choose the model with the best fit for the data. It examines the goodness-of-fit relative to the number of parameters that need to be derived. Models with the lowest values are deemed to be the best.

coefficient of determination: This is also called the R^2 value. It is a measure of overall fit in regression analysis and reflects the proportion of variance of the dependent variable that has been explained by the regression model. It varies from 0 to 1 (can also be expressed as a percentage), and the larger the value, the greater is the overall fit of the regression model.

dependent variable: When evaluating causal relationships, the dependent variable is the response variable, the consequence of events or processes that are characterized by the independent/predictor variables. In a regression equation, the dependent variable is typically denoted as Y, a function of the independent variable, X, and an error term.

homoscedasticity: When performing traditional regression analysis, a core assumption is that the error terms must have constant variance. Violation of the assumption results in a heteroscedastic model that could arise from the omission of an important predictor variable in the analysis. Heteroscedasticity increases the chances of committing a type I error; specifically, it leads to the underestimation of the standard error of the regression coefficients, inflating the t-values and leading one to conclude that the variables are statistically significant (rejecting H_O), when in reality they are not significant.

independent variable: When testing causal relationships, the independent variable plays the antecedent or causal role. It is the predictor variable in the relationship and is used to explain or predict the variability of the dependent variable, *Y*. In a regression equation, the independent variable is denoted as X.

joint *F*-statistics: A goodness-of-fit test that measures the overall fit and significance of the regression model in GWR. It essentially captures the proportion of explained variance relative to the unexplained variance in the dependent variable.

Jarque-Bera statistics: This is also a goodness-of-fit test in GWR. When it is statistically significant, it suggests that the model is biased, and the results are unreliable. A significant statistic could be caused by the omission of an important predictor variable in the regression model.

Moran's *I* coefficient: A useful and popular test of spatial autocorrelation. The measure consists of a value ranging between 0 and 1 signifying the strength of autocorrelation, a positive or negative sign denoting clustering or dispersal, and a related probability value for use in assessing the overall significance. This test can be run after a traditional regression analysis (OLS) to ensure that the residuals are not correlated. If they are, then a GWR is warranted.

multicollinearity: This is a statistical violation in multiple regression analysis that is caused by a high correlation between the predictor variables. This violation results in unstable regression coefficients, insignificant *t*-values, and overestimation of the overall fit of the model. A more severe condition called *singularity* arises when there is a perfect correlation between the predictor variables included in the analysis. Singularity results in a positive definite scenario that prevents the computation of the regression estimates.

Pearson's product moment correlation: A bivariate correlation measure that is used to assess the linear association between interval/ratio scaled variables. The measure consists of a value ranging between 0 and 1 signifying the strength of the association, a positive or negative sign denoting the direction of the relationship, and a related probability value for use in assessing the overall significance.

residual: This is the error term, and it is often denoted as *E* in a regression equation. It captures the portion of the dependent variable that has not been explained by the regression model.

Spearman's rank correlation: An alternate correlation measure that is used to assess the linear association between two ordinal scaled variables, or interval/ratio scaled variables that exhibit significant departures from normality. The measure also consists of a value ranging between 0 and 1 signifying the strength of the association, a positive or negative sign denoting the direction of the relationship, and a related probability value for use in assessing the overall significance.

spuriousness: This is a serious challenge that could arise when evaluating relationships between variables and could potentially lead to confounding results. A spurious variable is one that impacts both the response and predictor variables such that when it is controlled for, or removed from the analysis, the original relationship between the predictor and response variables diminishes or disappears.

time precedence: An important requirement in the testing of causal relationships is the need to avoid a temporal mismatch between the data compiled for the predictor variables and the response variables. Specifically, data generated for the predictor variables must either precede or be concurrent with the data compiled for the response variables.

variance inflation factor: A useful diagnostic measure for identifying collinear variables in the regression equation. The higher the variance inflation factor, the more difficult it is to establish the unique contributions of that variable in the regression analysis.

References

Burt, J.E., G.M. Barber, and D.L. Rigby. 2009. *Elementary Statistics for Geographers*, 3rd ed. New York, NY: Guilford Press.

Chambers, J., W. Cleveland, B. Kleiner, and P. Tukey. 1983. *Graphical Methods for Data Analysis*. Pacific Grove, CA: Wadsworth International Group.

Demšar, U., A.S. Fotheringham, and M. Charlton. 2008a. Combining geovisual analytics with spatial statistics: The example of geographically weighted regression. *Cartographic Journal* 45(3): 182–192.

Demšar, U., A.S. Fotheringham, and M. Charlton. 2008b. Exploring the spatiotemporal dynamics of geographical processes with geographically weighted regression and geovisual analytics. *Information Visualization* 7: 181–197.

Fotheringham, A.S., C. Brunsdon, and M. Charlton. 1996. The geography of parameter space: An investigation of spatial non-stationarity. *International Journal of Geographical Information Systems* 10: 605–627.

Geographically Weighted Regression (GWR) Software GWR 3.X and 4.x for fitting Poisson, Logistic and Gaussian GWR models. Spatial Analysis Research Center, Arizona State University. Accessed on December 9, 2014. Available at https://geodacenter.asu.edu/gwr_software or https://gwrtools.github.io/.

Harris, P., C. Brunsdon, and M. Charlton. 2011a. Geographically weighted principal components analysis. *International Journal of Geographical Information Science* 25(10):1717–1736.

Harris, P., C. Brunsdon, and A.S. Fotheringham. 2011b. Links, comparisons and extensions of the geographically weighted regression model when used as a spatial predictor. *Stochastic Environmental Research and Risk Assessment* 25(2): 123–138.

Kenny, D.A. 1979. *Correlation and Causality*. New York, NY: Wiley.

Nakaya, T. 2011. Evaluating socio-economic inequalities in cancer mortality by using areal statistics in Japan: A note on the relation between the municipal cancer mortality and the areal deprivation index. *Proceedings of the Institute of Statistical Mathematics* 59(2): 239–265.

Nakaya, T., S. Fotheringham, C. Brunsdon, and M. Charlton. 2005. Geographically weighted Poisson regression for disease associative mapping. *Statistics in Medicine* 24: 2695–2717.

Neter, J., M.H. Kutner, C.J. Nachtsheim, W. Wasserman. 1996. *Applied Linear Statistical Models*. 4th ed. Boston, MA: WCB/McGraw-Hill.

Tufte, E. 1983. *The Visual Display of Quantitative Information*. Cheshire, CT: Graphics Press.

Tukey, J. 1977. *Exploratory Data Analysis*. Reading, MA: Addison-Wesley.

6

Engaging in Point Pattern Analysis

Learning Objectives

1. Understand point patterns in a spatial distribution.
2. Explore attribute data using different weighting schemes.
3. Generate and interpret point pattern descriptors.
4. Detect and interpret clustering of spatial point patterns/events.
5. Explore and interpret space-time point patterns.

Introduction

The motivation to work with spatial data is partly driven by the need to gain a deep understanding of the spatial structure of a range of phenomena such as crime incidents, injuries, diseases, retail, or bird nesting sites that are represented by point features. Such features are amenable to point pattern analysis in which emphasis is placed on the complete set of observations as well as the location of each observation and its distance relative to others in the distribution. Although the analysis of the point distributions provides us with fundamental clues about the underlying spatial processes and relationships, the main focus is on the examination of any static evidence of spacing. This evidence is normally depicted as a random or nonrandom pattern. If the point pattern is identified as nonrandom, it can be further described as more clustered than random or more dispersed than random. Therefore, three basic pattern structures exist: random, clustered, or dispersed. These patterns are illustrated in Figure 6.1. In the upper panel, the spatial pattern is clustered and has a large variance. The middle panel is a randomly dispersed pattern, has a moderate variance, and is similar to a Poisson distribution. The lower panel is a dispersed/uniform pattern with no or little variance. The data depicted in this figure are based on the simulation of nesting sites of the African black coucals *(Centropus grillii)* in the Ssezibwa wetlands, north of the town of Kayunga, Uganda. A polygon layer

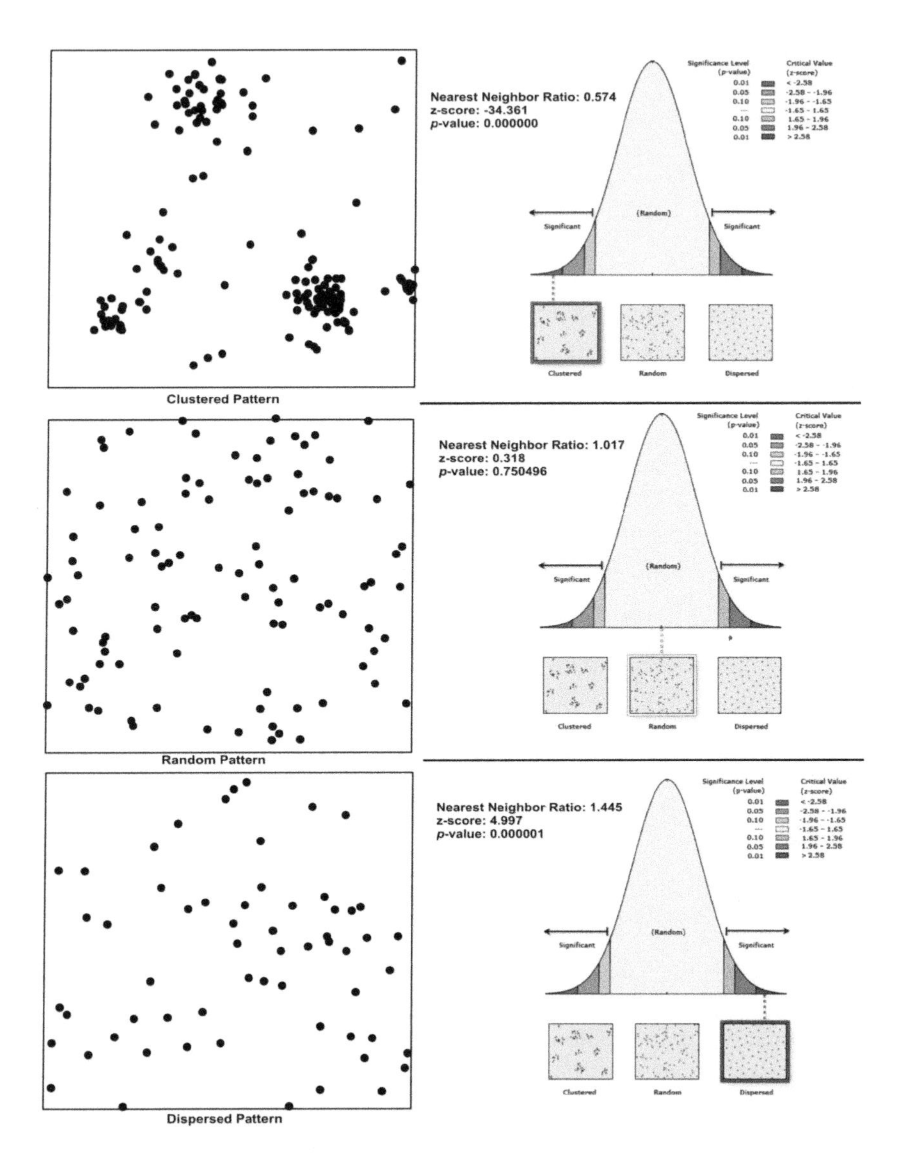

FIGURE 6.1
Three different spatial point patterns showing potential nest locations of the African black coucals in the Ssezibwa wetlands, north of the town of Kayunga, Uganda.

of distribution and habitats for the African black coucals from the International Union for Conservation of Nature's (IUCN) 2012 Red List of Threatened Species database was used to identify the potential nesting sites. Ancillary information compiled from the 2012 aerial and satellite images was used in identifying land cover with open, dense, marshy, or swampy grassland. The area was delineated into a rectangular shape of 1507 m × 1370 m so that all the nest sites were within the boundaries.

The purpose of this chapter is to explore the range of approaches that are used to analyze the distributional patterns of point features such as previously depicted bird nesting sites. Our focus is on quadrat analysis, nearest neighbor, Ripley's *K*-function, and Kernel estimation. Two other methods, Voronoi mapping and Kulldorff's spatial scan statistic, are also introduced at the end of the chapter. Using data drawn from previous research projects, we run through a series of tasks to illustrate the applications of these methods.

Rationale for Studying Point Patterns and Distributions

The statistical tests for studying point distributions rely on the comparison between an observed spatial pattern and a random theoretical pattern (i.e., Poisson distribution). The tests are used to determine the probability of the observed pattern, which may be equal to or more extreme than the critical value at a given significance level. In theory, the distribution of observation points throughout a given study region follows a homogenous Poisson process. The assumption behind this relates to two core principles that define complete spatial randomness (CSR): (1) each event has an equal probability of occurring at any position in the study region, and (2) the position of any event is independent of the position of any other. In framing a statistical test, our goal, therefore, is to test the null hypothesis that the observed pattern is random and is produced by the CSR process. However, there are challenges in meeting this assumption due to the nature of geographic data. First, if we were to explore the absolute locations of a spatial phenomenon, we are bound to encounter a first-order effect (no equal probability). Second, if we were to explore the interactions between locations, we are bound to encounter a second-order effect (no independence). In essence, point pattern descriptors are designed to take these effects into consideration under the CSR process.

Exploring Patterns, Distributions, and Trends Associated with Point Features

A variety of spatial techniques can be used to analyze spatial phenomena that possess discrete spatial properties represented as points on a map. In this chapter, we feature five of these methods and related measures: quadrat count, nearest neighbor, Ripley's *K*-function, kernel estimation, and spatial-time scan statistics. These measures are designed to determine the density of events or interaction between the locations that develop over space and time. For example, we can use the nearest neighbor approach to compute the relationships between pairs of the closest points assigned as neighboring locations (Clark and Evans 1954), or we can use the *K*-function to determine patterns across spatial ranges (Bailey and Gatrell 1995; Fotheringham and Zhan 1996; Gatrell et al. 1996). When studying these point distributions, one must be cognizant of the impact of scale (magnitude of study and extent) on the identified patterns. As described in the introductory chapters of the book, the modifiable areal unit problem (MAUP) problem is an inherent spatial data problem and is definitely one to look out for in point pattern analysis.

Quadrat Count

The quadrat count method determines the point distribution by examining its density over the study area. Analysis is based on subquadrats (or grid cells) that are constructed over a given study area (A). Again, given the MAUP problem, the size of each grid cell is critical and could influence the estimation of measures derived from the analysis. Also, while the most commonly used surfaces in quadrat analysis are square grids, it is important to note that other surfaces can be used depending on the analytical objectives of the study and the nature of the spatial phenomena under investigation. Once the surface is established over the study area, the next step is the quantification of the number of points per cell (subquadrat) and the frequency distribution of points in the entire quadrat. The end goal is to compare the observed distribution of points to a theoretical random pattern to assess whether it is clustered, dispersed, or random. If the results show that events in the population have a randomly dispersed pattern, this confirms that there are a random number of points in each subquadrat. If the results show that points in the population exhibit a dispersed spatial pattern, this confirms that there are a dispersed number of points in each subquadrat. If the results show that points in the population exhibit a clustered spatial pattern, this confirms that the points are concentrated in a few subquadrats, and many are empty. Following are the major steps in conducting a quadrat count analysis:

1. Divide a study area into a set of equal-area quadrats (grid cells). Ideally, the formula for dividing the area is as follows:

$$\sqrt{2}\,.\frac{A}{n}$$

   where A is the study area, and n is the number of points.

2. State the appropriate null hypothesis for the statistical test.
3. Count the number of events falling in each of the subquadrats to create the variable X, and compute the frequency distribution of X.
4. Calculate the observed and expected probability of the points for X.
5. Compute the variance and mean of the variable X.
6. Calculate a chi-square test of hypothesis and the variance mean ratio (VMR).
7. Use the observed statistics and critical value to confirm or deny the null hypothesis.

According to the frequency distribution results for a clustered pattern presented in Table 6.1, there are 68 subquadrats without any nests. Out of 110 subquadrats, 94 have less than 3 nests. The highest concentration of nests is 27, which is in one subquadrat; and anywhere from 8 to 25 nests are clumped in a few of the other subquadrats. The average number of nests per subquadrat is 2.1109. The chi-square statistic is 969.61, and it is statistically significant at $p < .05$.

$$\bar{X} = \frac{\sum X_j f_j}{\sum f_j} = \frac{232.2}{110} = 2.1109$$

$$\text{Chi-square statistic } \chi^2 = 2046.77/2.1109 = 969.61$$

$$P\text{-}value = 4.8248\text{E-}138$$

The frequency distribution results for a randomly dispersed pattern are presented in Table 6.2. In this table, there are 52 subquadrats without any nests. Thirty of the subquadrats have at least one nest, while the rest of the subquadrats have either two or three nests with the exception of one

TASK 6.1 SAMPLE DATA, SYNTHESIS, AND INTERPRETATION OF QUADRAT COUNT MEASURES

If we consider our data of potential nest sites in Figure 6.2, we can assume that a set of locations of these nest sites is represented by S with n events. Each event (nest) is represented by a pair of coordinates (X, Y) in a study area of 2,064,590 m^2. The entire study area was partitioned into 110 square quadrats each measuring 137 $\times$ 137 m. The results for the distribution pattern of nests throughout the landscape are presented in Tables 6.1 through 6.4.

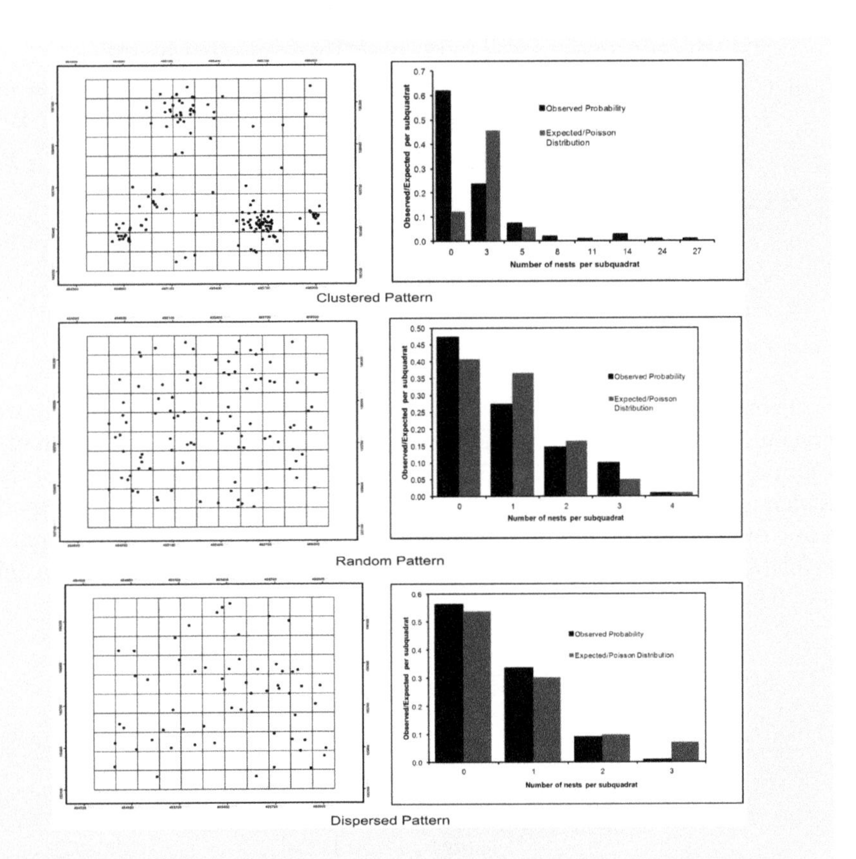

FIGURE 6.2
The spatial distributions of three dispersion patterns are presented as indicated earlier. On the left panels are rectangles with the entire quadrat. Each small 137 × 137 m square represents one subquadrat, and on the right panels are observed and expected quadrat events/nest sites of the African coucals.

TABLE 6.1

Worktable for Chi-Square Test and Nest Dispersions with Their Observed Probability and Expected (Poisson) Distribution Showing a Clustered Pattern

Number of Nests per Subquadrat (x_i)	Number of Subquadrats (f_i)	$f_i x_i$	Observed Probability	$P(x)$	$x_i-\mu$	$(x_i-\mu)^2$	$x(x_i-\mu)^2$
0.0	68.0	0.0	0.6182	0.1211	−2.11	4.46	303.00
2.7	26.0	70.2	0.2364	0.4553	0.59	0.35	9.02
5.4	8.0	43.2	0.0727	0.0570	3.29	10.82	86.55
8.1	2.0	16.2	0.0182	0.0013	5.99	35.87	71.74
10.8	1.0	10.8	0.0091	0.0001	8.69	75.50	75.50
13.5	3.0	40.5	0.0273	0.0000	11.39	129.71	389.13
24.3	1.0	24.3	0.0091	0.0000	22.19	492.36	492.36
27.0	1.0	27.0	0.0091	0.0000	24.89	619.47	619.47
	$\sum = 110$	$\sum = 232.2$					$\sum = 2046.77$

TABLE 6.2

Worktable for Chi-Square Test and Nest Dispersions with Their Observed Probability and Expected (Poisson) Distribution Showing a Randomly Dispersed Pattern

Number of Nests per Subquadrat (x_i)	Number of Subquadrats (f_i)	$f_i x_i$	Observed Probability	$P(x)$	$x_i-\mu$	$(x_i-\mu)^2$	$x(x_i-\mu)^2$
0	52	0	0.4727	0.4066	−0.90	0.81	42.12
1	30	30	0.2727	0.3659	0.10	0.01	0.30
2	16	32	0.1455	0.1647	1.10	1.21	19.36
3	11	33	0.1000	0.0494	2.10	4.41	48.51
4	1	4	0.0091	0.0111	3.10	9.61	9.61
	$\sum = 110$	$\sum = 99$					$\sum = 119.90$

TABLE 6.3

Worktable for Chi-Square Test and Nest Dispersions with Their Observed Probability and Expected (Poisson) Distribution Showing a Dispersed Pattern

Number of Nests per Subquadrat (x_i)	Number of Subquadrats (f_i)	$f_i x_i$	Observed Probability	$P(x)$	$x_i-\mu$	$(x_i-\mu)^2$	$x(x_i-\mu)^2$
0	62	0	0.5636	0.5379	−0.62	0.38	23.83
1.2	37	44.4	0.3364	0.3031	0.58	0.34	12.45
2.1	10	21	0.0909	0.0986	1.48	2.19	21.90
2.8	1	2.8	0.0091	0.0705	2.18	4.75	4.75
	$\sum = 110$	$\sum = 68.2$					$\sum = 62.94$

TABLE 6.4

Worktable for Variance Mean Ratio (VMR) for Potential Nesting Sites

	Mean	Standard Deviation	Variance	VMR
Clustered	2.11	4.31	18.61	8.81
Dispersed	0.62	0.75	0.57	0.92
Random	0.90	1.04	1.09	1.21

subquadrat that has four nests. In general, the distribution pattern among the subquadrats is somewhat random with an average number of 0.9 nests per subquadrat throughout the landscape. The chi-square test statistic is 133.22, and it is not significant ($p > .05$).

$$\bar{X} = \frac{\sum X_j f_j}{\sum f_j} = \frac{99}{110} = 0.90$$

$$\text{Chi-square statistic } \chi^2 = 119.90/0.9 = 133.22$$

$$P\text{-}value = 0.057389389$$

According to the frequency distribution results for a dispersed pattern presented in Table 6.1, there are 62 subquadrats without any nests. Most of the subquadrats have one to three nests that are spatially dispersed throughout the landscape. The average number of nests is 0.62 per subquadrat. The chi-square test statistic is 101.51 and is statistically insignificant.

$$\bar{X} = \frac{\sum X_j f_j}{\sum f_j} = \frac{68.2}{110} = 0.62$$

$$\text{Chi-square statistic } \chi^2 = 62.94/0.62 = 101.51$$

$$P\text{-}value = 0.682243941$$

Chi-square tests were used to determine whether the distribution of nesting sites occurs randomly throughout the landscape. These tests compared observed distributions of nesting sites to Poisson distributions; if we were to find the patterns to be random, then we would conclude that these were produced by CSR. The chi-square test was conducted at a 95% significance level and 109 degrees of freedom, so that there was only a 5% chance of committing a type I error if we were to incorrectly reject the null hypothesis.

According to the chi-square result (see Figure 6.2 and Table 6.1), we reject the null hypothesis that potential nesting sites occur randomly throughout the landscape. Also, the VMR is 8.81, which is significantly greater than 1. This confirms the pattern is clustered. We, therefore, conclude that potential nesting sites were not an outcome of the CSR process.

Based on the statistical results generated from the other distributions and the chi-square tests (see Tables 6.2 and 6.3), we do not reject the null hypotheses that potential nesting sites occur randomly throughout the landscape. The VMR was 0.92 and 1.21 for the dispersed and randomly distributed point patterns, respectively. Both of them are below 1 or barely over 1, suggesting that the point patterns exhibit more randomness than nonrandomness. The chi-square tests were statistically insignificant, implying that the observed distribution patterns were similar to Poisson distributions (see Figure 6.2). This lends further credence to the fact that the patterns are an outcome of the CSR process.

The interpretation of quadrat count results was based on the frequency distribution comparisons. For the VMR results, we expected a VMR that was close to 0 to yield a dispersed distribution, around 1 to yield a random distribution, and greater than 1 to yield a clustered distribution. Although the patterns for the nesting sites reflected this, it should be noted that the dispersed pattern had a VMR that was close to 1. This is not a completely uniform distribution.

Nearest Neighbor Approach

The nearest neighbor approach compares the distances between nearest points and distances that would be expected on the basis of chance or simply measures the distance between an individual point and its nearest neighbor (Clark and Evans 1954). The approach computes the average distance between nearest neighbors in a point distribution (observed distance) and compares it to that of a theoretical pattern (expected distance). This approach assumes that observation points represent a sample in a two- or more-dimensional Euclidean space. Relationships between neighboring points are derived under the Poisson distribution assumption, such that if points are randomly distributed, then they can be used to detect the presence of nonrandomness for any given pattern (Clark and Evans 1954; Bailey and Gatrell 1995; Fotheringham and Zhan 1996; Gatrell et al. 1996). The Euclidean space between two or more objects, or distance, captures neighboring relations, which enables different orders of neighbors to be quantified when studying any given neighboring points. Different-ordered neighbor statistics, first-ordered, second-ordered, and other higher-ordered neighbors can be derived.

In a study region, we have a set of events (N) in a population. Each of the events has a nearest neighbor, which can be represented by r. The observed distances (r_i) defined as $r_1, r_2, r_3, r_4 \ldots r_n$ represent the distance between each item and its closest neighbor in an area (A). The values are expressed using similar units of measurement. To compute the nearest neighbor, we divide the sum of r_i by N to get the mean observed distance (r_o) for all points and compare it with the expected mean distance (r_e). The formulas to complete this analysis are as follows.

The density of points, p, is given by number of points (N) per study area (A):

$$p = \frac{N}{A}$$

Mean observed distance (r_o) is given by

$$r_0 = \frac{\sum_{i=1}^{N} r_i}{N}$$

The expected value of the nearest neighbor distance (r_e) in a hypothetical random pattern is

$$r_e = 0.5\sqrt{\frac{A}{N}} + \left(0.0514 + \frac{0.041}{\sqrt{N}}\right) \times \frac{B}{N}$$

where B is the length of the perimeter of the study area.

The nearest neighbor ratio, R, measures the degree to which the observed distribution departs from the expectation in a random pattern:

$$R = \frac{r_o}{r_e}$$

The divergence from randomness along the R scale is interpreted as follows: When R is equal to 1, the distribution of events in the study region is perfectly random. When R is equal to 0, the distribution of events is completely clustered, and if R is greater than 1, the distribution of events tends toward uniformity. The R scale shows different dispersion patterns and ranges from 0 to 2.149. Small or large divergences are indicative of the underlying processes that are producing a dispersion pattern.

A statistical test of significance is conducted by looking at the difference in the observed and expected mean distance of the nearest neighbor, divided by the standard error:

$$z = \frac{r_o - r_e}{\sigma_r} = \frac{r_o - r_e}{.261/\sqrt{np}}$$

The resulting quantity is a standard normal variable (z) that can be used to evaluate the null hypothesis of randomness.

To summarize, there are six major steps in conducting a nearest neighbor analysis:

1. Calculate the density of points in an area.
2. Derive observed average distances.
3. Determine the hypothetical random pattern.
4. Compute R statistic and perform statistical test.
5. Interpret R statistic (when r_0 is less than r_e, then more clustered patterns are associated with smaller R values; when r_0 is greater than r_e, then more dispersed patterns are associated with larger R values).
6. Calculate the z-scores and use the appropriate critical values to confirm or deny the null hypothesis.

TASK 6.2 SAMPLE DATA, SYNTHESIS, AND INTERPRETATION OF NEAREST NEIGHBOR MEASURES

We use our data of potential nest sites in Figure 6.2. However, the dispersed distribution has been simulated to give a near-perfect uniform pattern. In this analysis, we can assume that a set of locations of these nest sites is located in a study region. Each of the points represented by a pair of coordinates (X, Y) has a closest neighbor that is represented by r in a predefined study area of 2,064,590 m^2. The density of points is 5.38e-5 per square meter.

Results for Three Basic Distributions. The R values in Table 6.5 provide an estimate of the degree of clustering for three distributions, while the test of significance (p value) assumes that the statistical distributions of observed distances were approximately normal.

The observed mean distance, expected mean distance, and z-score for distributions are presented in the same table. The R value for the clustered distribution is less than 1; for the dispersed distribution, it is greater than 1; for the randomly distributed pattern, it is close to 1. The z-score result for the clustered distribution is –10.55, and this is far below –1.96 (the critical value observed at a 0.05 significance level). For the dispersed distribution, the z-score is 15.08, and this too is above the critical value of +1.96 at a significance level of 0.05. Based on these z-scores, we are 95% confident that the two spatial patterns are not randomly distributed. However, when examining the random distribution, the z-score is 0.396, which is below +1.96, so the null hypothesis of spatial randomness cannot be rejected. Overall, based on these statistical results, we can make the following observations regarding the spatial distribution of potential nesting sites of African black coucals (Figure 6.3). The distribution listed in the first row of Table 6.5 shows a significant degree of clustering, implying that potential nesting sites are more clustered than random, and the pattern is not due to the CSR process. In the second row, potential nesting sites are more dispersed than random; thus, the spatial distribution shows a significant level of regularity. In the third row, nesting sites are randomly dispersed and are essentially produced by the CSR process.

TABLE 6.5

Worktable for Nearest Neighbor Analysis for Potential Nesting Sites Showing Results for Three Basic Distributions

	Observed Mean Distance	Expected Mean Distance	Nearest Neighbor Ratio (R)	z-score	p-value
Clustered	31.671	54.30851	0.58316	−10.549206	0.00000
Dispersed	120	68.5	1.751825	15.08495	0.00000
Random	95.231	92.749438	1.026753	0.396444	0.691778

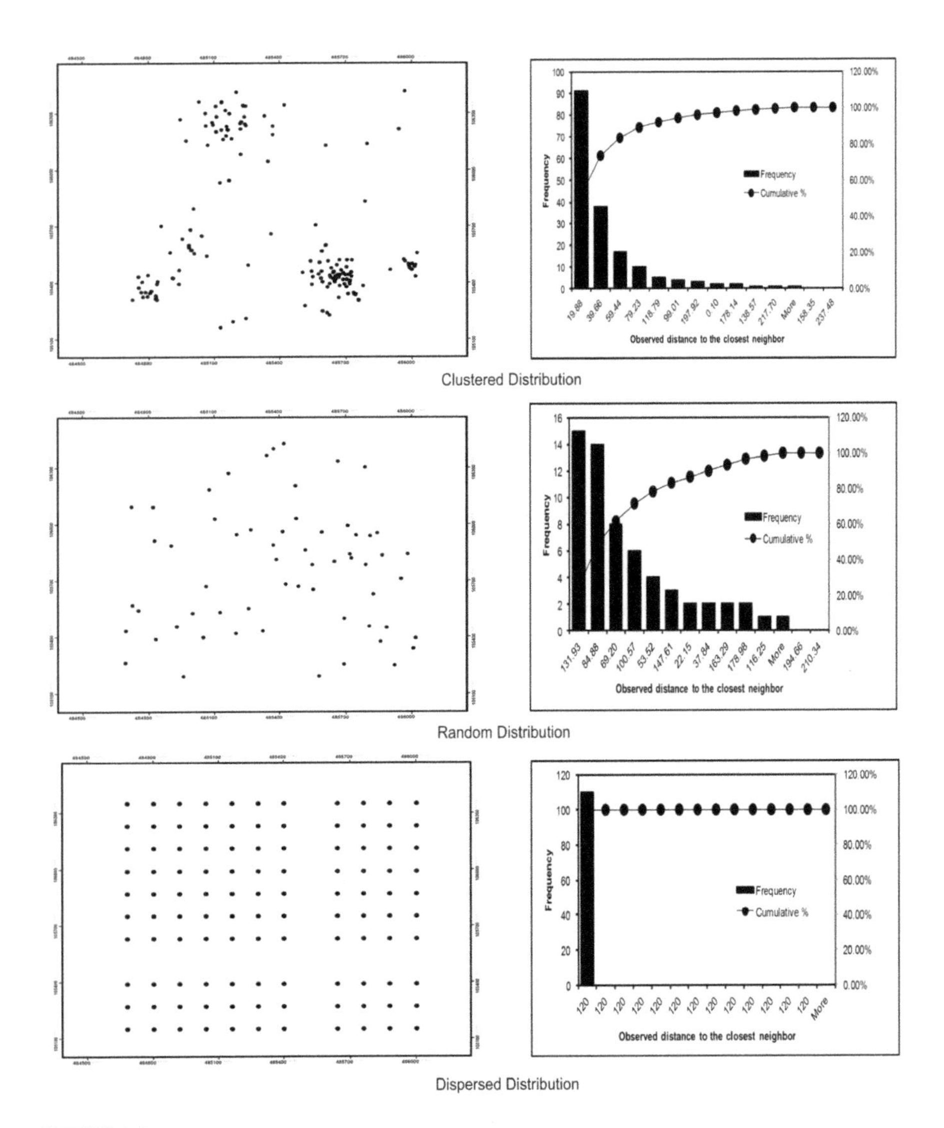

FIGURE 6.3
The spatial distribution of three dispersion patterns and a plot of the observed distance to the closest neighbor.

K-Function Approach

Unlike the nearest neighbor method, which relies on distances only to the closest events, the *K*-function approach explores a spatial pattern across a range of spatial scales (Bailey and Gatrell 1995; Fotheringham and Zhan 1996; Gatrell et al. 1996). It is based on all interevent distances between observation points and provides another way to summarize and fit the models that best describe spatial patterns in a given study region. The *K*-function is given as

$$K(h) = \frac{A}{n^2} \Sigma_{i=1}^{N} \Sigma_{j=1, i \neq j}^{N} \frac{I_h(d_{ij})}{w_{ij}}$$

where *K*(*h*) is the expected number of events inside the radius (*h*), *A* is area, *n* is the number of observed events, and d_{ij} is the distance between events *i* and *j*. $I_h(d_{ij})$ is an indicator function, which is 1 if $d_{ij} \leq h$ and 0 otherwise. The w_{ij} are weights associated with edge correction, which is most often taken as the proportion of the circumference of a circle with radius *h* centered at a point that is contained within the study area. *K*(*h*) is normally graphed against the distances to reveal if any clustering occurs at certain distances. *K*(*h*) should be transformed into a square root function to make it linear *L*(*d*) under a Poisson distribution to a value of zero with clumped alternatives being positive and regular alternatives being negative. This is done by applying Besag's (1977) zero benchmark to normalize the *K*(*h*):

$$L(d) = \sqrt{\frac{K(h)}{\pi}} - h$$

The normalization of *K*(*h*) to *L*(*d*) enables fast computation and simple interpretation of the result. We can evaluate different *K*(*h*) models using simulated confidence envelopes. For example, when *L*(*d*) is equal to zero, the process is considered to be random.

Just like in the previous point pattern methods, the basis for conducting a *K*-function is by comparing the expected and observed distributions. So for any distance, if the observed *L*(*d*) is less or greater than the expected *L*(*d*), the null hypothesis of CSR is rejected at a specified significance level. With this in mind, the interpretation of the *K*-function is as follows: (1) an observed *L*(*d*) greater than the upper limit of the simulations indicates clustering in concentration, (2) an observed *L*(*d*) lesser than the lower limit of the simulations indicates dispersion, and (3) an observed *L*(*d*) in between the lower and upper limit of the simulations indicates a random distribution.

There are six major steps in conducting a *K*-function:

1. Determine/compare the observed and expected *K*. The observed *K* is obtained through the construction of a circle around each point event (*i*), counting the number of other events (*j*) within the radius (*h*) of the circle, and repeating the same process for all other events (*i*).
2. Next, determine the average number of events within successive distance bands. Find the overall point density for the study area. The

observed K is the ratio of the numerator to the density of events. This can then be compared to the expected K, which is a random pattern, $K(h) = \pi h^2$.

3. Transform $K(h)$ estimates into a square root function to make it linear $L(d)$.
4. Determine the confidence envelope by estimating min $L(d)$ and max $L(d)$ values from several simulations at $\alpha = 0.05$ under the null hypothesis of random distribution.
5. Plot $L(d)$ estimates on a graph to reveal if any clustering occurs at certain distances.
6. Interpret the results.

TASK 6.3 SAMPLE DATA, SYNTHESIS, AND INTERPRETATION OF *K*-FUNCTION MEASURES

In this example, we use two datasets drawn from ecological and medical domains to illustrate the application of the *K*-function methods. For the ecological application, we use the bird nesting dataset introduced earlier in this chapter. The null hypothesis is that the distribution of nesting sites is random (nonhomogenous) throughout the landscape under a Poisson distribution and is statistically significant at $\alpha=0.05$. A plot of the results generated for $L(d)$ is shown in Figure 6.4. In the upper panel, the observed distance is above the min $L(d)$ and max $L(d)$, suggesting a clustered distribution. In the middle panel, the observed distance is in between the min $L(d)$ and max $L(d)$, indicating the distribution is random. Finally, in the lower panel, the observed distance is below the min $L(d)$ and max $L(d)$ suggesting a dispersed distribution.

For the medical application, we examine injury location data drawn from the city of Syracuse, New York region. The data were originally obtained from the Department of Emergency Medicine, University of Buffalo. The data consisted of 911 reported calls for all patients transported directly to the trauma center from the scene of injury, incident location, and travel time to trauma center, covering a 6-year study period (1993–1998). Eighty-one percent of 750 incident locations were successfully geocoded to create an injury location database. We use the data to determine whether the distribution of injury locations (and the influence of prehospital travel time from the scene of injury to trauma center) exhibits a random pattern under a Poisson distribution. A plot of the results for $L(d)$ is shown in Figure 6.5. In the upper panel, the observed distance is above the min $L(d)$ and max $L(d)$ that is outside of the confidence envelope. We can therefore conclude that the spatial distributions of injury locations were more clustered than random throughout the Syracuse region. However,

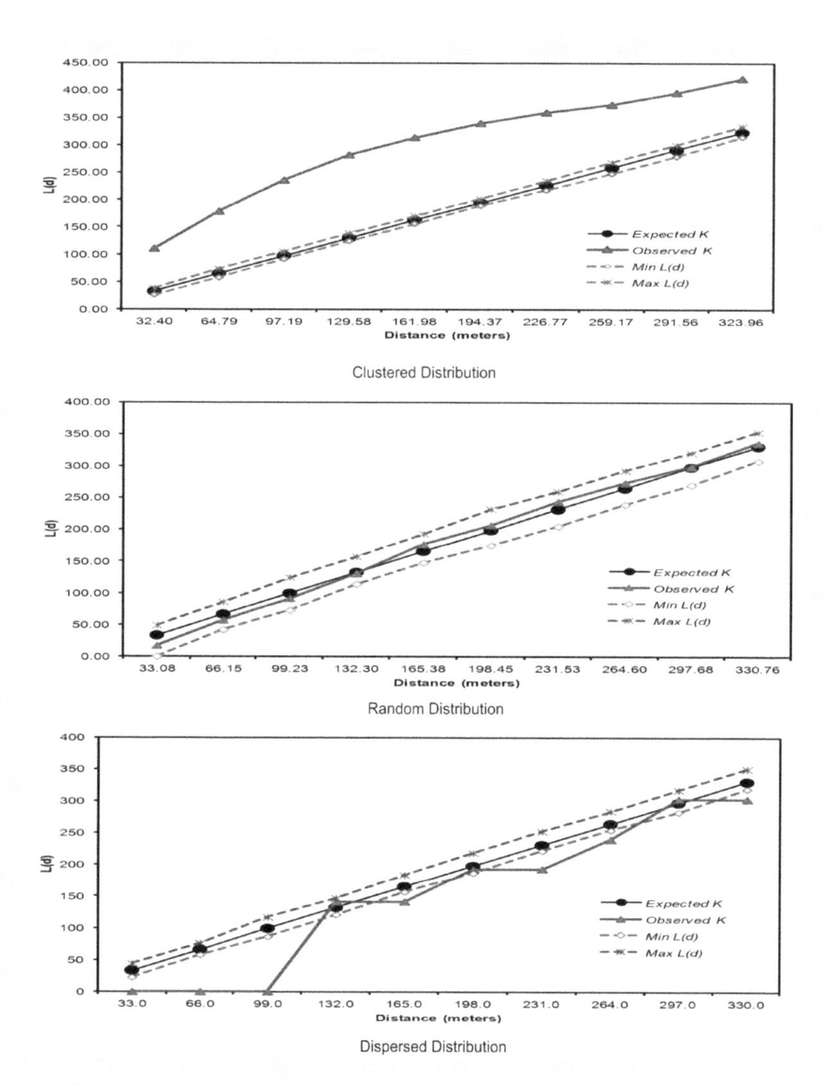

FIGURE 6.4
Plots of $L(d)$ values for three dispersion patterns of an ecological study obtained from the K-function analysis. The findings were generated on the basis of 99 simulations under the null hypothesis of random distribution.

upon weighting with prehospital travel time, a distributional change was observed as depicted in the lower panel. From a distance of 13,115 to 65,576 m, the observed distance is below the min $L(d)$ and max $L(d)$, which is outside of the confidence envelope. This finding suggests that the distribution of injury locations is more dispersed than random. For remaining distances occurring after 65,576 m, the observed distance is in between the min $L(d)$ and max $L(d)$, suggesting that the distribution is more random than dispersed and is due to the CSR process. There is a significant change in $L(d)$ values when prehospital travel time is considered in the distribution analysis. The mixed distribution pattern reminds us of the need to proceed with caution when doing dispersion studies. In this medical example, the results seem to suggest that prehospital travel time is not associated with injury locations.

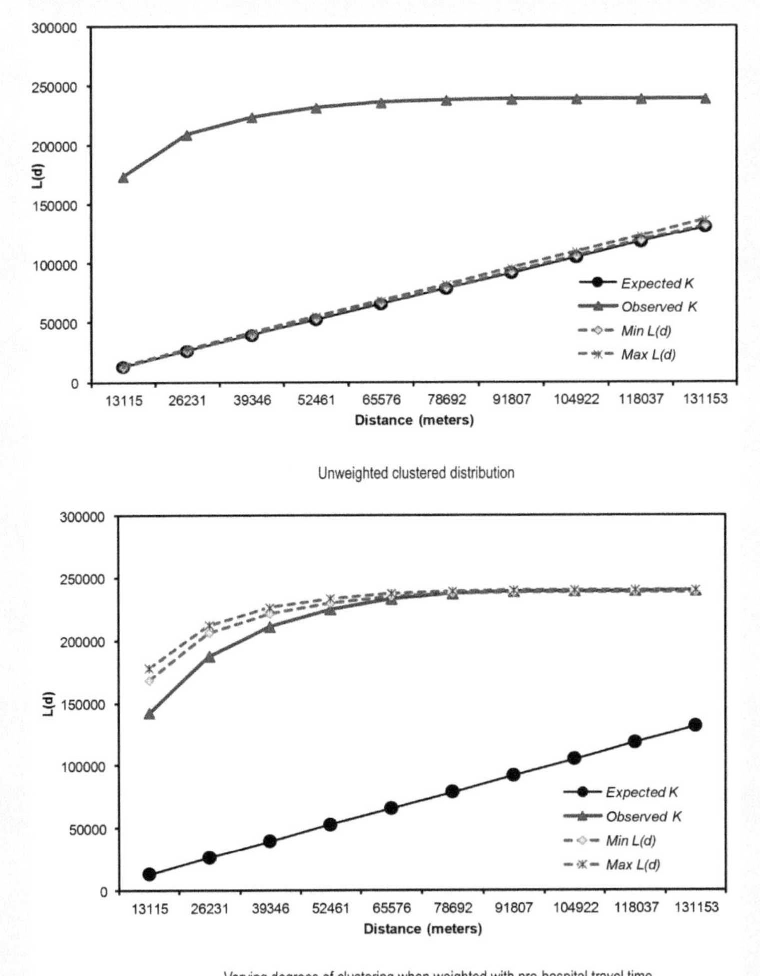

FIGURE 6.5
A plot of $L(d)$ values for injury locations of a medical study obtained from the *K*-function analysis. The findings were generated on the basis of 99 simulations under the null hypothesis of random distribution.

Kernel Estimation Approach

This approach utilizes kernel functions to estimate the density surface of events within a specified radius (bandwidth) around each event in a study region. From a statistical perspective, it is basically a nonparametric method that estimates the probability density function of a random variable. We can apply this method to study the density of events in a study region (*R*) by using a moving two- or more-dimensional function (the kernel).

Each of the events lies in a specified location, which can be represented as *s*. Let s_1, s_2, $\ldots s_n$ be the location of a set of *n* events in a study region, *R*. We can derive the surface intensity of *n* events using this equation (Bailey and Gatrell 1995):

$$\widehat{\lambda_\tau}(s) = \sum_{i=1}^{n} \frac{1}{\tau^2} k\left(\frac{s - s_i}{\tau}\right)$$

where τ is the bandwidth (a smoothing parameter, i.e., radius of the circle), $k(\cdot)$ is the kernel, and $s - s_i$ is the distance between two events (point *s* and s_i).

For an event to be incorporated in a density surface estimate, a suitable kernel function needs to be applied to spread its effect across space (Figure 6.6). What follows is a smoothened density surface, which is produced after summing all the individual kernels across the study region. The influence of an event at s_i to all point events can be adjusted by scaling the kernel function. This function provides an appropriate interpolation technique for generalizing individual-level events in a given location to the entire study region. The kernel density estimation results can be displayed by either using surface maps or contour maps. It is also possible to construct a histogram of a kernel density estimate.

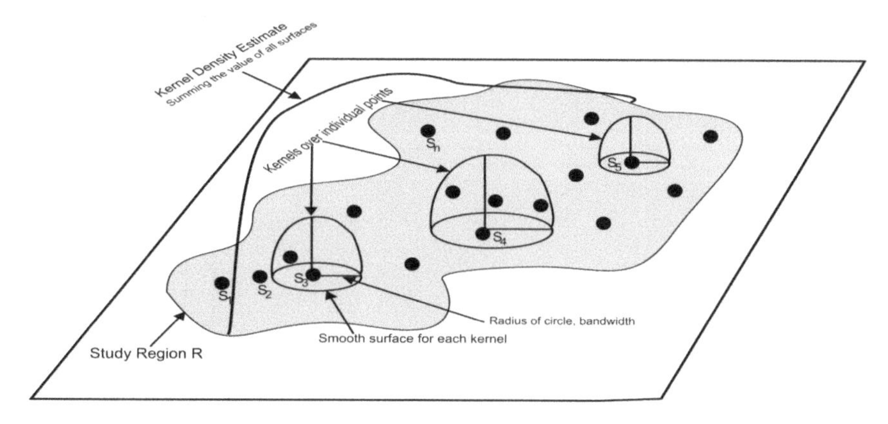

FIGURE 6.6
The kernel estimation method applied to study region *R*.

TASK 6.4 SAMPLE DATA, SYNTHESIS, AND INTERPRETATION OF THE KERNEL DENSITY MEASURES

In this example, we apply the kernel estimation method on the injury location dataset that was previously described for the *K*-function. The maps from the kernel estimation method are presented in Figure 6.7. Two continuous injury density surfaces with population field as prehospital travel time are given in Maps B (5000 m bandwidth) and C (10,000 m bandwidth). Although Maps B and C have similar spatial patterns in terms of their density surface, the 10,000 m bandwidth provides a more generalized density pattern and spatial extent than the 5000 m bandwidth. This is because the 10,000 m bandwidth includes a larger number of point events in its calculations.

In both Maps B and C, the intensity of injury locations was apparent in the central portion of the study region. The highest density surface values were located within the center surrounding the trauma center, and because of distance decay, the surface values gradually level off. There is a diminishing of density surface values as one moves farther away, implying that there were more observed injury locations that fell in this neighborhood than farther away. The results offer confidence in the spatial patterning of injury locations and the role of the trauma center.

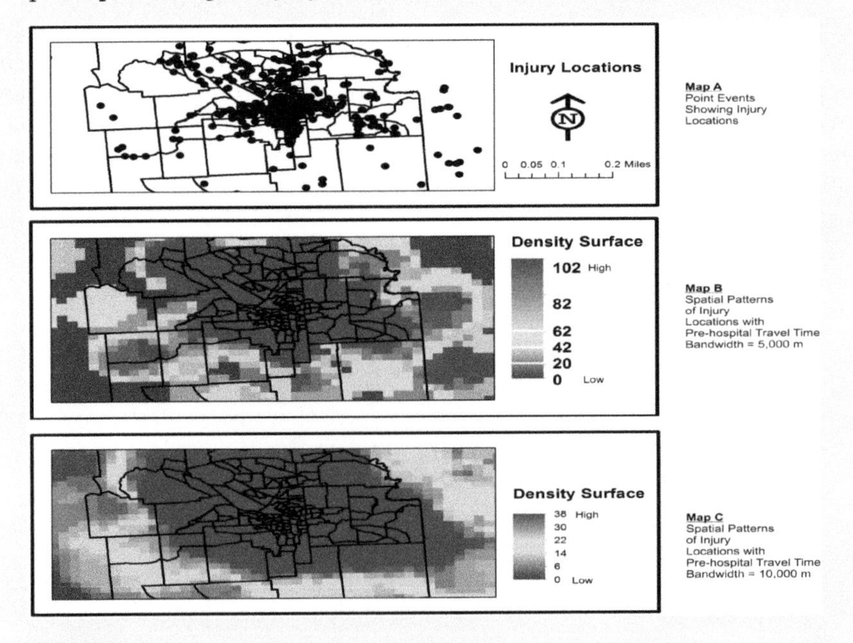

FIGURE 6.7
Intensity patterns of injury location relative to prehospital travel time in Syracuse. Kernel density interpolation of injury location estimates with prehospital travel time as a population field.

The kernel function calculates the probability density of an event at a specified distance using an observed reference point. However, the event intensity of spatial point patterns is contingent upon kernel type and bandwidth. There are different kernel types including normal, uniform, triangular, quartic, and Gaussian. The kernel density estimation is useful for characterizing spatial patterns of point events and is normally employed in many spatial applications, including population density, housing density, crime, ecology, and health. Let us work through Task 6.4 to illustrate this application.

Constructing a Voronoi Map from Point Features

Along with the methods described in the preceding section on exploring patterns, distributions, and trends associated with point features, Voronoi maps are fundamental tools for uncovering the geometric structures that underlie spatial data. They have been used in applications that draw from the location of point features to delineate space into so-called "spheres of influence," such as trade areas in the retail industry, in hospital service areas, and more. The Voronoi method offers an excellent example of how spatial analysis builds upon the synergies between multiple analytical domains including mathematics, computational geometry, and geographic information system (GIS). The technique offers a computational means to partition a plane (space) using a set of individual points into convex polygons (Klein 1989).

TASK 6.5 SAMPLE DATA, SYNTHESIS, AND INTERPRETATION OF A VORONOI MAP

In this example, we create a Voronoi map for the bird nesting locations of the African black coucals dataset that was used in earlier sections of this chapter. Figure 6.8 presents Voronoi maps for three spatial patterns for nesting locations and comparison histograms of areas of proximity polygons. In the upper panel, the area of proximity polygons is very tight; most nests are in close proximity to each other, frequency distribution is heavily skewed, and there are five potential spatial clusters of nesting sites encoded with yellow color. In the middle panel, most of the individual nests fall in their own polygon (Voronoi cell), and this pattern was repeated throughout the study region. The frequency distribution of nesting sites is skewed with most observations occurring within three sets of areas of proximity polygons. In the lower

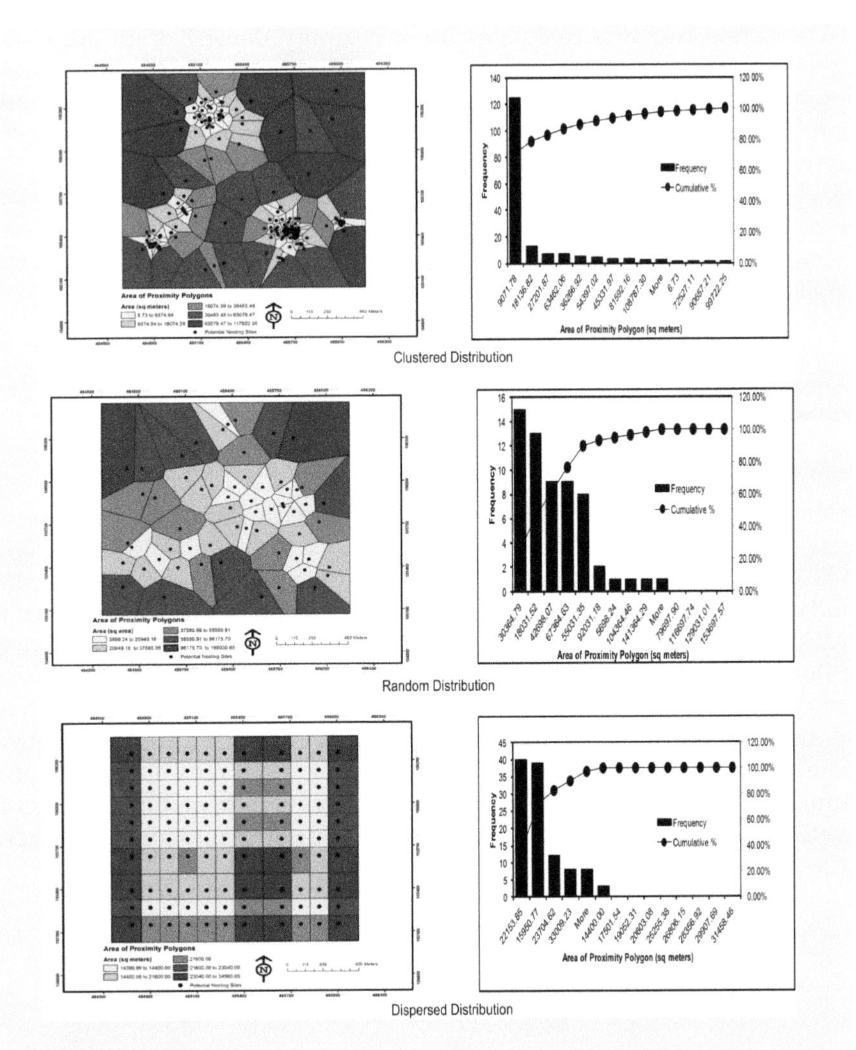

FIGURE 6.8
Voronoi maps illustrating three-point distributions for potential nesting sites in relation to area and comparison histogram of area of proximity polygons.

panel, individual nesting sites are uniformly distributed and dispersed throughout the study region. There is one nest site in each polygon (Voronoi cell). We can draw conclusions about the type of spatial patterns of potential nesting sites of African black coucals based on these Voronoi maps. Overall, the upper panel shows a clustered spatial pattern, the middle panel shows a random spatial pattern, and the lower panel shows a dispersed spatial pattern.

Let us assume we have a set of points, n, with the following points $\{v_1, v_2, \ldots, v_n\}$ in a Euclidean space. Each site v_k is simply a point and has a corresponding Voronoi cell, R_K, which consists of every point whose distance is less than or equal to any other site. We can use a Voronoi diagram to measure the proximity of a polygon area to a particular event or investigate a source of concern.

Exploring Space-Time Patterns

While several of the applications introduced so far in this chapter are based on point patterns that are rooted in space, it is important to point out that there are several geographic problems that call for and entail the use of space-time applications. The visualization and analysis of space-time patterns were inspired by the groundbreaking work of Hägerstrand (1970). Hägerstrand's idea focused on how we can better understand human spatial activity through the space-time path concept. He identified two types of human activities that could be examined using this concept: fixed and flexible activities. A fixed activity entails core aspect of an individual's schedule that occur at a defined location, while a flexible activity represents any secondary activity an individual would schedule or engage in. Flexible activity may occur around a fixed activity. It is important to note that individual activities a person may engage in are typically constrained by spatial and temporal factors. An individual's travel activities have origin and destination locations with start and end times.

Decades later, Miller (1991) illustrated how Hägerstrand's space-time path concept could be extended into new areas. Harvey not only modeled individuals' accessibility to an environment using space-time prism concepts but also advocated for its widespread use in spatial modeling and analysis.

Motivated by this prior work, GIS/spatial analysts now apply these perspectives to study different spatial phenomena, including incidents of crimes and diseases, tweet movements on the Twitter network, and consumer activities on social media and online shopping sites. The new knowledge that is derived can be useful for synthesizing life trajectories and the development of superior study hypotheses for more in-depth studies. Specifically, the space-time perspectives can be used to construct three-dimensional (3D) distributional characteristics (space-time path, space-time prism, and potential path area [PPA] space-time) of any human activity or moving objects provided that they are constrained by physical or virtual spaces. We explore PPA and activity space (AS) concepts using R.

Although many space-time methods are available (Groff 2007), one that has been used frequently to analyze crime as well as disease incidents is Kulldorff's space-time scan statistic. In an effort to find statistically significant clusters, Kulldorff's space-time method employs an elliptic search window to determine whether the point process is purely random or if any potential clusters exist in the study area under the homogenous Poisson distribution. Within each search window, the method assigns a likelihood function to a potential cluster, which is then compared with a randomly generated theoretical pattern. To compute the spatial scan statistic, a circular window is imposed on the map, and the center of the circle is allowed to move flexibly over the area to include different neighborhood positions within each search window (Kulldorff 2001). A likelihood value (Kulldorff 2001) is then calculated for each window using this formula:

$$S = \frac{\max_{z}\{L(Z)\}}{L_0} = \max_{z}\left\{\frac{L(Z)}{L_0}\right\}$$

Given a total number of observed incidents, *N*, the definition of the spatial scan statistic *S* is the maximum likelihood ratio over all possible circles *Z*. *L(Z)* is a measure of how likely it is that the observed data (in our example, it is the rate of crime incidents) within the window are different from out of the window. The maximum likelihood ratio test statistic (L_0) is calculated under the null hypothesis of no spatial heterogeneity in the spatial distribution of observations.

A single *p*-value is generated for the test of null hypothesis through Monte Carlo simulations, and the theoretical pattern reflects the number of random replications on the basis of a number of simulations (at least 999 to ensure excellent power). The theoretical patterns are compared with the observations, and if the observations are among the highest 5%, then the test is significant at the 0.05 level (Kulldorff 1997; Kulldorff et al. 2006; Dai and Oyana 2008). Based

TASK 6.6 SAMPLE DATA, SYNTHESIS, AND INTERPRETATION OF KULLDORFF'S SCAN STATISTIC

For this task, we utilize crime data covering a 5-year study period (2008–2012) for the city of Spokane, Washington. The data were obtained from the city's GIS website (http://www.spokanecity.org/services/gis/). The crime dataset provides individual-level incident location information for different types of crimes. In this case study, we chose to examine the overall crime patterns and dynamics, and we also focused on detecting the clusters for two crimes in particular—theft and burglary crimes. We applied the space-time prospective scan statistic to study crime incidents over a 5-year period. The crime incidents were analyzed at two spatial levels: individual and group.

In 2010, Spokane had an estimated population of 210,000 with most people living in the north and south. Very few people lived in the central and western portions of the city. The city has 27 neighborhoods and covers an area of 156 km^2. Figure 6.9 provides a 3D representation of crime rate distribution by neighborhood during the study period. In this figure, it appears that crime rates were highest in the Riverside, West Central, Cliff/Cannon, and Bemiss neighborhoods. However, the rates are misleading because these neighborhoods have a low population (a small number problem), thus inflating the crime rates. One must exercise caution in interpreting the crime rates and to the extent possible, conduct a more detailed spatial analysis such as one shown in Figure 6.10. In this figure, the space-time clusters of crime incidents are far more realistic than the observed trends in Figure 6.9.

The space-time prospective analyses detected four sets of clusters of crime incidents shown in Figure 6.10. The life trajectory of overall crime incidents in Spokane has a start date of January 1, 2011, to December 31, 2012 (see Figure 6.10: Map B). The biggest set was detected in Bemiss, Hillyard, Chief Garry Park, and Minnehaha. The second set was located in Brownes Addition, Peaceful Valley, and Riverside. The third set was detected in West Hills, while the fourth set was a big cluster that overlapped in six neighborhoods, including Latah Valley, Comstock, Lincoln Height, Southgate, Rockwood, and Manito/Cannon Hill. Burglary had a similar spatial pattern to overall crime incidents (see Figure 6.10: Map C). However, the spatial pattern for theft was slightly different from burglary; three sets of theft clusters were detected in Bemiss, Hillyard, Chief Garry Park, and Minnehaha; Brownes Addition; and West Hills (see Figure 6.10: Map D).

Five life trajectories were built from the space-time prospective analyses of crime incidents. The most important life trajectory was the one that overlapped in four neighborhoods. Most of the detected clusters in 2011–2012 were common in three analyses, suggesting a consistent finding among the different types of crime incidents in the city of Spokane.

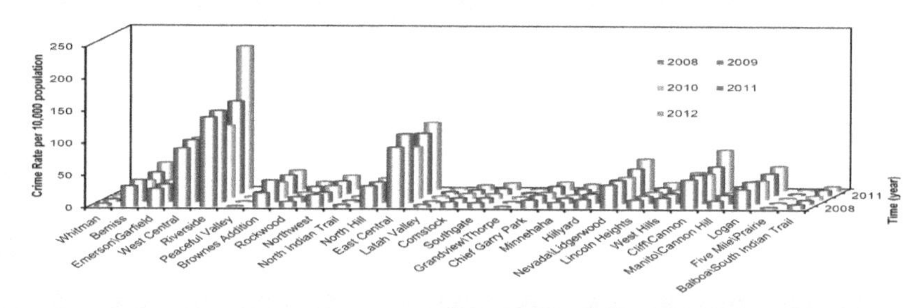

FIGURE 6.9
Three-dimensional representation of crime rate per 10,000 people between 2008 and 2012.

FIGURE 6.10
Crime incident patterns and dynamics in the city of Spokane. Map A provides the place names of the neighborhood; Map B provides four space-time clusters derived from a spatial space-time scan statistic; Map C provides four space-time clusters derived from a spatial space-time scan statistic; and Map D provides three space-time clusters derived from a spatial space-time scan statistic.

on these statistics, we can reject the null hypothesis and specify the estimated location of the most likely space-time cluster of the events.

Conclusions

One of the most fundamental applications in spatial analysis is point pattern analysis. In this chapter, we explored a number of approaches for quantifying the pattern of these distributions from the most basic using quadrat analysis to more complex approaches that use circular/cylindrical windows to characterize the events in space and time. For each of the techniques, we have provided the analytical steps and examples to help you learn how to compute and synthesize the results. Following are more examples and sample exercises.

Worked Examples in R and Stay One Step Ahead with Challenge Assignments

Explore Potential Path Area and Activity Space Concepts

Create two points to model an individual's travel routine with eight different activities. Suppose this individual is a college student at Makerere University Main Campus, Kampala, Uganda, with a daily journey to lectures and then back to the hostel. This journey to attend lectures and then back to the hostel is done by two other students. Specific geographic coordinates have been created to represent these students' single-trip activity to attend lectures and return to a nearby hostel with following specific daily activities based on a 24-hour time stamp (Figure 6.11).

1. Wake up
2. Start to leave for campus
3. Eat breakfast in a nearby student restaurant
4. Arrive at campus

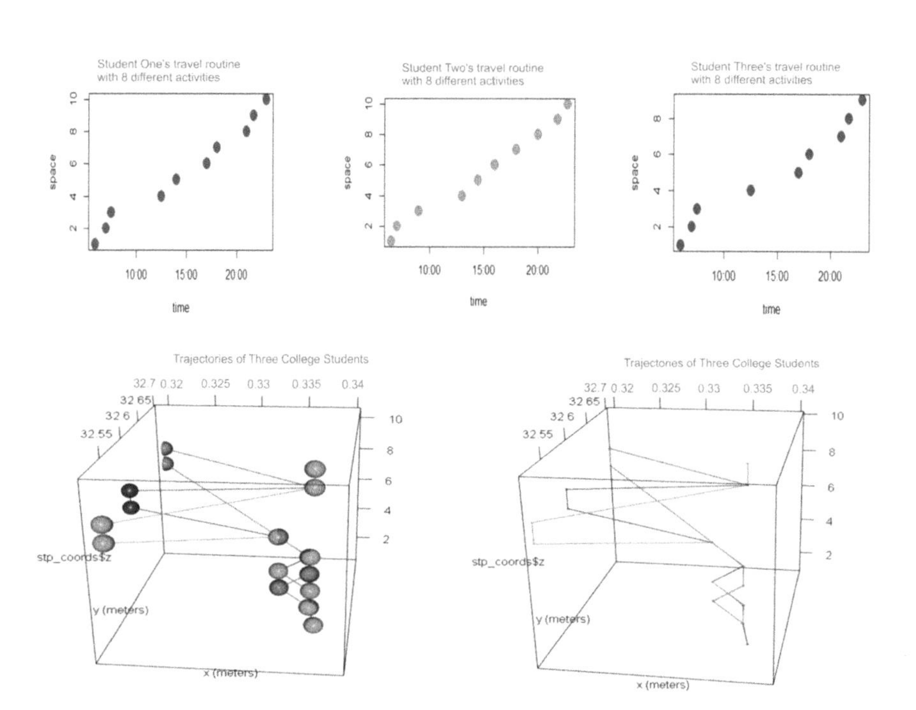

FIGURE 6.11
Representations of potential path area and activity space for three college students' daily routines (generated using the R Code in explore PPA and activity space section).

5. Leave for another destination
6. Arrive at another destination
7. Start to leave for home
8. Arrive at home

PPA and AS Script of Three Makerere University Students

```
## Load packages
# Ensure you are running this code with R Version 3.6.2 and
RStudio Version 1.2.5033
library(sp)
library(spacetime)
library(knitr)
library(rgl)
library(xts)
library(rgeos)
require(rgdal)
knit_hooks$set(webgl = hook_webgl)
library(STPtrajectories)
library(trajectories)
library(zoo)
library(digest)

#POSIXct is the number of seconds from a specific date and
time.
#Student One
#Generate the start time of each activity
t1 <- as.POSIXct("27/01/2020 23:00:00″, "%d/%m/%Y %H:%M:%S",
origin = "1960-01-01", tz = "UTC") # sleeping, Wake up
t2 <- as.POSIXct("27/01/2020 06:00:00", "%d/%m/%Y %H:%M:%S",
origin = "1960-01-01",tz = "UTC") # Starts to leave for campus
t3 <- as.POSIXct("27/01/2020 07:00:00", "%d/%m/%Y %H:%M:%S",
origin = "1960-01-01",tz = "UTC") # Eats breakfast in a nearby
student restaurant
t4 <- as.POSIXct("27/01/2020 07:30:00", "%d/%m/%Y %H:%M:%S",
origin = "1960-01-01",tz = "UTC") # Arrives at campus and start
morning lectures
t5 <- as.POSIXct("27/01/2020 17:00:00", "%d/%m/%Y %H:%M:%S",
origin = "1960-01-01",tz = "UTC") # start to leave for another
destination
t6 <- as.POSIXct("27/01/2020 18:00:00", "%d/%m/%Y %H:%M:%S",
origin = "1960-01-01",tz = "UTC") # Arrives at another
destination
t7 <- as.POSIXct("27/01/2020 21:00:00","%d/%m/%Y %H:%M:%S",
origin = "1960-01-01",tz = "UTC") # Starts to leave for hostel
t8 <- as.POSIXct("27/01/2020 21:45:00", "%d/%m/%Y %H:%M:%S",
origin = "1960-01-01",tz = "UTC") # Arrives at hostel
t9 <- as.POSIXct("27/01/2020 12:30:00", "%d/%m/%Y %H:%M:%S",
origin = "1960-01-01",tz = "UTC") # goes for lunch,
```

```
t10 <- as.POSIXct("27/01/2020 14:00:00", "%d/%m/%Y %H:%M:%S",
origin="1960-01-01",tz="UTC") # start afternoon lectures,
start_time <- c(t1,t2,t3,t4,t9,t10,t5,t6,t7,t8)

#Generate the end time of each activity
e1 <- as.POSIXct("27/01/2020 06:00:00", "%d/%m/%Y %H:%M:%S",
origin="1960-01-01",tz="UTC") # wake up and preparation to
leave for campus
e2 <- as.POSIXct("27/01/2020 07:00:00", "%d/%m/%Y %H:%M:%S",
origin="1960-01-01",tz="UTC") # Finish Break Fast
e3 <- as.POSIXct("27/01/2020 07:30:00", "%d/%m/%Y %H:%M:%S",
origin="1960-01-01",tz="UTC") # Arrive at campus
e4 <- as.POSIXct("27/01/2020 17:00:00", "%d/%m/%Y %H:%M:%S",
origin="1960-01-01",tz="UTC") # Leave for another destination
e5 <- as.POSIXct("27/01/2020 18:00:00", "%d/%m/%Y %H:%M:%S",
origin="1960-01-01",tz="UTC") # Arrive at destination
e6 <- as.POSIXct("27/01/2020 21:00:00", "%d/%m/%Y %H:%M:%S",
origin="1960-01-01",tz="UTC") # Do personal matters and
leave for hostel
e7 <- as.POSIXct("27/01/2020 21:45:00", "%d/%m/%Y %H:%M:%S",
origin="1960-01-01",tz="UTC") # Arrives at hostel
e8 <- as.POSIXct("27/01/2020 23:00:00", "%d/%m/%Y %H:%M:%S",
origin="1960-01-01",tz="UTC") # End of the day,
e9 <- as.POSIXct("27/01/2020 12:30:00", "%d/%m/%Y %H:%M:%S",
origin="1960-01-01",tz="UTC") # morning lectures ends,
e10 <- as.POSIXct("27/01/2020 14:00:00", "%d/%m/%Y %H:%M:%S",
origin="1960-01-01",tz="UTC") # lunch ends,
end_time <- c(e1,e2,e3,e9,e10,e4,e5,e6,e7,e8)

#Student two
#Generate the start time of each activity
st1 <- as.POSIXct("27/01/2020 22:40:00", "%d/%m/%Y %H:%M:%S",
origin="1960-01-01",tz="UTC") # sleeping, Wake up
st2 <- as.POSIXct("27/01/2020 06:30:00", "%d/%m/%Y %H:%M:%S",
origin="1960-01-01",tz="UTC") # Starts to leave for campus
st3 <- as.POSIXct("27/01/2020 07:00:00", "%d/%m/%Y %H:%M:%S",
origin="1960-01-01",tz="UTC") # Eats breakfast in a nearby
student restaurant
st4 <- as.POSIXct("27/01/2020 09:00:00", "%d/%m/%Y %H:%M:%S",
origin="1960-01-01",tz="UTC") # Arrives at campus
st5 <- as.POSIXct("27/01/2020 16:00:00", "%d/%m/%Y %H:%M:%S",
origin="1960-01-01",tz="UTC") # start to leave for another
destination
st6 <- as.POSIXct("27/01/2020 18:00:00", "%d/%m/%Y %H:%M:%S",
origin="1960-01-01",tz="UTC") # Arrives at another destination
st7 <- as.POSIXct("27/01/2020 20:00:00","%d/%m/%Y %H:%M:%S",
origin="1960-01-01",tz="UTC") # Starts to leave for hostel
st8 <- as.POSIXct("27/01/2020 21:45:00", "%d/%m/%Y %H:%M:%S",
origin="1960-01-01",tz="UTC") # Arrives at hostel
```

```
st9 <- as.POSIXct("27/01/2020 13:00:00", "%d/%m/%Y %H:%M:%S",
origin = "1960-01-01",tz = "UTC")# goes for lunch,
st10 <- as.POSIXct("27/01/2020 14:30:00", "%d/%m/%Y %H:%M:%S",
origin = "1960-01-01",tz = "UTC") # Start afternoon lectures,
sstart_time <- c(st1,st2,st3,st9,st10,st4,st5,st6,st7,st8)

#Generate the end time of each activity
se1 <- as.POSIXct("27/01/2020 06:30:00", "%d/%m/%Y %H:%M:%S",
origin = "1960-01-01",tz = "UTC") # wake up and preparation to
leave for campus
se2 <- as.POSIXct("27/01/2020 07:00:00", "%d/%m/%Y %H:%M:%S",
origin = "1960-01-01",tz = "UTC") # Finish Break Fast
se3 <- as.POSIXct("27/01/2020 09:00:00", "%d/%m/%Y %H:%M:%S",
origin = "1960-01-01",tz = "UTC") # Arrive at campus
se4 <- as.POSIXct("27/01/2020 16:00:00", "%d/%m/%Y %H:%M:%S",
origin = "1960-01-01",tz = "UTC") # Leave for another
destination
se5 <- as.POSIXct("27/01/2020 18:00:00", "%d/%m/%Y %H:%M:%S",
origin = "1960-01-01",tz = "UTC") # Arrive at destination
se6 <- as.POSIXct("27/01/2020 20:00:00", "%d/%m/%Y %H:%M:%S",
origin = "1960-01-01",tz = "UTC") # Do personal matters and
leave for hostel
se7 <- as.POSIXct("27/01/2020 21:45:00", "%d/%m/%Y %H:%M:%S",
origin = "1960-01-01",tz = "UTC") # Arrives at hostel
se8 <- as.POSIXct("27/01/2020 22:40:00", "%d/%m/%Y %H:%M:%S",
origin = "1960-01-01",tz = "UTC") # End of the day,
se9 <- as.POSIXct("27/01/2020 13:00:00", "%d/%m/%Y %H:%M:%S",
origin = "1960-01-01",tz = "UTC") # end morning lectures
se10 <- as.POSIXct("27/01/2020 14:30:00", "%d/%m/%Y %H:%M:%S",
origin = "1960-01-01",tz = "UTC") # Comes back for lunch,
send_time <- c(se1,se2,se3,se9,se4,se10,se5,se6,se7,se8)

#Student Three
#Generate the start time of each activity
tt1 <- as.POSIXct("27/01/2020 23:40:00", "%d/%m/%Y %H:%M:%S",
origin = "1960-01-01",tz = "UTC") # sleeping, Wake up
tt2 <- as.POSIXct("27/01/2020 06:45:00", "%d/%m/%Y %H:%M:%S",
origin = "1960-01-01",tz = "UTC") # Starts to leave for campus
tt3 <- as.POSIXct("27/01/2020 07:00:00", "%d/%m/%Y %H:%M:%S",
origin = "1960-01-01",tz = "UTC") # Eats breakfast in a nearby
student restaurant
tt4 <- as.POSIXct("27/01/2020 08:00:00", "%d/%m/%Y %H:%M:%S",
origin = "1960-01-01",tz = "UTC") # Arrives at campus
tt5 <- as.POSIXct("27/01/2020 15:00:00", "%d/%m/%Y %H:%M:%S",
origin = "1960-01-01",tz = "UTC") # start to leave for another
destination
tt6 <- as.POSIXct("27/01/2020 18:00:00", "%d/%m/%Y %H:%M:%S",
origin = "1960-01-01",tz = "UTC") # Arrives at another destination
tt7 <- as.POSIXct("27/01/2020 21:00:00","%d/%m/%Y %H:%M:%S",
origin = "1960-01-01",tz = "UTC") # Starts to leave for hostel
```

```
tt8 <- as.POSIXct("27/01/2020 21:30:00", "%d/%m/%Y %H:%M:%S",
origin = "1960-01-01",tz = "UTC") # Arrives at hostel
tt9 <- as.POSIXct("27/01/2020 12:30:00", "%d/%m/%Y %H:%M:%S",
origin = "1960-01-01",tz = "UTC") # goes for lunch,
tt10 <- as.POSIXct("27/01/2020 14:00:00", "%d/%m/%Y %H:%M:%S",
origin = "1960-01-01",tz = "UTC") # start afternoon lecs,
tstart_time <- c(t1,t2,t3,t9,t4,t5,t6,t7,t8)

#Generate the end time of each activity
te1 <- as.POSIXct("27/01/2020 06:45:00", "%d/%m/%Y %H:%M:%S",
origin = "1960-01-01",tz = "UTC") # wake up and preparation to
leave for campus
te2 <- as.POSIXct("27/01/2020 07:00:00", "%d/%m/%Y %H:%M:%S",
origin = "1960-01-01",tz = "UTC") # Finish Break Fast
te3 <- as.POSIXct("27/01/2020 08:00:00", "%d/%m/%Y %H:%M:%S",
origin = "1960-01-01",tz = "UTC") # Arrive at campus
te4 <- as.POSIXct("27/01/2020 15:00:00", "%d/%m/%Y
%H:%M:%S", origin = "1960-01-01",tz = "UTC") # Leave for
another destination
te5 <- as.POSIXct("27/01/2020 18:00:00", "%d/%m/%Y %H:%M:%S",
origin = "1960-01-01",tz = "UTC") # Arrive at destination
te6 <- as.POSIXct("27/01/2020 21:00:00", "%d/%m/%Y %H:%M:%S",
origin = "1960-01-01",tz = "UTC") # Do personal matters and
leave for hostel
te7 <- as.POSIXct("27/01/2020 21:30:00", "%d/%m/%Y %H:%M:%S",
origin = "1960-01-01",tz = "UTC") # Arrives at hostel
te8 <- as.POSIXct("27/01/2020 23:40:00", "%d/%m/%Y %H:%M:%S",
origin = "1960-01-01",tz = "UTC") # End of the day,
te9 <- as.POSIXct("27/01/2020 14:00:00", "%d/%m/%Y %H:%M:%S",
origin = "1960-01-01",tz = "UTC") # Comes back for lunch,
te10 <- as.POSIXct("27/01/2020 12:30:00", "%d/%m/%Y %H:%M:%S",
origin = "1960-01-01",tz = "UTC") # end of morning lectures,
tend_time <- c(te1,te2,te3,te10,te9,te4,te5,te6,te7,te8)

#Student One
## X coordinates of each activity place
x <- c(
0.337087, # Hostel
0.337087, # Hostel (time to leave)
0.336674, # break fast
0.333793, # campus 32.56851912
0.336674,
0.336674,
0.333793, # campus (time to leave for another destination)
0.319626, # Another destination (private matter)
0.319626, # Another destination (time to leave for Hostel)
0.337087 # Hostel (night time until sleep)
)
```

```
# count of x coordinates
n = length(x)

## Y coordinates of each activity place
y <- c(
32.561918,
32.561918,
32.564176,
32.567961,
32.564176,
32.564176,
32.567961,
32.694195,
32.694195,
32.561918
)

#Student two
## X coordinates of each activity place
sx <- c(0.337100,0.337100,0.336600,0.333750,0.336740,0.336740,
0.333750,0.319800,0.319800,0.337100)

# count of x coordinates
sn = length(sx)

## Y coordinates of each activity place
sy <- c(32.561900,32.561900,32.564150,32.567930,32.564190,
32.564190,32.567930,32.514105,32.514105,32.561900)

#Student three
## X coordinates of each activity place
ssx <- c(0.337120,0.337120,0.336650,0.333750,0.336640,0.336640,
0.333750,0.319700,0.319700,0.337120)

# count of x coordinates
ssn = length(sx)

## Y coordinates of each activity place
ssy <- c(32.561500,32.561500,32.564250,32.567930,32.564130,
32.564130,32.567930,32.584155,32.584155,32.561500)

## CRS
proj = CRS("+init = epsg:28992")

## Build STIDF class object
#student one
stidf_data <- STIDF(SpatialPoints(cbind(x,y),
proj4string = proj),start_time,
  data.frame(x = rnorm(n),y = rnorm(n)),end_time)
```

```
#student two
stidf_data_s <- STIDF(SpatialPoints(cbind(sx,sy),
proj4string = proj),sstart_time,
  data.frame(x = rnorm(sn),y = rnorm(sn)))

#student three
stidf_data_ss <- STIDF(SpatialPoints(cbind(ssx,ssy),
proj4string = proj),tstart_time,
  data.frame(x = rnorm(ssn),y = rnorm(ssn)))

#visualize
#student one
plot(stidf_data, col="purple",pch=16,cex=2)
#output

#student two
plot(stidf_data_s, col="green",pch=16,cex=2)
#output

#student three
plot(stidf_data_ss, col="brown",pch=16,cex=2)
#output

#### Build track class object

#student one
track <- Track(stidf_data)
#student two
track_s <- Track(stidf_data_s)
#student three
track_ss <- Track(stidf_data_ss)

## Convert meters into kilometers for distance, speed and
duration
#It's not really necessary, we normally do it to increase
velocity. In this case, we only need it for student three.

#student three
track_ss@connections$distance <- track_ss@
connections$distance/1000

## Convert velocity m/s into km/h (same for above
explanation)

#student three
track_ss@connections$duration <- track_ss@
connections$duration/3600

#### Build STP_track class object
```

```
#student one
vmax_bear1 <- track@connections$speed * 3.6
activity_time = track@connections$distance * 360

#student two
vmax_bear1_s <- track_s@connections$speed * 3.5
activity_time_s = track_s@connections$distance * 360

#student three
vmax_bear1_ss <- track_ss@connections$speed
activity_time_ss = track_ss@connections$distance * 360

## STP_track class objects
#student one
location_uncertainty <- mean(x)
stp <- STP_Track(track,vmax_bear1, activity_time,
location_uncertainty)
#student two
stp_s <- STP_Track(track_s,vmax_bear1_s, activity_time_s)

#student three
stp_ss <- STP_Track(track_ss,vmax_bear1_ss, activity_time_ss)

#create xyz coords of the track

#student one
stp_coords <- xyz.coords(data.frame(stp))

#student two
stps_coords <- xyz.coords(data.frame(stp_s))

#student three
stpss_coords <- xyz.coords(data.frame(stp_ss))

###plotting
xrange <- c(0.319500,0.340000) # Manually set to see too
large PAA
yrange <- c(32.561500,32.702000) # Manually set to see too
large PAA
#type s
plot3d(stp_coords, type='s",pch=16,cex=0.8, col="red",
grid=TRUE,xlim=xrange,ylim=yrange,
 xlab="x (meters)",ylab="y (meters)",main="Tragectories for
the College Students")
lines3d(stp_coords, col = "red")
plot3d(stps_coords, type='s",pch=16,cex=2, col="green", add=T)
lines3d(stps_coords, col = "green")
plot3d(stpss_coords, type='s",pch=16,cex=2, col="blue",
add=T)
lines3d(stpss_coords, col = "blue")
```

```
#type p
plot3d(stp_coords, type='p',pch=16,cex=0.8, col="red",xlim=x
range,ylim=yrange,grid=TRUE,
 xlab="x (meters)",ylab="y (meters)",main="Tragectories for
the College Students")
lines3d(stp_coords, col = "red")
plot3d(stps_coords, type='p',pch=16,cex=2, col="green",
add=T)
lines3d(stps_coords, col = "green")
plot3d(stpss_coords, type='p',pch=16,cex=2, col="blue",
add=T)
lines3d(stpss_coords, col = "blue")
#output

#type b
plot3d(stp_coords, type='b',pch=16,cex=0.8, col="red",xlim=x
range,ylim=yrange,
 xlab="x (meters)",ylab="y (meters)",main="Tragectories for
the College Students")
lines3d(stp_coords, col = "red")
plot3d(stps_coords, type='b',pch=16,cex=2, col="green",
add=T)
lines3d(stps_coords, col = "green")
plot3d(stpss_coords, type='b',pch=16,cex=2, col="blue",
add=T)
lines3d(stpss_coords, col = "blue")
legend('topright',c('Student One','Student two','Student
three'),pch = 16,col=c('red','green','blue'))

#Calculate Estimated PPA, RTG and Area
# get new vmax for student one
stp@connections$vmax<- vmax_bear1 + 0.10

# calculate estimated Potential Path Area (PPA)
stp_ppa <- PPA(stp, time = start_time)

#PPA Area
ppa.area <- gArea(stp_ppa)
ppa.area

#plotting
plot(stp, type='p',pch=16,cex=0.8,
col="#003300",ylim=c(-3000,4000),xlim=c(-3000,4000),
 xlab="x (meters)",ylab="y (meters)",main="Activity Space
time for the College Students")
plot(stp_ppa, type='p',col="#FFFFFF",border="purple",yli
m=c(-3000,4000),
 xlim=c(-4000,7000), add=T)
plot(stp_s, type='p',pch=16,cex=2, col="#000099", add=T)
plot(stp_ss, type='p',pch=16,cex=2, col="#000066", add=T)
```

```
legend('topright',c('PPA Area -',ppa.
area),pch = 20,col=c('red','red'))

#output

#random trajectory generator(RTG)
#Adding 2 random pints between two existing points
#random trajectory generator(RTG)
set.seed(10)
rtg_bear1 <- RTG(stp,n_points = 2, max_time_interval = 20,
quadsegs = 12,iter = 4)
rtg_bear1
#output

> rtg_bear1
bbox:
   min max
x -250.9967 350.8386
y -147.8564 555.8966
```

Stay One Step Ahead with Challenge Assignments

Concept: Recall in this chapter and Task 6.3, we discussed the *K*-function and other spatial measures for point (representing events) patterns and processes. Just like we did before in Chapters 2 through 5, in this task, we explore derived *K*-function results. The *K*-function spatial technique has a powerful capability to account for interevent distances and a range of spatial scales in a spatial point process.

Task: Interpret different spatial patterns of food stores in Chicago using *K*-function *K*-function.

Question 1: Using Figure 6.12, accept or reject the following null hypothesis: The spatial pattern of food stores is not significantly more clustered than the underlying pattern of those observed weighted food stores in Chicago. Explain why.

Question 2: Using Figure 6.13, accept or reject the following null hypothesis: The spatial pattern of food stores is not significantly more clustered than the underlying pattern of those observed weighted food stores in Chicago. Explain why.

Question 3: Using Figure 6.14, accept or reject the following null hypothesis: The spatial pattern of food stores is not significantly more clustered than the underlying pattern of those observed unweighted food stores in Chicago. Explain why.

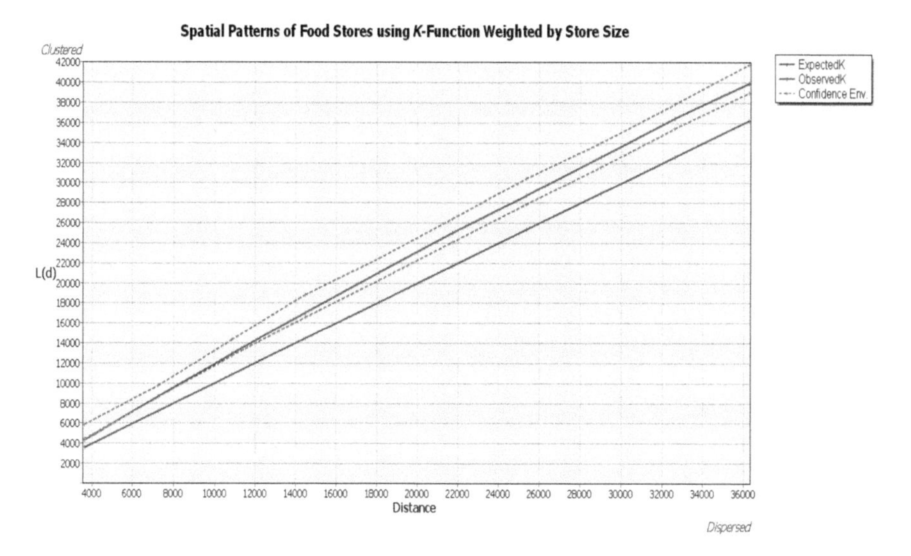

FIGURE 6.12
At spatial range from 3600 to 9200 feet (at smaller distances), the spatial patterns of observed weighted food store touch the lower limits (initially dispersed), but at more than 9200 feet (at larger distances), observed *K* completely falls within the confidence envelope.

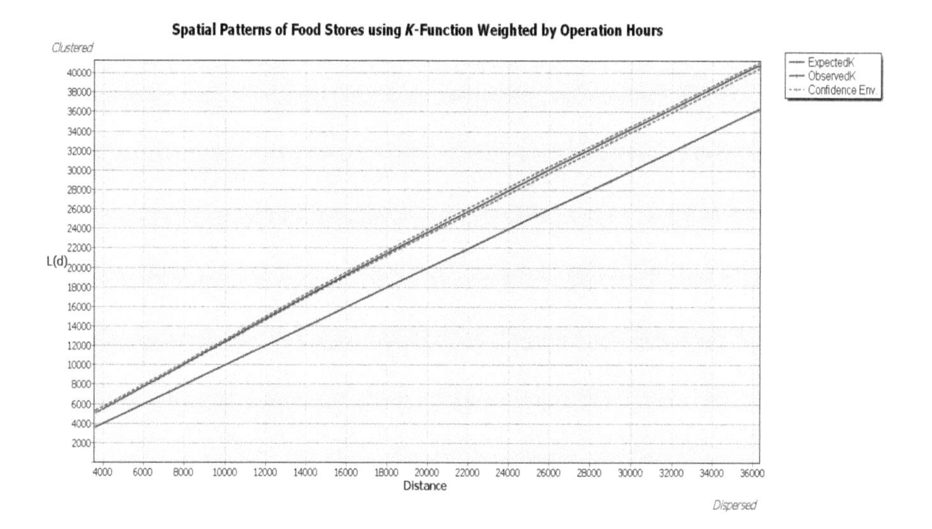

FIGURE 6.13
At spatial range from 3600 to 17,600 feet (at smaller distances), the spatial patterns of observed weighted food store touch the lower limits (initially dispersed), but at more than 17,600 feet (at larger distances), observed *K* completely falls within the confidence envelope.

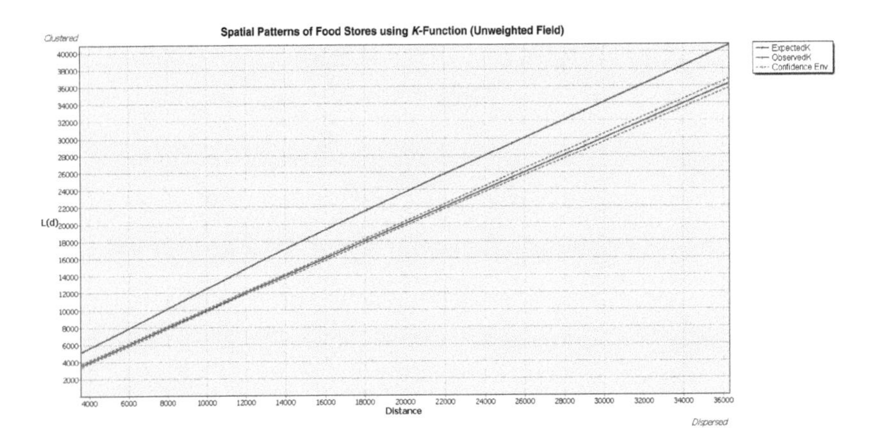

FIGURE 6.14
The spatial patterns of observed unweighted food store (observed *K*) completely fall outside of the confidence envelope, way above its upper limits.

Approach/strategy: Use grocery store results to complete the task. The grocery stores in Chicago are mapped at their true geographic positions as points. The point location data representing grocery stores was projected using NAD 1983 State Plane Illinois East FIPS 1201 Feet before spatial analysis. The *K*-function analysis was conducted on the basis of 99 simulations with one distance vector under the null hypothesis of random distribution. Use your newly acquired knowledge and skills from this chapter to interpret the graphical *K*-function results presented earlier.

Possible solution: Review this chapter, especially Task 6.3, for guidance on how to interpret the *K*-function results.

TASK 6.7 GENERATE AND INTERPRET POINT PATTERN DESCRIPTORS AND STATISTICS

1. One way of overcoming the limitations of quadrat analysis is to compare the observed distribution of points with the Poisson random model for a whole range of different distance radii using Ripley's *K*-function. We can conduct a Ripley's *K*-test on our data very simply with the spatstat package using the kest function. Refer to Chapter6-R-Code for earlier commands that will assist in completing this task. The data for completing this challenge assignment is located in Chapter6_Data_Folder.

```
K <- Kest(OSA.Events.ppp, correction="none")
plot(K)
```

TASK 6.8 EXPLORE AND INTERPRET SPACE-TIME POINT PATTERNS AND STATISTICS

1. There are two ways of conceptualizing and modeling the complex patterns of spatiotemporal dynamics: (1) continuous space and time models and (2) discrete space-time models. In this example, we learn how to conduct a basic time series analysis of noise-level data spanning a 7-year (2004–2010) study period. The data have been split into four categories/subsets using MS Excel: Tier 1 (80th percentile and above), Tier 2 (between 50th and 80th percentiles), Tier 3 (20th and 50th percentiles), and Tier 4 (20th percentile and below). The average for the four categories is presented in Table 6.6.
   a. Plot the temporal trends using a line chart for the four categories.
   b. Describe the temporal trends among the four categories.
   c. Test the following hypothesis using a one-way analysis of variance (ANOVA).
      i. The null hypothesis is that the decibels' means over the study period are equal: $\mathbf{H_O}$: *Tier1_Mean=Tier2_Mean=Tier3_Mean=Tier4_Mean*
      ii. The alternate hypothesis is that the decibels' means over the study period are not equal: $\mathbf{H_A}$: *Tier1_Mean <> Tier2_Mean <> Tier3_Mean <> Tier4_Mean*
      iii. Is the result statistically significant? What does this mean? Explain.
2. Use name places within the attribute table together with a general map of O'Hare International Airport (e.g., from Google Maps) to identify specific neighborhoods and any other identifiable characteristics from each of the four tiers. Perhaps after doing Task 6.5 you will have additional information to effectively respond to this question.
3. Describe the spatiotemporal patterns of noise levels in the study region.

TABLE 6.6

A Summary Showing Averages of Four Categories of Day/Night Sound Levels

Levels	2004	2005	2006	2007	2008	2009	2010
Tier1	69.29755	69.4	70.70509	69.26111	69.25972	66.72942	67.61667
Tier2	62.34501	62.86818	63.00463	61.67189	61.11852	59.0463	59.92407
Tier3	58.52424	57.9871	59.01704	58.80031	58.37204	57.11894	56.12129
Tier4	56.67273	56.04306	55.86667	55.90972	55.60417	54.66959	54.40238

Review and Study Questions

1. What are the three common distributions encountered in point pattern analysis? Using a variable with point features drawn from your area of interest, speculate on the observed distributional pattern of this variable. What technique would you use to confirm or deny your claim?
2. What is the theory of CSR? Describe the various ways in which this can be violated in spatial analysis.
3. Point pattern measures are all based on the comparison of observed and expected distributions. Choosing two of the approaches introduced in this chapter, first explain how the expected distributions are derived. Then explain the statistical measures that are used to confirm or deny the null hypothesis of CSR.
4. Compare and contrast the measures derived from Kulldorff's Scan Statistics with two other measures introduced in this chapter.

Glossary of Key Terms

clustered pattern: When point features (events) are detected to be spatially concentrated in a specific location of a study region.

complete spatial randomness: This principle states that each event has an equal probability of occurring at any position in the study region, and the position of any event is independent of the position of any other.

kernel estimation: This method is used to estimate the density surface of events within a specified radius (bandwidth) around each event in a study region.

nearest neighbor analysis: This method compares the distances between nearest points (events) and distances that would be expected on the basis of chance.

point pattern analysis: This is a means through which we describe or examine a complete set of observations as well as the location of each observation and its distance relative to others in a distribution.

quadrat count method: This method is used to determine the frequency of point distribution by measuring the density of points (events) over the study region.

Ripley's *K*-function: This method is used to describe spatial patterns across a range of spatial scales.

space-time scan statistic: This statistic is used to describe the distribution of spatial, temporal intervals, or spatiotemporal patterns of events in a study region.

spatial scan statistic: This statistic measures the maximum likelihood ratio over all possible search radii.

variance mean ratio: This is a normalized measure of the dispersion of a probability distribution.

Voronoi map: This is used to delineate or represent space, and it helps to computationally determine the "spheres of influence." A Voronoi map is derived using the complex geometric and topologic structures of the underlying spatial data.

References

Bailey, T.C. and A.C. Gatrell. 1995. *Interactive Spatial Data Analysis*. England, UK: Longman.

Besag, J. E. 1977. Comment on "Modelling Spatial Patterns" by B.J Ripley. *Journal of the Royal Statistical Society B* 39(2): 193–195.

Clark, P.J. and F.C. Evans. 1954. Distance to nearest neighbor as a measure of spatial relationships in populations. *Ecology* 35(4): 445–453. http://dx.doi.org/10.2307/1931034

Dai, D. and T.J. Oyana. 2008. Spatial variations in the incidence of breast cancer and potential risks associated with soil dioxin contamination in Midland, Saginaw, and Bay Counties, Michigan, USA. *Environmental Health* 7(1): 49.

Fotheringham, A.S. and F.B. Zhan. 1996. A comparison of three exploratory methods for cluster detection in spatial point patterns. *Geographical Analysis* 28: 200–218.

Gatrell, A.C., T.B. Bailey, P.J. Diggle, and B.S. Rowlingson. 1996. Spatial point pattern analysis and its application in geographical epidemiology. *Transactions of the Institute of British Geographers* 21(1): 256–274.

Groff, E.R. 2007. "Situating" simulation to model human spatio-temporal interactions: An example using crime events. *Transactions in GIS* 11(4): 507–530.

Hägerstrand, T. 1970, What about people in regional science? *Papers of the Regional Science Association* 24(1): 6–21.

Klein, R. 1989. *Concrete and Abstract Voronoi Diagrams. Lecture Notes in Computer Science Series* 400. Berlin-Heidelberg, Germany: Springer.

Kulldorff, M. 2001. Prospective time periodic geographical disease using a scan statistic. *Journal of the Royal Statistical Society, Series A (Statistics in Society)* 164(1): 61–72.

Kulldorff, M. 1997. A spatial scan statistic. *Communications in Statistics: Theory and Methods* 26: 1481–1496.

Kulldorff, M., L. Huang, L. Pickle, and L. Duczmal. 2006. An elliptic spatial scan statistic. *Statistics in Medicine* 25: 3943–3943.

Miller, H.J. 1991. Modeling accessibility using space-time prism concepts within geographical information systems. *International Journal of Geographical Information Systems*, 5(3): 287–301.

7

Engaging in Areal Pattern Analysis Using Global and Local Statistics

Learning Objectives

1. Construct and use different spatial weights for areal pattern descriptors.
2. Generate and interpret clustering of values of areal patterns at a global level.
3. Generate and interpret clustering of values of areal patterns at a local level.
4. Identify clustering of areal pattern values using different spatial weights.
5. Explore, analyze, and interpret areal patterns based on advanced spatial analysis.

Rationale for Studying Areal Patterns

As we embark on the analysis of group-level spatial datasets, a number of questions come to mind. Specifically, what are some of the most robust spatial methods for analyzing areal units? What role do measurement scales play in the selection of these methods? What is the significance of spatial weights, and what are the implications of using these weights on the analysis? And what are the benefits of using either global or local statistics to uncover the spatial areal patterns? Along with these questions, it is important to note that changes in spatial patterns over time are often the result of underlying spatial processes. Therefore, when exploring spatial patterns, we need to focus not only on the spatial patterns but also on the spatial processes. As we discovered in Chapter 6, the patterns can be clustered, dispersed, or random. Our task is to formulate a statistical hypothesis of complete spatial randomness and then validate this based on the empirical observations derived from the analysis. If the pattern is nonrandom, we then proceed to uncover the processes that underlie the observed pattern.

The Notion of Spatial Relationships

Spatial statistics does not simply mean the application of statistical methods to data that just happens to be spatial, encompassing *x*- and *y*-coordinates. Rather, it entails the integration of space and spatial relationships (area, distance, length, etc.) directly into the analysis. In Chapter 1, we discussed the notion of spatial dependency as a principal characteristic of geographic data. Knowledge of this basic characteristic lies at the core of a successful spatial analysis. The existence of spatial dependency requires more attention to avoid biased estimates of spatial effects (Armhein 1995; Nakaya 2000; Getis 2010). Handling geographic data involves a systematic examination of spatial dependency and then figuring out how to incorporate the true spatial structure in the spatial analysis. If this is successfully accomplished, we are more likely to generate unbiased estimates and possibly identify influential factors that may explain spatial patterns and processes underlying any phenomenon.

In spatial analysis, we model spatial relationships based on the principle of spatial neighbors, best captured by Tobler's first law of geography: "everything is related to everything else, but near things are more related than distant things." Spatial autocorrelation can be measured for both point and areal spatial patterns. Strong spatial autocorrelation means that attribute values of adjacent geographic objects are strongly related (whether positively or negatively). Results of this type of analysis often lead to further inquiry of how the spatial patterns change from the past to the present, or estimates of how the spatial patterns will change from the present to the future.

In addition, the study of spatial autocorrelation has significant implications for the use of statistical techniques in analyzing spatial data. For many classical statistics, including various regression models, a fundamental assumption is that observations are randomly selected or independent of each other. Unfortunately, when spatial data are analyzed, this assumption of independence is often violated because most spatial data have certain degrees of spatial autocorrelation (Anselin and Griffith 1988; Nakaya 2000; Getis 2010), as stated in Tobler's law. This often prompts the use of alternative techniques such as geographically weighted regression to accommodate these attributes of spatial data.

As a data scientist, it is good practice to start out by examining the degree of spatial autocorrelation in the aggregated spatial data, following which one can decide on the next steps. In practice, the two spatial neighbors that are commonly used are contiguity-based neighbors (the adjacency of boundaries) and distance-based neighbors (critical distance thresholds). We assume that the influence of spatial neighbors between n spatial units can be quantified using a spatial weight; this is reflected in the way we summarize the spatial structure using a spatial weight matrix (mathematical terms). A spatial weight matrix is a representation of the spatial structure of the dataset. It is a quantification of the spatial relationships that exist among the features within the dataset. The primary weights are conceptualized in terms of spatial contiguity or adjacency (Rook's or Queen's) and the distance between two events. If you are measuring clustering of events/values that depict an inverse relationship, then the inverse distance is probably most appropriate. However, if you are assessing the geographic distribution of commuting patterns, for example, in a city, then travel time or travel cost is a better choice. It is therefore incumbent upon anyone conducting a spatial analysis to determine the best weighting scheme for computing a spatial relationship, because it is highly consequential for tests of spatial autocorrelation. Furthermore, some spatial units may have no spatial neighbors. If the selection of weighting scheme is done correctly, then we are likely to capture the effects of spatial autocorrelation.

In some cases, contiguity-based and distance-based neighbors are combined to create a spatial weighting scheme that is reflective of conceptualized spatial relationships (Cliff and Ord 1969; Griffith 1995; Getis and Aldstadt 2004; Kelejian and Prucha 2010). Figures 7.1 and 7.2 present visual schematic representations of spatial neighbors for computing the effects of spatial autocorrelation. We normalize spatial weights to remove dependence on irrelevant scale factors using either row or scalar standardizations.

Quantifying Spatial Autocorrelation Effects in Areal Patterns

A statistical test is applied to determine whether there is a match between locational and attribute similarity. The effects of spatial autocorrelation are commonly quantified using Moran's I index (Moran 1948, 1950) and Geary's C ratio, both of which are statistical in nature. In this chapter, we examine a variety of existing methods. The methods use a measure known as the spatial autocorrelation coefficient, which statistically tests how clustered/dispersed features lie in space with respect to their attribute values. The measure examines whether an observed attribute of a variable at one location is independent of values of that variable at neighboring locations. If these values are similar and statistically significant, then we can conclude that positive spatial autocorrelation is evident in the spatial distribution. However, in a case when values in the neighboring location exhibit different characteristics

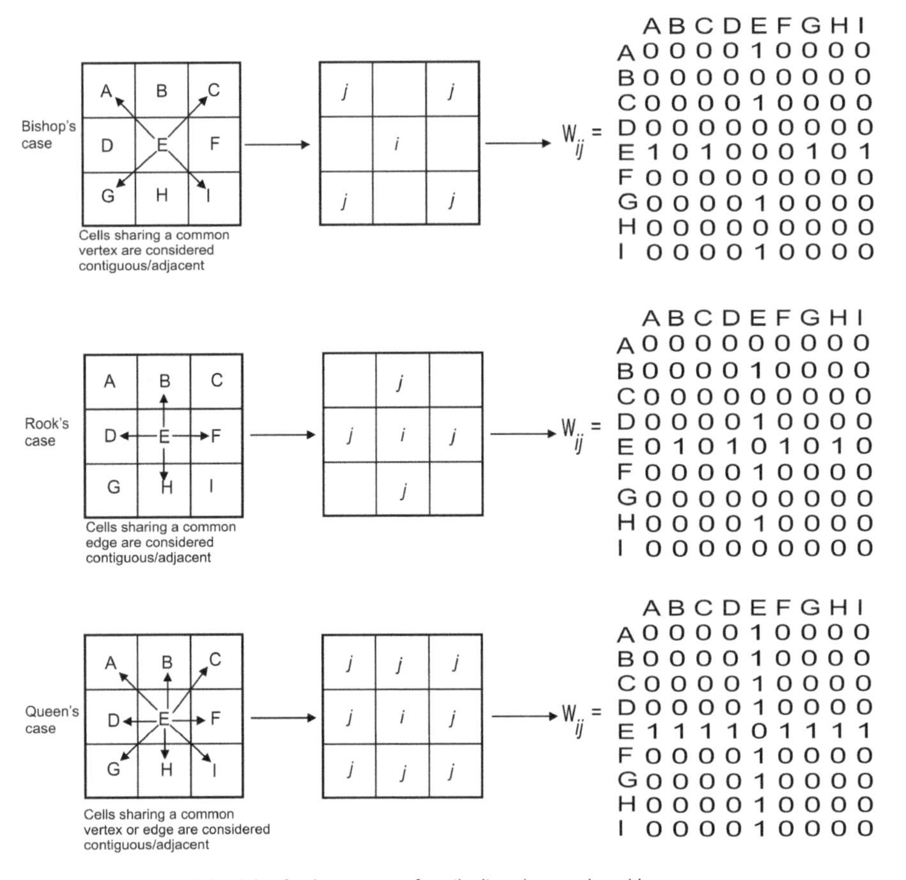

We quantify the degree of spatial influence for the three cases using a connectivity matrix.

A connectivity matrix C is given by $m \times m$,

where $i = \{1, 2, \dots n\}$ and $j = \{1, 2, \dots n\}$

$C_{ij} = 1$ if the two spatial units $i \neq j$ are considered connected, and $C_{ij} = 0$ if they are not

FIGURE 7.1
Spatial neighbors using a contiguity-based weighting scheme. Contiguity cells defining Bishop's case include A, C, G, I; Rook's case, B, D, F, H; and Queen's case, A, B, C, D, F, H, I.

(are dissimilar), then we can conclude that spatial autocorrelation is weak or nonexistent in the spatial distribution. In Figure 7.3, the map in the upper panel shows clustered/positive spatial autocorrelation, with adjacent or nearby polygons having similar values; the map in the middle panel exhibits a random/independent spatial autocorrelation; and the map in the lower panel shows a dispersed pattern/negative spatial autocorrelation, with changes in shade often occurring between adjacent polygons.

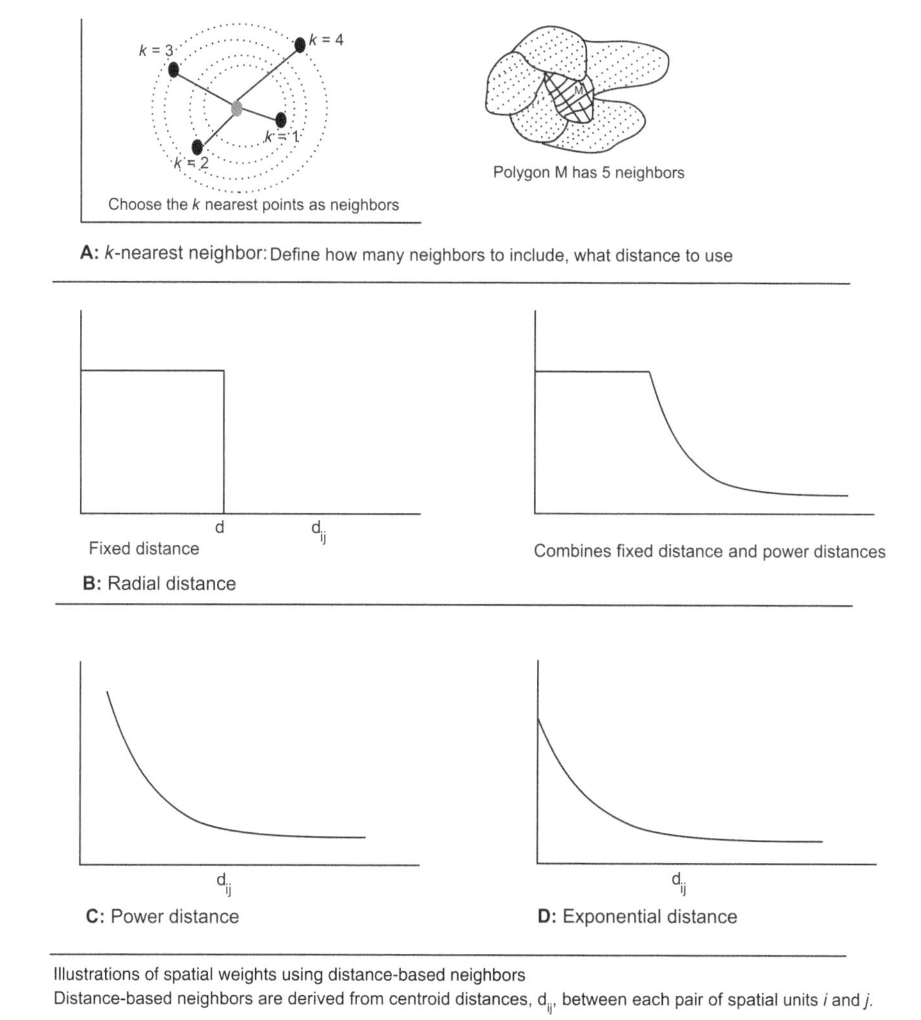

FIGURE 7.2
Spatial neighbors for distance-based weighting scheme: (A) *k*-nearest neighbor, (B) radial distance, (C) power distance, and (D) exponential distance.

Join Count Statistics

This is a basic method that quantitatively determines the degree of clustering or dispersion among a set of spatially adjacent polygons (Cliff and Ord 1973; Goodchild 1986). It is used for binary nominal data such as 1/0, yes/no, arable/nonarable lands, and urban/rural counties. The method measures the spatial relationships between similar or dissimilar attributes in adjacent areas. The binary variable is denoted by two colors, black (B) and white (W)

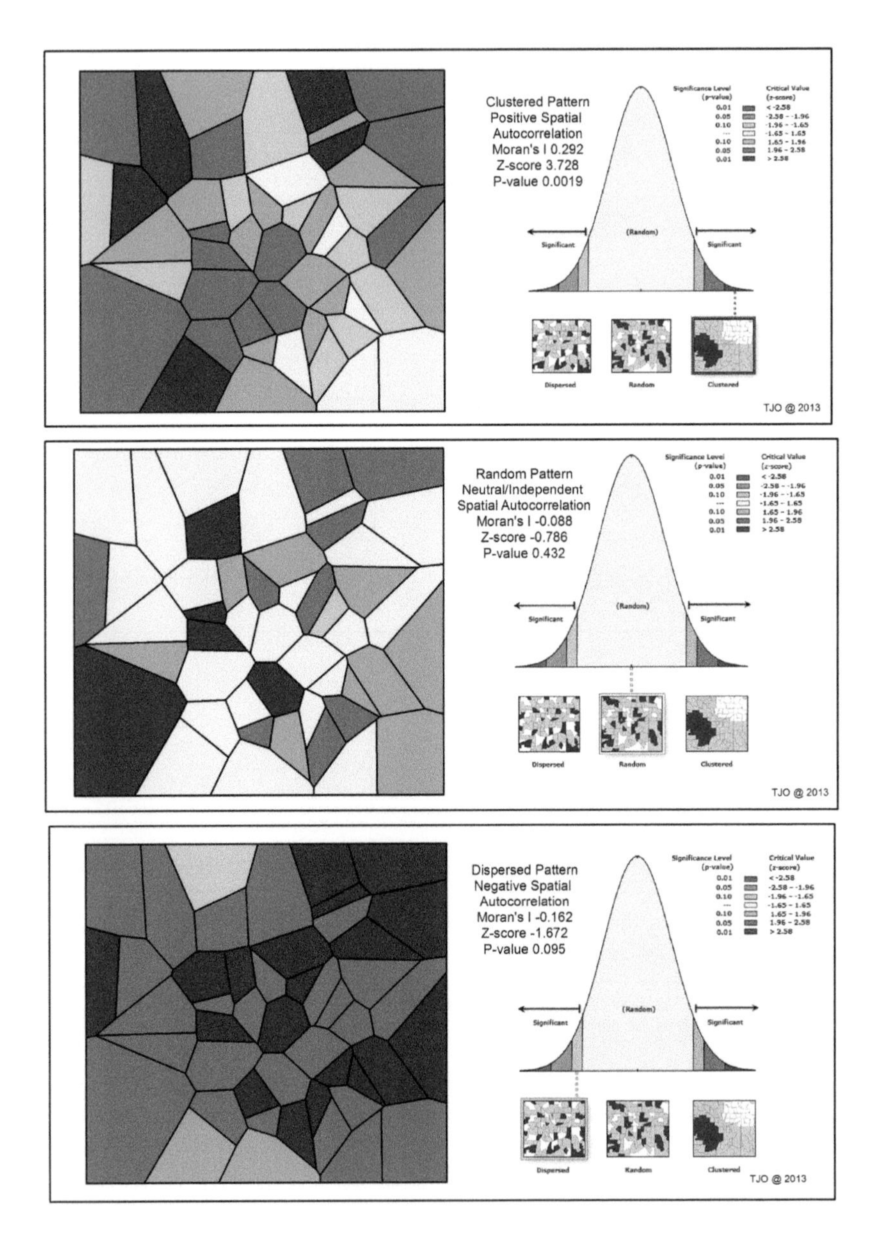

FIGURE 7.3
Three different spatial areal patterns showing tree height near a residential neighborhood in Chicago.

(Figure 7.4). If a given attribute of 1 occurs in an area, then the area will be assigned B. If it does not and has an attribute of 0, then it will be assigned W. If two neighboring areas share a common boundary, they are conceptualized as joined.

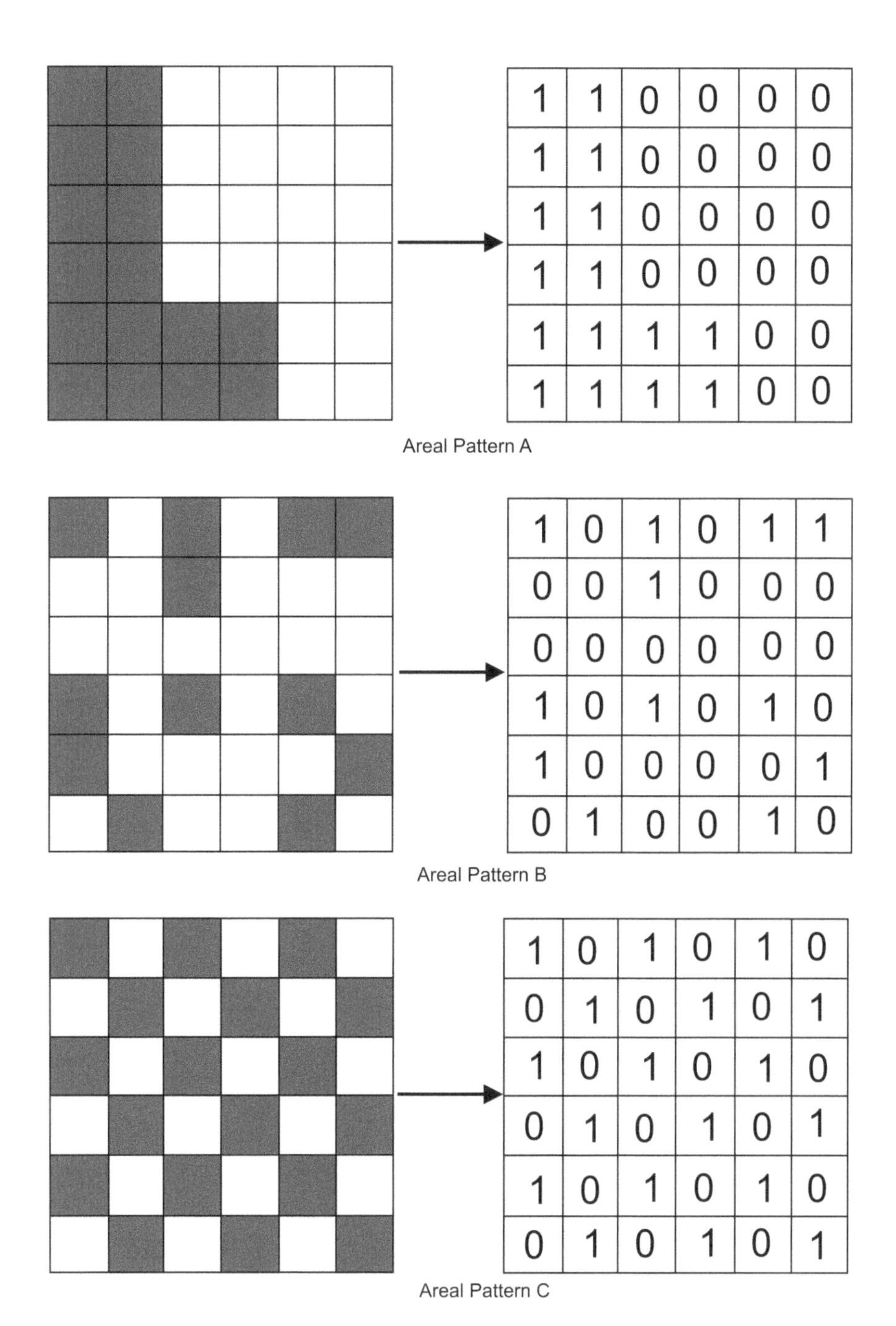

FIGURE 7.4
Three different areal spatial patterns (A) through (C). Spatial autocorrelation for these maps can be calculated using join count statistics where each of the shaded cells (B) is assigned a value of 1, while each of the nonshaded cells (W) is assigned a value of 0.

There are three possible types of joins, black–black (BB), two B neighboring areas; white–white (WW), two W neighboring areas; and black–white (BW), B and W neighboring areas. Join counts tally the numbers of black–black, white–white, and black–white joins in the study area. Observed join counts are derived as follows:

BB (black–black) joins:

$$BB = \frac{1}{2}\sum_i \sum_j w_{ij} x_i x_j$$

BW (black–white) joins:

$$BW = \frac{1}{2}\sum_i \sum_j w_{ij} (x_i - x_j)^2$$

WW (white–white) joins:

$$WW = \frac{1}{2}\sum_i \sum_j w_{ij} (1 - x_i)(1 - x_j)$$

where x_i is the observer value for variant X_i, $x_i = 1$ when the ith area is B, and $x_i = 0$ when the ith area is W, and w_{ij} is weight for each pair of objects i and j.

We use the observed patterns of join counts to compare whether it is different from a random/expected pattern under the null hypothesis of no spatial autocorrelation. Each of the null hypotheses for the three types of joins determines whether the compared differences are statistically significant at p-value less than 0.05. This is done by calculating the Z-test for each join and deciding whether the null hypothesis is true. The Z-test is calculated as

$$Z = \frac{\text{Observed} - \text{Expected}}{\sigma_{\text{Expected}}}$$

A z-score for each of the joins is calculated in the following example.

TASK 7.1 EXAMINING LAND USE PATTERNS OF A FARMLAND

Let us look at a hypothetical case of an area of farmland (Figure 7.5). Within the farm area, we may assign white to areas or cells representing nonarable and black to areas or cells for arable land. Spatial autocorrelation for these maps can be calculated using Join Count Statistics, where each of the filled cells (B) is assigned a value of 1 while each of the nonshaded cells is assigned a value of 0.

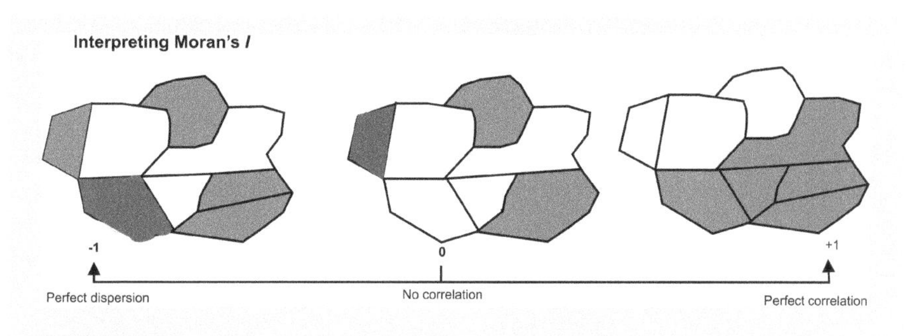

FIGURE 7.5
Resultant values of Moran's *I*.

Interpreting the Join Count Statistics and Methodological Flaws

Statistical results for the join counts are presented in Table 7.1 based on Rook's case, Bishop's case, and Queen's case. Areal Pattern A shows more arable/arable land joins (Rook's case observed = 30, Bishop's case observed = 23, Queen's case observed = 53) that would be expected under Rook's case (18.1), Bishop's case (15.1), and Queen's case (33.2), implying the presence of positive spatial autocorrelation in land use patterns. A similar observation is evident for the nonarable/nonarable land joins. There are far fewer arable/nonarable land joins (Rook's case observed = 8, Bishop's case observed = 12, Queen's case observed = 20) that would be expected under Rook's case (30.5), Bishop's case (25.4), and Queen's case (55.9), implying the presence of positive spatial autocorrelation in land use patterns.

TABLE 7.1

Worktable for Deriving Join Count Statistics for Three Cases of Contiguity-Based Spatial Neighbors

	Areal Pattern A			Areal Pattern B			Areal Pattern C		
Case	Rook's	Bishop's	Queen's	Rook's	Bishop's	Queen's	Rook's	Bishop's	Queen's
BB	30	23	53	26	25	51	0	25	25
BW	8	12	20	31	22	53	60	0	60
WW	22	15	37	3	3	6	0	25	25
n	36	36	36	36	36	36	36	36	36
B	20	20	20	24	24	24	18	18	18
W	16	16	16	12	12	12	18	18	18
Total	60	50	110	60	50	110	60	50	110
E_{BB}	18.1	15.1	33.2	26.3	21.9	48.2	14.6	12.1	26.7
E_{BW}	30.5	25.4	55.9	27.4	22.9	50.3	30.9	25.7	56.6
E_{WW}	11.4	9.52	20.9	6.3	5.2	11.5	14.6	12.1	26.7
Z_{BB}	5.47	3.09	5.15	−0.14	1.16	0.69	−6.80	5.28	−0.47
Z_{BW}	−6.02	−3.92	−7.49	1.05	−0.26	0.59	7.72	−7.47	0.71
Z_{WW}	5.12	2.41	4.76	−1.85	−1.21	−2.05	−6.80	5.28	−0.47

Areal Pattern B shows no strong clustering evidence for arable/arable land joins except in Bishop's and Queen's cases. Also, more arable/nonarable land joins exist in both Rook's and Bishop's cases except in Queen's case. Areal Pattern C shows negative spatial autocorrelation in all the cases except Bishop's. However, in all three cases of Areal Pattern B, there are far fewer nonarable/nonarable land joins, suggesting there is no spatial autocorrelation.

It must be emphasized that join count statistics offer an easy way to represent spatial distribution. However, it can only be applied to nominal data and does not provide a simple summary index that is similar to Geary's *C* or Moran's *I*. Caution is therefore required when classifying continuous variables into binary variables, because the aggregation of the data could lead to loss of information and biased estimates (Goodchild 1986; Odland 1988).

Global Moran's *I* Coefficient of Spatial Autocorrelation

Moran's *I* measures the degree of spatial autocorrelation (Moran 1950) in ordinal- and interval-measured data. It is one of the widely used indices that evaluates the extent of spatial autocorrelation between a set of n cells = $\{x_i\}$ located in neighboring areas, where x_i is either the rank of the ith cell (ordinal data) or the value of X in the ith cell (interval data). The computation of Moran's *I* is achieved by dividing the spatial covariation by the total variation. The resultant values range from approximately −1 (perfect dispersion) to 1 (perfect correlation). The positive sign represents positive spatial autocorrelation, while the converse is true for the negative sign, and a zero result represents no spatial autocorrelation (Figure 7.5).

Suppose we have a study region, *R*, which is subdivided into n cells, where each cell is identified with a spatial feature. Moran's *I* is calculated as follows:

$$I = \frac{\sum_i \sum_j w_{ij} c_{ij}}{s^2 \sum_i \sum_j w_{ij}}$$

where $w_{ij} = 1$ if cells i and j are neighbors, $w_{ij} = 0$ otherwise; $c_{ij} = (X_i - \bar{X})(X_j - \bar{X})$, X_i and X_j are variables at a particular and another location, respectively:

$$s^2 = \sum_{i=1}^{n} \frac{(X_i - \bar{X})^2}{n}$$

The average of all the n cells is the mean ($\bar{X}$), which is used to compute (s^2) based on the differences that each X value has from the mean ($\bar{X}$).

TASK 7.2 SPATIAL DISTRIBUTION OF LOW BIRTH WEIGHT RATES IN A STUDY REGION A

Figure 7.6 depicts the rates of low birth weight per 1000 children in hypothetical region A. The values in the upper left corner represent the unique identifier for the enumeration spatial units, and values in the center represent the low birth weight rates. Using Moran's *I*, we can determine the type of areal pattern in this figure. Table 7.2 presents a worktable and results for Moran's *I*.

$n = 4 \; X_1 = 3 \; w_{ij} = 0\,1\,1\,0$

$X_2 = 2\,1\,0\,0\,1$

$X_3 = 4\,1\,0\,0\,1$

$X_4 = 7\,0\,1\,1\,0$

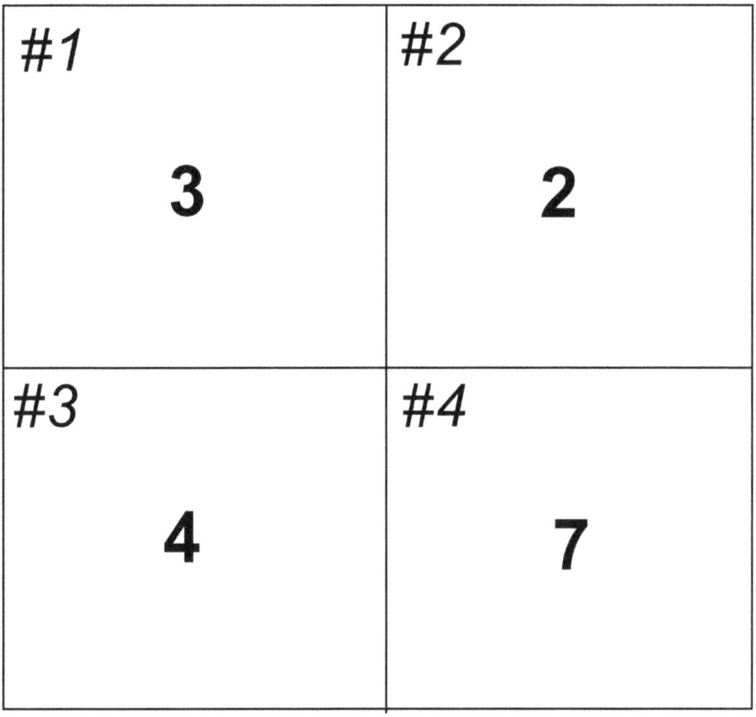

#1 3	#2 2
#3 4	#4 7

FIGURE 7.6
A regular grid/spatial units of low birth weight example.

TABLE 7.2

Worktable for Deriving Global Moran's *I* Coefficient for Low Birth Weight Rates

i	j	w_{ij}[a]	$(X_i - \bar{X})(X_j - \bar{X})$	$\sum\sum w_{ij}c_{ij}$
1	2	1	(3 – 4)(2 – 4)	2
1	3	1	(3 – 4)(4 – 4)	0
1	4	0	(3 – 4)(7 – 4)	0
2	1	1	(2 – 4)(3 – 4)	2
2	3	0	(2 – 4)(4 – 4)	0
2	4	1	(2 – 4)(7 – 4)	–6
3	1	1	(4 – 4)(3 – 4)	0
3	2	0	(4 – 4)(2 – 4)	0
3	4	1	(4 – 4)(7 – 4)	0
4	1	0	(7 – 4)(3 – 4)	0
4	2	1	(7 – 4)(2 – 4)	–6
4	3	1	(7 – 4)(4 – 4)	0
		$\sum w_{ij}$=8		=–8

[a] Weighting scheme based on Rook's case.

$$\bar{X} = \frac{\sum_{i=1}^{n} X_i}{n} = \frac{16}{4} = 4$$

$$s1 = (3 - 4) * (3^* - 4) = 1$$
$$s2 = (2 - 4) * (2 - 4) = 4$$
$$s3 = (4 - 4) * (4 - 4) = 0$$
$$s4 = (7 - 4) * (7 - 4) = 9$$

$$\sum_i \sum_j w_{ij} c_{ij} = -8$$

$$\sum_i \sum_j w_{ij} = 8$$

$$s^2 = \sum_{i=1}^{n} (X_i - \bar{X})^2 / n = 14/4 = 3.5$$

$$I = \frac{(-8)}{8 * 3.5} = -0.2857$$

Interpreting Moran's *I* and Methodological Flaws

Having computed the Moran I statistic, one can proceed to evaluate the statistical significance of the test statistic. As noted earlier, the null hypothesis is one of spatial randomness, meaning that the spatial autocorrelation of the given variable is zero. The statistical significance of Moran's I is based on the normal frequency distribution (Z-score):

$$z = \frac{I - E(I)}{S_{\text{error}(I)}}$$

where I is the computed Moran's I value, $E(I)$ is the expected Moran's I under the null hypothesis of spatial randomness, and S is the standard error of the Moran's I value.

Thus, given a Moran's I value of -0.286 with a Z-score of 0.1597, we fail to reject the null hypothesis and conclude that the areal pattern for low birth rates is statistically insignificant with a weak negative spatial autocorrelation.

It is important to keep in mind that the Moran's I statistic only provides a measure of spatial autocorrelation for spatial data measured at ordinal and interval scales, and may be sensitive to extreme values in a positive or negative correlation. In some cases, Moran's I may not be useful due to its sensitivity to spatial patterning and spatial weight selection.

Global Geary's *C* Coefficient of Spatial Autocorrelation

Geary's C is an alternative measure of spatial autocorrelation. It determines the degree of spatial association using the sum of squared differences between pairs of data values as its measure of covariation (Goodchild 1986). The computation of Geary's C results in a value within the range of 0 to +2 (Figure 7.7). When we obtain a zero value, it is interpreted as a strong positive spatial autocorrelation (perfect correlation), a value of 1 indicates a random spatial pattern (no autocorrelation), and a value between 1 and 2 represents a negative spatial autocorrelation (2 is a perfect dispersion).

Suppose we have a study region, R, that is subdivided into n cells, where each cell is identified with a spatial feature. Geary's C can be computed by

$$C = \frac{\sum_i \sum_j w_{ij} c_{ij}}{(2\sum_{ij} w_{ij} s^2)}$$

where $w_{ij} = 1$ if cells i and j are neighbors, $w_{ij} = 0$ otherwise; $c_{ij} = (X_i - X_j)^2$;

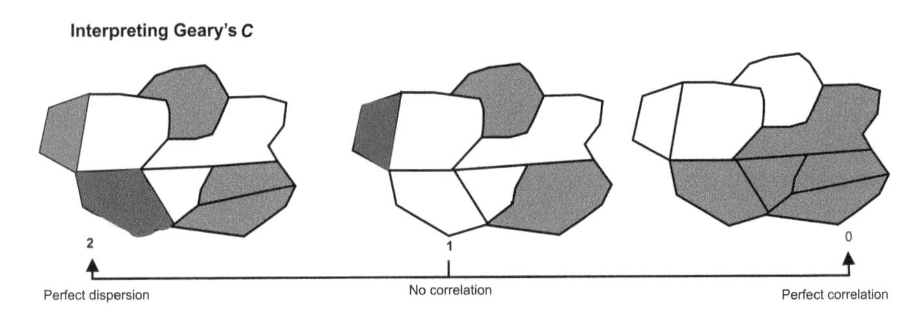

FIGURE 7.7
Resultant values of Geary's C.

$$s^2 = \sum_{i=1}^{n} \frac{(X_i - \bar{X})^2}{n-1}$$

The average of all n cells is the mean ($\bar{X}$), which is used to compute (s^2) based on the differences that each X value has from the mean ($\bar{X}$).

TASK 7.3 COMPUTING GEARY'S *C* FOR LOW BIRTH WEIGHTS

We now examine the low birth weight rates used earlier to compute Moran's *I*. Table 7.3 presents a worktable and results for Geary's C.

TABLE 7.3
Worktable for Deriving Geary's Coefficient for Low Birth Rates

i	j	w_{ij}[a]	$(X_i - X_j)(X_i - X_j)$	$\sum\sum w_{ij}c_{ij}$
1	2	1	(3 – 2)(3 – 2)	1
1	3	1	(3 – 4)(3 – 4)	1
1	4	0	(3 – 7)(3 – 7)	0
2	1	1	(2 – 3)(2 – 3)	1
2	3	0	(2 – 4)(2 – 4)	0
2	4	1	(2 – 7)(2 – 7)	25
3	1	1	(4 – 3)(4 – 3)	1
3	2	0	(4 – 2)(4 – 2)	0
3	4	1	(4 – 7)(4 – 7)	9
4	1	0	(7 – 3)(7 – 3)	0
4	2	1	(7 – 2)(7 – 2)	25
4	3	1	(7 – 4)(7 – 4)	9
		$\sum w_{ij}=8$		=72

[a] Weighting scheme based on Rook's case.

$N = 4\ x_1 = 3\ w_{ij} = 0\,1\,1\,0$

$x_2 = 2\,1\,0\,0\,1$

$x_3 = 4\,1\,0\,0\,1$

$x_4 = 7\,0\,1\,1\,0$

$$\bar{x} = \frac{\sum_{i=1}^{n} x_i}{n} = 16/4 = 4$$

$s_1 = (3 - 4) * (3 - 4) = 1$

$s_2 = (2 - 4) * (2 - 4) = 4$

$s_3 = (4 - 4) * (4 - 4) = 0$

$s_4 = (7 - 4) * (7 - 4) = 9$

$$C = \frac{72}{2 * 8 * 4.6667} = 0.964$$

Interpreting Geary's *C* and Methodological Flaws

Geary's *C* also requires the formulation of a null hypothesis of spatial randomness, which holds true when the spatial autocorrelation of a variable is 1. The statistical significance of Geary's *C* is also based on the normal frequency distribution (*Z*-score). For the previous example, Geary's *C* is 0.964 with a *Z*-score of 0.1597. Therefore, we do not reject the null hypothesis and conclude that the areal pattern for low birth rates shows a spatial autocorrelation that is statistically random.

Both Moran's *I* and Geary's *C* only detect spatial patterns (clusters) of an entire region and are unable to distinguish local patterns. Geary's *C* is less arranged; therefore, the extremes are less likely to correspond to the positive or negative correlation. Although their calculations are quite similar, Moran's *I* is based on the cross-product of deviations from the mean for variables at a particular cell and another neighboring cell (location), while Geary's *C* is a cross-product of actual values of a variable at a particular location and another neighboring cell.

Getis-Ord *G* Statistics

G(*d*) statistics is an alternative index among a family of conventional global spatial autocorrelation measures. Unlike Moran's *I* and Geary's *C*, which are unable to discriminate whether spatial patterns are due to high or low values, the $G(d)$ method is able to discern between hot spots and cold spots over the entire study region. This method can be used to identify spatial concentrations of particular phenomena, such as particulate matter, birth rates, crime rates, or poverty rates, as well as, account for spatial autocorrelation (Nakaya 2000; Getis 2010).

Assume we have a study region, R, that is subdivided into n cells, where each cell is identified with a spatial feature. We can measure the degree of spatial association in study region R, by computing the spatial concentration of weighted point feature values within a radius of distance d (neighbor distance),

TASK 7.4 EXAMINING THE SPATIAL DISTRIBUTION OF NITROGEN OXIDES

Let us look at a hypothetical case of the spatial distribution of nitrogen oxides in a study region; the number in each cell in Figure 7.8 represents emissions in tons per year. We will derive $G(d)$ statistics for this case. Table 7.4 presents a worktable and results for $G(d)$.

$$G(d) = \frac{\sum_{i=1}^{n}\sum_{j=1}^{n} w_{ij}(d)x_i x_j}{\sum_{i=1}^{n}\sum_{j=1}^{n} x_i x_j} = \frac{10000}{36100} = 0.277$$

a 20	*b* 20	*c* 30
d 20	*e* 10	*f* 30
g 20	*h* 20	*i* 20

FIGURE 7.8
A regular grid/spatial units of nitrogen oxide emissions in tons per year.

TABLE 7.4

Worktable for Deriving ***G(d)*** Statistics for Nitrogen Oxides

	a	b	c	d	e	f	g	h	i	
a	400	400	600	400	200	600	400	400	400	
b	400	400	600	400	200	600	400	400	400	
c	600	600	900	600	300	900	600	600	600	
d	400	400	600	400	200	600	400	400	400	
e	200	200	300	200	100	300	200	200	200	
f	600	600	900	600	300	900	600	600	600	
g	400	400	600	400	200	600	400	400	400	
h	400	400	600	400	200	600	400	400	400	
i	400	400	600	400	200	600	400	400	400	
	=3800	**=3800**	**=5700**	**=3800**	**=1900**	**=5700**	**=3800**	**=3800**	**=3800**	$\sum x_i x_j$ = **36100**

w^{a}_{ij}	a	b	c	d	e	f	g	h	i
a	0	1	0	1	0	0	0	0	0
b	1	0	1	0	1	0	0	0	0
c	0	1	0	0	0	1	0	0	0
d	1	0	0	0	1	0	1	0	0
e	0	1	0	1	0	1	0	1	0
f	0	0	1	0	1	0	0	0	1
g	0	0	0	1	0	0	0	1	0
h	0	0	0	0	1	0	1	0	1
i	0	0	0	0	0	1	0	1	0

0	**400**	**0**	**400**	**0**	**0**	**0**	**0**	**0**	
400	0	600	0	200	0	0	0	0	
0	600	0	0	0	900	0	0	0	
400	0	0	0	200	0	400	0	0	
0	200	0	200	0	300	0	200	0	
0	0	900	0	300	0	0	0	600	
0	0	0	400	0	0	0	400	0	
0	0	0	0	200	0	400	0	400	
0	0	0	0	0	600	0	400	0	
=800	**=1200**	**=1500**	**=1000**	**=900**	**=1800**	**=800**	**=1000**	**=1000**	$\sum_{i=1}^{n}\sum_{j=1}^{n} w_{ij}(d)x_i x_j$ = **10000**

[a] The 9×9 matrix grid is labeled by cell locations; when two cells are adjacent, it is assigned a value of 1, and when they are not, a value of 0 is assigned. However, diagonal neighbors were excluded. Although the contiguity spatial weight has been used to illustrate the calculations, centroid distances among the 9×9 cells is more reflective of proximity relationships.

where we expect a cluster to occur. Getis-Ord's G statistics (Getis and Ord 1992; Ord and Getis 1995) for study region R can be derived as follows:

$$G(d) = \frac{\sum_{i=1}^{n}\sum_{j=1}^{n} w_{ij}(d)x_i x_j}{\sum_{i=1}^{n}\sum_{j=1}^{n} x_i x_j} \quad j \neq i$$

where $w_{ij}=1$ if cell j is within distance d from cell i, or $w_{ij}=0$ if it is outside. $G(d)$ statistics is interpreted relative to its expected value. If, for example, high values are clustered together, then $G(d)$ is relatively large. This means $G(d)$ is greater than the expected value ($G[d]$ > expected value), suggesting a potential hot spot. However, if low values are clustered together, then $G(d)$ is relatively small. This means that $G(d)$ is smaller than the expected value ($G[d]$ < expected value), suggesting a potential cold spot.

Interpretation of Getis-Ord G and Methodological Flaws

When using the Getis-Ord G statistic, the null hypothesis is that for the phenomenon under investigation, there is no clustering of high or low values in a given location or in its neighborhood. The alternative hypothesis is that for the phenomenon under study, the spatial distribution may exhibit a significantly more clustered than random pattern. In the case study provided (nitrogen oxides), since the $G(d)$ is relatively small (0.277), we reject the null hypothesis and conclude that the low values or below-average values of nitrogen oxides may be clustered in the study region.

When using the $G(d)$ statistics, there might be some difficulty in distinguishing between a random pattern and one in which there is little deviation from the mean (Getis and Ord 1992). Further, although the $G(d)$ statistic has gained wide acceptance for determining spatial concentrations at local scales, it can only be applied to analyze ratio-scale data having a natural zero. The $G(d)$ statistic evaluates the total concentration or lack thereof among all pairs of (x_i, x_j), where i and j are within d distance. If x values change in proportion, $G(d)$ remains the same. To attain a deeper understanding of spatial patterns, it is recommended that the $G(d)$ statistic be used in conjunction with Moran's I (Getis and Ord 1992).

Local Moran's I

Local Moran's I determines the degree of spatial association at the location-specific level. It belongs to a family of Local Indicators of Spatial Association (LISA) (Anselin 1995) that is used to identify clusters among individual spatial units. LISA statistics measure the degree to which one areal unit is autocorrelated relative to its neighbors (Figure 7.9). Following the analysis, a Moran's scatterplot can be used to identify the leverage points and spatial outliers. The plot has four

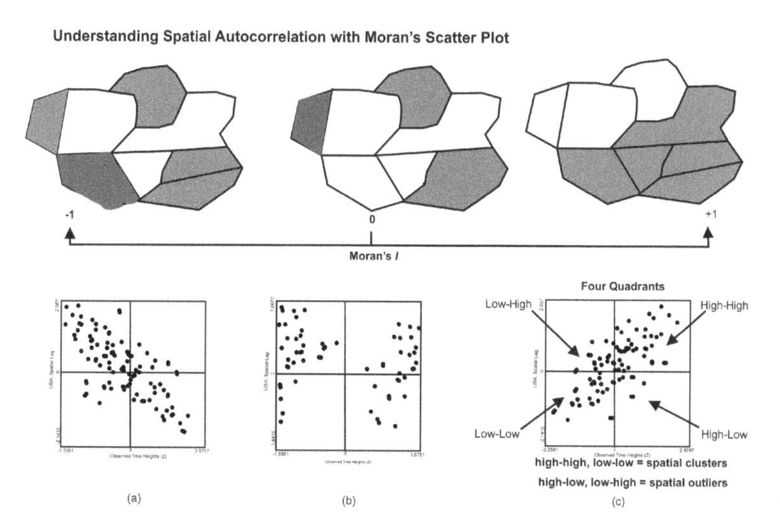

FIGURE 7.9
Moran's *I* and the four quadrants of Local Indicators of Spatial Association statistics shown in (a) through (c).

quadrants: high–high, high–low, low–high, and low–low. High–high denotes the presence of spatial clustering of neighbors with high values surrounded by those with similar values, low–low denotes spatial clustering of neighbors with low values surrounded by those with similar values, high–low or low–high represents spatial outliers or neighbors with values that are statistically insignificant. In line with Anselin's (1995) suggestions, there are two notable aspects of these statistics: (1) the LISA for each observation gives an indication of the extent of significant spatial clustering of similar values around that observation and (2) the sum of LISAs for all observations is in proportion to a global statistic of spatial association. There are several local versions of global statistics such as Moran's *I*, Geary's *C*, and Getis-Ord's *G*. These measures serve four principal aims: (1) provide a finer-grained analysis at the local level, (2) identify spatial patterns at the local level or hot spots, (3) measure spatial autocorrelation at the local level, and (4) detect spatial clusters or spatial outliers at the local level. Using working examples, we next describe how each index is derived and how to interpret the statistical results.

A Local Moran's *I* for an observation *i* is defined as

$$L_i = z_i \sum_j w_{ij} z_j$$

$$\mathrm{I} = \sum_i I_i$$

where w_{ij} are spatial weights matrix; the observations z_i, z_j are the deviations from the mean; and L_i is the summation of the spatial weights matrix multiplied by z_j, z_i. Deriving the mean deviations for each of the observations is similar to how we calculate a Z-score.

The L_i values and related statistics for each of the spatial units are given in Table 7.5. Five of the spatial units (*a, b, c, d,* and *h*) have an L_i value of zero. Positive values of L_i indicate spatial clustering of similar values (high or low), while negative values indicate a spatial clustering of dissimilar values (high–low or low–high). Three negative L_i values fall within the low–high quadrant of Moran scatterplot implying that there are spatial outliers. It is evident from

TASK 7.5 SPATIAL DISTRIBUTION OF SMARTPHONES IN A SMALL TOWN

Let us illustrate the calculation of Local Moran's *I* by using a hypothetical example of the number of smartphones per 1000 people in a regular grid/spatial units (Figure 7.10). The number in each cell represents the number of smartphones per 1000 people in each neighborhood. Table 7.5 provides a worktable and results for the Local Moran's *I*.

$n = 9, \bar{x} = 180/9 = 20$, $\sigma = 9.43$, $z_j = (x_j - \bar{x})/\sigma$ derived in Table 7.5 under Z-score

$$I = \sum_i I_i = [(0 + 0 + 0 + 0 + (-4.5) + (-3.375) + 1.125 + 0 + (-1.125))] = -7.875$$

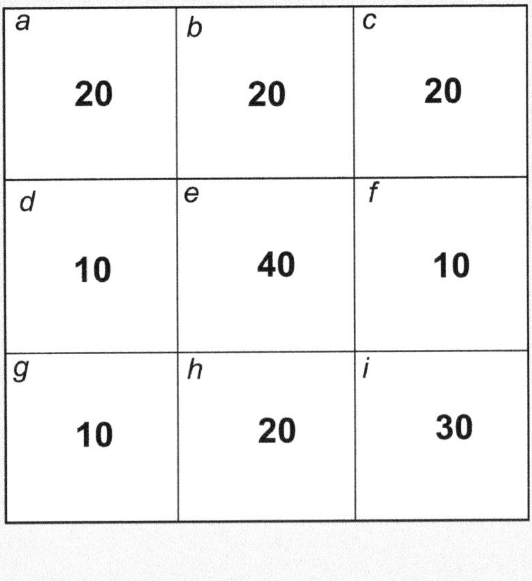

a 20	*b* 20	*c* 20
d 10	*e* 40	*f* 10
g 10	*h* 20	*i* 30

FIGURE 7.10
A regular grid/spatial units of the number of smartphones per 1000 people in a small town.

TABLE 7.5

Worktable for Deriving Local Moran's I and Related Local Indicators of Spatial Association Statistics

$w^{a}_{ij}=$	a	b	c	d	e	f	g	h	i
A	0	1	0	1	0	0	0	0	0
B	1	0	1	0	1	0	0	0	0
C	0	1	0	0	0	1	0	0	0
D	1	0	0	0	1	0	1	0	0
E	0	1	0	1	0	1	0	1	0
F	0	0	1	0	1	0	0	0	1
G	0	0	0	1	0	0	0	1	0
H	0	0	0	0	1	0	1	0	1
I	0	0	0	0	0	1	0	1	0

Deriving the L_i value for each observation

$L_i = z_i \sum_j w_{ij} z_j$		a	b	c	d	e	f	g	h	i		
	Z-score	0	0	0	−1.061	2.121	−1.061	−1.061	0	1.061	$\sum_j w_{ij} z_j$	L_i[b]
A	0	0	0	0	−1.061	0	0	0	0	0	−1.061	0
B	0	0	0	0	0	2.121	0	0	0	0	2.121	0
C	0	0	0	0	0	0	−1.061	0	0	0	−1.061	0
D	−1.061	0	0	0	0	2.121	0	−1.061	0	0	1.061	0
E	2.121	0	0	0	−1.061	0	−1.061	0	0	0	−2.121	−4.5
F	−1.061	0	0	0	0	2.121	0	0	0	1.061	3.182	−3.375
G	−1.061	0	0	0	−1.061	0	0	0	0	0	−1.061	1.125
H	0	0	0	0	0	2.121	0	−1.061	0	1.061	2.121	0
I	1.061	0	0	0	0	0	−1.061	0	0	0	−1.061	−1.125

[a] The 9×9 matrix grid is labeled by cell locations; when two cells are adjacent, it is assigned a value of 1, and when they are not, a value of 0 is assigned. However, diagonal neighbors were excluded.

[b] Extreme L_i values indicate outliers, in this example, −4.5 is such a value.

this statistical data that there is no clear spatial clustering of local values of smartphones, with the L_i values suggesting strong evidence of the presence of dissimilar values that require further scrutiny.

When compared to the global statistics Moran's *I*, Local Moran's *I* statistics contribute to local spatial association and overcome the local instabilities of spatial observations (Anselin 1995). The LISA method is especially applicable for spatial data that are heterogeneous among areas, as they are able to compute subregions of the datasets at a local scale (Boots and Okabe 2007).

Local *G*-Statistic

The Local *G*-statistic is designed to measure specific spatial associations that may not be obvious when using global statistics. It is based on Getis-Ord General's *G* $G_{i,i}{}^*$ (Getis and Ord 1992) and determines the effects of individual locations (including detecting extremes) on the scale of global statistics (Anselin 1995; Ord and Getis 1995).

The local *G*-statistic is a standard variant that is calculated by taking general *G*-statistics minus its expectation $E(G_i)$ and dividing this by the square root of its variance. *G*-statistics is given by

$$G_i(d) = \frac{\sum_j w_{ij}(d)x_j - W_i\bar{x}(i)}{s(i)\left\{\left[(n-1)S_{1i}) - W_i^2\right]/(n-2)\right\}^{1/2}} \; j \neq i.$$

$$W_i = \sum_{j \neq i} w_{ij}(d), S_{1i} = \sum_{j \neq i} w_{ij}^2$$

$$G_i^*(d) = \frac{\sum_j w_{ij}(d)x_j - W_i^*\bar{x}}{s\left\{\left[(nS_{1i}^*) - W_i^{*2}\right]/(n-1)\right\}1/2} \quad all\, j$$

$$W_i^* = W_i + w_{ii}(w_{ii} \neq 0), S_{1i}^* = \sum_j w_{ij}^2(allj),$$

where $G_i(d)$ is a proportion of the sum of all x_j values that are within distance (d) of i; x_j is the variable of interest in a given study region; $\bar{X}$ and s^2 denote the sample mean and variance, where $w_{ij}=0$ is the spatial weight between neighbors i and j, and $\sum_j w_{ij}{}^2$ is the sum of squared weights.

Under the null hypothesis, we use Local *G*-statistics to determine whether there is evidence of spatial clustering of high or low values around each spatial unit of fast food restaurants. From the calculation, we observe that some G_i values are both negative and positive. Positive values of G_i indicate a spatial clustering of high values, while negative values of G_i indicate a spatial clustering of low values. However, to interpret Local *G*-statistics, we need to derive the *Z*-score. This has been done in the empirical examples that are presented later in the section on using scatterplots to synthesize and interpret LISA statistics. Although Local *G*-statistics are a more flexible form of LISA statistics, they do not have a natural origin. The use of nonbinary weight matrices also makes them more appealing for understanding spatial relationships.

TASK 7.6 SPATIAL DISTRIBUTION OF FAST FOOD RESTAURANTS IN A STUDY REGION

Let us illustrate the calculation of Local *G*-statistics by using a hypothetical example of fast food restaurants (Figure 7.11). The numbers in parentheses represent the identifying number of fast food restaurants. The locations of fast food restaurants are given in x and y centroid coordinates of an areal unit. We can use the set of fast food restaurants to calculate possible clustering of high or low values in the vicinity of point 5 at three critical threshold distances of 10 miles, 20 miles, and 30 miles from point 5, respectively. Getis and Ord's (1995) Local *G*-Statistics method is applied to find a solution.

Point 5 is not included:

$$\bar{x}(5) = \frac{\sum_j x_j}{n-1}$$

$$= \frac{(1+2+3+1)}{5-1} = 1.75, s^2(5) = \frac{\sum_j x_j^2}{n-1} - [\bar{x}(5)]^2 = 0.6875 =, s = 0.8292$$

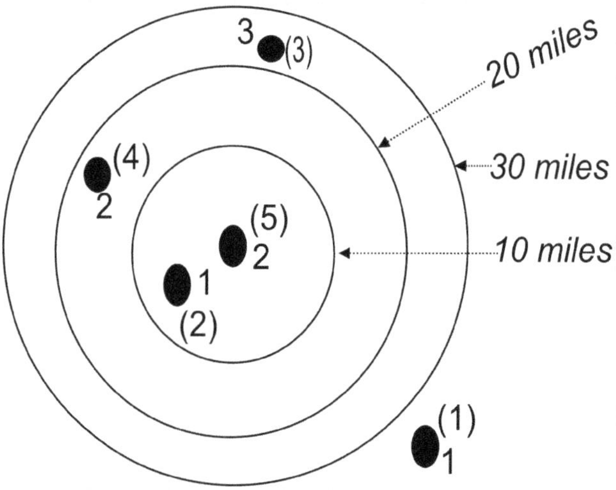

FIGURE 7.11
Location of fast food restaurants in a study region.

$$G_5(10) = \frac{1 - 1 * 1.75}{0.8292 * [(4 * 1 - 1)/3]^{1/2}} = -0.905$$

$$G_5(20) = \frac{3 - 2 * 1.75}{0.8292 * [(4 * 2 - 4)/3]^{1/2}} = -0.522$$

$$G_5(30) = \frac{6 - 3 * 1.75}{0.8292 * [(4 * 3 - 9)/3]^{1/2}} = 0.905$$

Point 5 is included:

$$\bar{x} = \frac{\sum_j x_i}{n} = \frac{9}{5} = 1.8, s^2 = \frac{\sum_j (x_i - \bar{x})^2}{n - 1} = \frac{2.8}{4} = 0.7, s = 0.837$$

$$G_5^*(10) = \frac{3 - 2 * 1.8}{0.837 * [(5 * 2 - 4)/4]^{1/2}} = -0.585$$

$$G_5^*(20) = \frac{5 - 3 * 1.8}{0.837 * [(5 * 3 - 9)/4]^{1/2}} = -0.390$$

$$G_5^*(30) = \frac{8 - 4 * 1.8}{0.837 * [(5 * 4 - 16)/4]^{1/2}} = 0.956$$

Under the null hypothesis, we use Local *G*-statistics to determine whether there is evidence of spatial clustering of high or low values around each spatial unit of fast food restaurants. From the calculation, we observe that some G_i values are both negative and positive. Positive values of G_i indicate a spatial clustering of high values, while negative values of G_i indicate a spatial clustering of low values. However, to interpret Local *G*-statistics, we need to derive the *Z*-score. This has been done in the empirical examples that are presented later in this chapter. Although Local *G*-statistics are a more flexible form of LISA statistics, they do not have a natural origin. The use of nonbinary weight matrices also makes them more appealing for understanding spatial relationships.

Local Geary

Local Geary measures local patterns of spatial association. Local Geary for each observation *i*:

$$c_i = \sum_j w_{ij}(z_i - z_j)^2$$

where w_{ij} are spatial weights matrix, and the observations z_i and z_j are in deviation from the mean. Local Geary, $c_{i,}$ is the summation of THE spatial weights matrix, which is then multiplied by the squared differences in *Z*-score $(z_{i,} z_j)$ of each observation and its neighboring cell. Deriving the mean deviations for each of the observations is similar to how we calculate the *Z*-score.

TASK 7.7 UNDERSTANDING THE SPATIAL DISTRIBUTION OF CAR ACCIDENTS IN A SMALL TOWN

Let us illustrate the Local Geary statistics by using a hypothetical case of incidents of car accidents in a small town (Figure 7.12). The number in each cell represents the number of car accidents for each neighborhood. We can calculate the Local Geary index for each of the neighbors. Table 7.6 gives a worktable and results for Local Geary statistics.

a 20	*b* 30	*c* 10
d 10	*e* 20	*f* 20
g 40	*h* 10	*i* 20

FIGURE 7.12
A regular grid/spatial units of incidents of car accidents in a small town.

TABLE 7.6

Worktable for Deriving Local Geary and Related Statistics

		w^{a}_{ij}=	a	b	c	d	e	f	g	h	i
		A	0	1	0	1	0	0	0	0	0
		B	1	0	1	0	1	0	0	0	0
		C	0	1	0	0	0	1	0	0	0
		D	1	0	0	0	1	0	1	0	0
		E	0	1	0	1	0	1	0	1	0
		F	0	0	1	0	1	0	0	0	1
		G	0	0	0	1	0	0	0	1	0
		H	0	0	0	0	1	0	1	0	1
		I	0	0	0	0	0	1	0	1	0
Deriving Mean Deviations			**Matrix for Neighboring Cells**								
X_j	*Z-score*	Spatial units	AA[b]	AB	AC	AD	AE	AF	AG	AH	AI
20	0	a	BA	BB	BC	BD	BE	BF	BG	BH	BI
30	1.061	b	CA	CB	CC	CD	CE	CF	CG	CH	CI
10	−1.061	c	DA	DB	DC	DD	DE	DF	DG	DH	DI
10	−1.061	d	EA	EB	EC	ED	EE	EF	EG	EH	EI
20	0	e	FA	FB	FC	FD	FE	FF	FG	FH	FI
20	0	f	GA	GB	GC	GD	GE	GF	GG	GH	GI
40	2.121	g	HA	HB	HC	HD	HE	HF	HG	HH	HI
10	−1.061	h	IA	IB	IC	ID	IE	IF	IG	IH	II
20	0	i									

$n = 9, \sigma = 9.428, \bar{x} = 20$

(Continued)

TABLE 7.6 (Continued)

Worktable for Deriving Local Geary and Related Statistics

Deriving c_i for each observations (spatial units), an example is provided in a previous row										Local Geary c_i
$c_i = \sum_j w_{ij}(z_i - z_j)^2$	0	1.125	0	1.125	0	0	0	0	0	2.25
	1.125	0	4.5	0	1.125	0	0	0	0	6.750
	0	4.5	0	0	0	1.125	10.125	0	0	5.625
	1.125	0	0	0	1.125	0	0	0	0	12.375
	0	1.125	0	1.125	0	0	0	1.125	0	3.375
	0	0	1.125	0	0	0	0	0	0	1.125
	0	0	0	1.125	0	0	0	10.125	0	20.25
	0	0	0	0	1.125	0	10.125	0	1.125	12.375
	0	0	0	0	0	0	0	1.125	0	1.125

[a] The 9×9 matrix grid is labeled by cell locations; when two cells are adjacent, it is assigned a value of 1, and when it is not, a value of 0 is assigned. However, diagonal neighbors were excluded.

[b] Example AA = weight $a_i a_j$*((a_i z-score - a_j z-score)[2])

For interpretative purposes, we need to derive p values to be able to meaningfully interpret Local Geary statistics.

Using Scatterplots to Synthesize and Interpret Local Indicators of Spatial Association Statistics

In this final section of the chapter, let us review two empirical examples of LISA statistics that are illustrated using Moran scatterplots (Figures 7.13 and 7.14) and LISA maps (Figures 7.14 and 7.15). The first empirical example consists of the 2013 blood lead levels (BLL) prevalence data for children (aged 5 years or younger) residing within the city of Chicago. This information was extracted from the Lead Poisoning Testing and Prevention Program database of the Chicago Department of Public Health (Oyana and Margai 2007, 2010). The second empirical example consists of crime incident data averaged over a 5-year period that was described in Chapter 6 (Figures 6.9 and 6.10). The two sets of data were first conditionally randomized 9999 times with Queen's spatial weights set at the nearest five neighbors for the BLL data and the nearest two neighbors (first order) for crime incident data.

Figure 7.13 presents both the adjusted (plot A) and nonadjusted (plot B) prevalence rates for BLL. Figure 7.15 shows local spatial clustering of BLL prevalence in the city of Chicago using Local Moran's I and Local G-statistics. In Map A (Local Moran's I), there are three major sets of spatial clusters of BLL depicting neighbors with high values surrounded by those with similar values. These neighborhoods are located in the west side, south side, and far south. However, there is a minor cluster located within the downtown area. Although Map B (Local G-statistics) shows similar locations to those detected in Map A, the sets of contiguous locations in Map A are much bigger in spatial extent than those in Map B. The hot spots of BLL identified by both methods should invite further in-depth scrutiny.

Figure 7.14 and Table 7.7 depict the local clusters of crime incidents in the city of Spokane, Washington, using Local Moran's I, Local G-statistics, and Moran scatterplot. In reviewing both the figure (see map and Moran scatterplot) and the table, one can see that there are two statistically significant L_i and G_i values obtained from each test. Local Moran's I detects the neighborhoods of Emerson/Garfield and Riverside to have high values of crime incidents, whereas Local G-statistics identifies Emerson/Garfield and Logan as having high rings. These results suggest that the two neighborhoods have a local mean that is higher than the regional mean. Emerson/Garfield is evident in both tests.

In the two sets of empirical examples presented, the adjusted rates are more stable and reliable than the unadjusted rates and therefore should be used in both spatial analysis and spatial modeling. The LISA statistics show that certain neighborhoods have a disproportionate BLL prevalence and crime incidents in comparison with the surrounding neighborhoods.

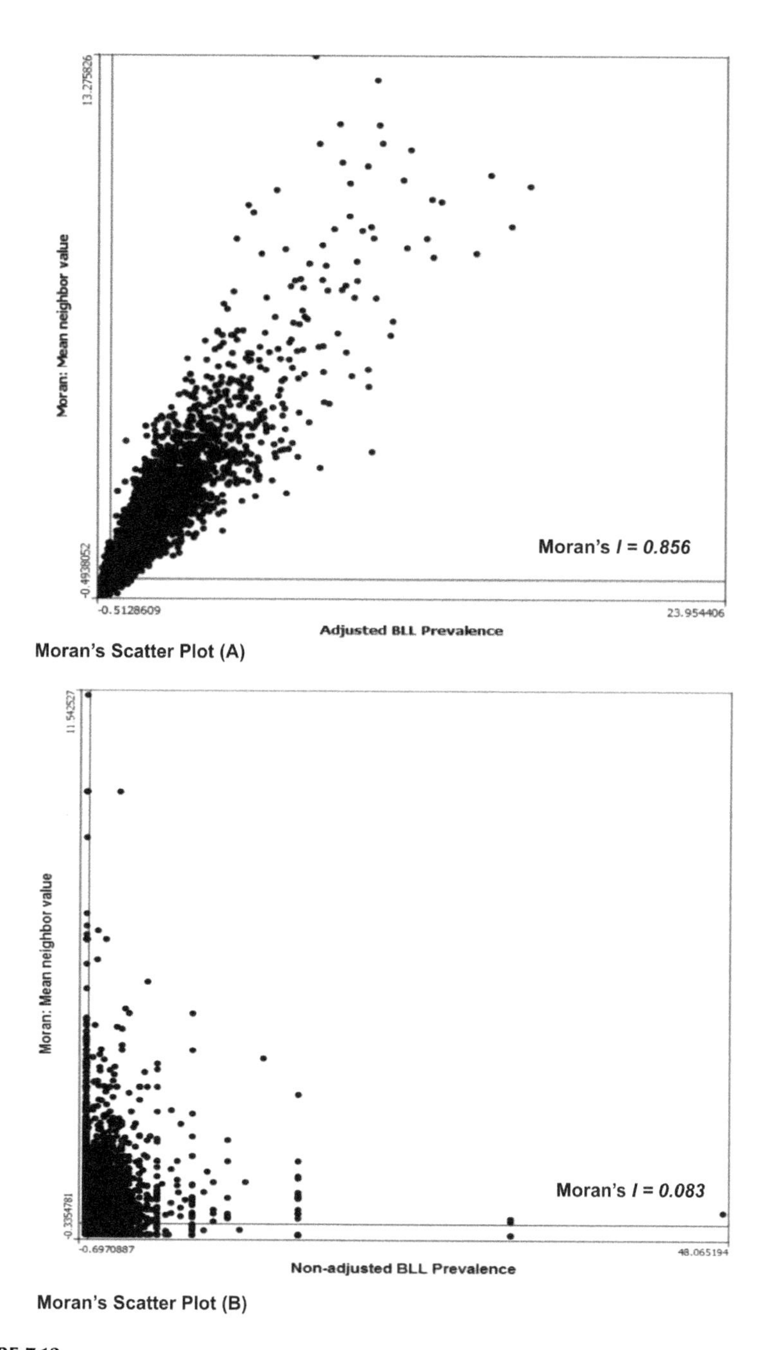

FIGURE 7.13
Plots of (A) filtered/adjusted and (B) unfiltered/nonadjusted blood lead level (BLL) prevalence in the city of Chicago. The value for Moran's I for adjusted rates is close to 1 and is linear because we have accounted for uncertainties. This plot gives unbiased estimates rather than nonadjusted rates.

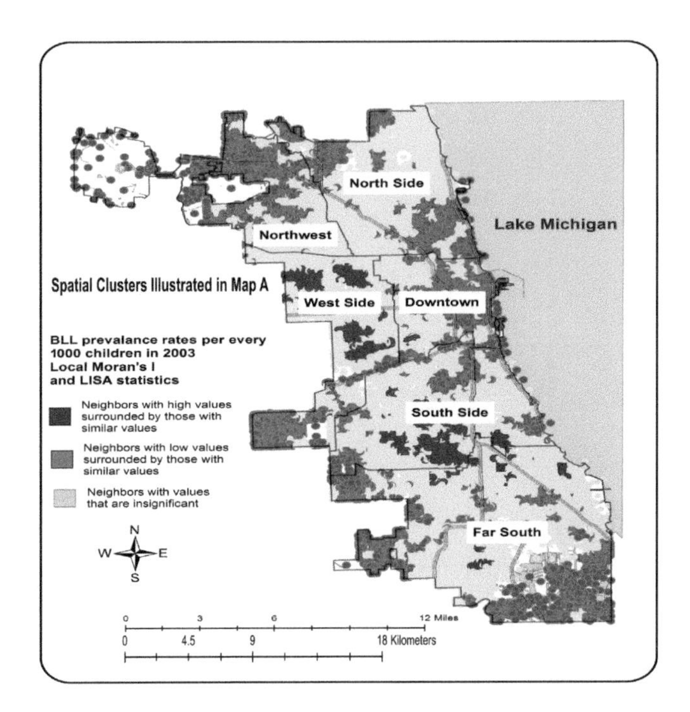

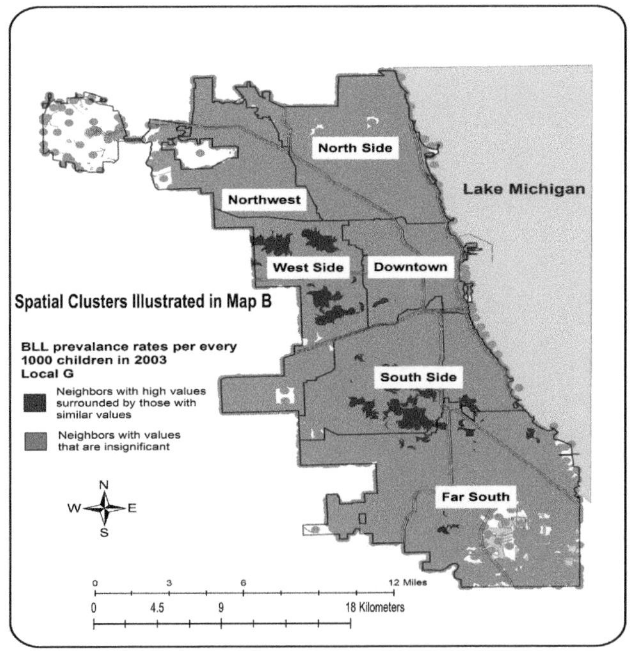

FIGURE 7.14
Identified local clusters of blood lead level (BLL) prevalence in the city of Chicago using Local Moran's I and Local G statistics.

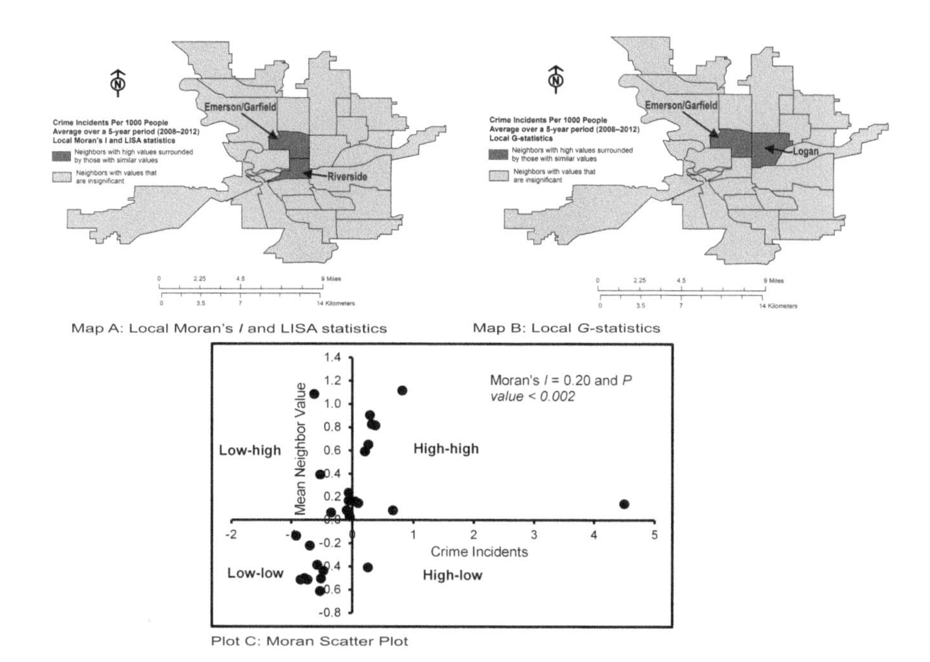

FIGURE 7.15
Identified local clusters of crime incidents in the city of Spokane, Washington, using Local Moran's *I*, Local *G*-statistics, and Moran scatter plot.

Conclusions

At the heart of spatial analysis is the notion of spatial dependency, and perhaps one of the best ways to demonstrate the relevance of this concept is through the analysis of areal data. We have done this through the use of several techniques that generate measures of spatial dependency and autocorrelation at both the global and localized levels. As illustrated in the examples, the global statistics are based on the entire dataset and seek to produce a single measure that reflects the average value (of spatial autocorrelation) for the entire study area. Although the global statistics provide a valuable first step in confirming the presence or absence of autocorrelation, the localized techniques are capable of pinpointing the locations of spatial outliers and notable hot spots that require further evaluation. The local statistics focus on each observational unit rather than the entire study area. These local measures are based on the assumption that different processes may underlie the existence of the geographic patterns that are observed in each area. The end result is a unique value or statistic that is produced for each spatial unit and can be used to delineate neighborhood clusters and other spatial anomalies. Using examples throughout the chapter, we demonstrated the practical applications of these techniques by working through the

TABLE 7.7

Worktable for Deriving Local Moran's I and Local G-Statistics

ObjectId	Name	Average_08_12	Crime_ Rate08_12	Pop_2011	Z-Score	Mean Neighbor Value	L_i[a]	High-Low	P-value	G_i[a]	P-value	G_iRing
27	Whitman	383.4	115.45	3321	−0.68	0.027	—0.02	NS	0.5	0.41	0.500	NS
2	Bemiss	1019	136.10	7487	0.144	−0.269	—0.04	NS	0.5	0.39	0.500	NS
7	Emerson\ Garfield	1647.6	161.01	10233	0.135	1.069	0.145	HH	0.035[a]	2.46	0.045[a]	High ring
17	West Central	1403	172.66	8126	2.500	0.422	1.055	NS	0.299	1.99	0.194	NS
23	Riverside	2144.2	714.02	3003	3.198	0.502	1.607	HH	0.042[a]	2.17	0.159	NS
14	Peaceful Valley	25	41.19	607	−0.759	0.942	—0.72	NS	0.13	2.57	0.054	NS
12	Brownes Addition	388.6	230.76	1684	0.038	0.576	0.022	NS	0.398	2.06	0.183	NS
3	Rockwood	194	31.82	6096	−0.515	0.234	—0.12	NS	0.5	—0.60	0.500	NS
26	Northwest	1753.2	77.62	22586	−0.154	0.396	—0.06	NS	0.5	0.12	0.500	NS
16	North Indian Trail	182.2	26.88	6779	−0.726	−0.629	0.457	NS	0.466	—0.54	0.465	NS
4	Logan	1684.2	165.02	10206	0.211	0.814	0.172	NS	0.085	2.496	0.019[a]	High ring
13	East Central	2013	157.75	12761	2.030	0.426	0.865	NS	0.283	1.78	0.430	NS
21	Latah Valley	165.2	54.11	3053	−0.713	0.126	—0.090	NS	0.5	1.22	0.500	NS
25	Comstock	377	53.88	6997	−0.545	−0.535	0.293	NS	0.29	—1.52	0.5	NS

[a] L_i and G_i values enable one to infer the statistical significance of the pattern at each location. Extreme values indicate the presence of outliers in the distribution. Statistically significant at $p < .05$.

computational steps. We also discussed the methodological limitations and the need for sound knowledge of analytical assumptions and criteria to ensure a reliable and robust spatial analysis of the data. By following these guidelines, we are able to account for the underlying spatial structure in our datasets, thus enabling a better understanding of spatial relationships.

Worked Examples in R and Stay One Step Ahead with Challenge Assignments

TASK 7.8 STAY ONE STEP AHEAD WITH CHALLENGE ASSIGNMENTS: CONCEPTS AND APPLICATIONS

We learned how to use a variety of exploratory data analysis techniques to search, characterize, and describe the spatial distribution of group-level data. We also examined the notion of spatial associations and the methods that are widely used to characterize these patterns at both the global and local levels. We now use R to demonstrate the concept of modeling spatial relationships using two spatial weight schemes, and second, quantifying spatial autocorrelation effects in areal patterns. Creating spatial weights is a crucial step in using areal data as a means to check that there is no remaining spatial patterning in the residuals. The first step is to define which relationships between observations are to be given a nonzero weight (i.e., choose the neighbor criterion to be used); the second is to assign weights to the identified neighbor links.

For this particular task, we use two spatial weight schemes: contiguity and distance-based neighbor schemes. It is important to note that creating the neighbors and weights is, however, not easy to do, and so a number of functions are included in the **spdep** package to help. In the **spdep** package, neighbor relationships between n observations are represented by an object of class **nb**. It is a list of length n with the index numbers of neighbors of each component recorded as an integer vector. If any observation has no neighbors, the component contains an integer zero. Also, while working with spatial data, it is better to use projected data to enhance distance accuracy.

CONTIGUITY-BASED NEIGHBORS

The contiguity-based neighbors consider neighboring polygons to be those that "touch" another polygon. We derive this using **spdep** package with the **poly2nb** function. We can use this function to generate either a queen's-case neighborhood object or a rook's-case neighborhood object.

```
setwd("~/ChapterData/Chapter7_Data_Folder")
list.files()

library(raster)
library(sp)
library(spdep)
repos='https://nowosad.github.io/drat/',

#Read in the areal spatial dataset. Since our dataset is
already in a projected coordinate system, there is no
need for transformation.
MMA_Region <- shapefile("MMA_Region.shp")
  plot(MMA_Region)
```

#Create a Queen's case neighborhood object.

```
nb1 <- poly2nb(MMA_Region, row.names=MMA_Region$ObjectID,
queen=TRUE)
class(nb1)
## [1] "nb"
  summary(nb1)
```

#Plot the links between the polygons

```
plot(MMA_Region, col='gray', border='blue')
xy <- coordinates(MMA_Region)
plot(nb1, xy, col='red', lwd=2, add=TRUE)
```

We already have the first one (Queen's-case neighborhood object, plotted previously). The rook's-case neighborhood object can be generated using the following command line. Instead of queen=True, we use False.

```
nb2 <- poly2nb(MMA_Region, row.names=MMA_Region$ObjectID,
queen=FALSE)
```

Now try plotting the links between the polygons for the Rook's-case neighborhood scheme.

Distance-Based Neighbors

Distance-based neighbors are those within a given proximity threshold to another polygon; distances are measured between Polygon centroids. The function; knn2nb(knearneigh(coordinates(spoly),k=n)) creates an object of class nb that retains the *n* nearest neighbors for each polygon. You can also use a fixed distance band with the function, dnearneigh(coordinates(spoly), d1=n1, d2=n2), where n1 is the minimum distance threshold (commonly set to 0) and n2 is the maximum distance at which polygons will be considered neighbors.

Create nearest neighbors as follows:

```
xy <- coordinates(MMA_Region)
k1 <- knn2nb(knearneigh(xy, k=1),row.
names=MMA_Region$ObjectID)
k1

k2 <- knn2nb(knearneigh(xy, k=2),row.
names=MMA_Region$ObjectID)
k2

k4 <- knn2nb(knearneigh(xy, k=4),row.
names=MMA_Region$ObjectID)
k4

Now, plot the neighbor relationships for k=1, 2 & 4.

par(mfrow=c(1,3))
#Panel 1
plot(MMA_Region, col='gray', border='white')
xy <- coordinates(MMA_Region)
plot(k1, xy, col='black', lwd=1, pch=20, add=TRUE)
```

For Panels 2 and 3, R-Code is located in Chapter7-R-Code.doc, and the respective datasets are in Chapter Seven Folder.

Quiz

1. Individual-level data at national level. Is this high resolution? *True* or *False*
2. Group-level data at census tract, county, and state levels. Which has a better resolution?
   a. Census tract
   b. County
   c. State
3. In 2 above, which has a modifiable areal unit problem (MAUP)?
4. Individual-level data at local level. Is this high resolution? *True* or *False*
5. Here is a list of different types of methods. What is the right measurement scale for each of them?
   a. Joint Count
   b. Moran's I
   c. Geary's C
   d. General G
6. Name the two major types of spatial weights.

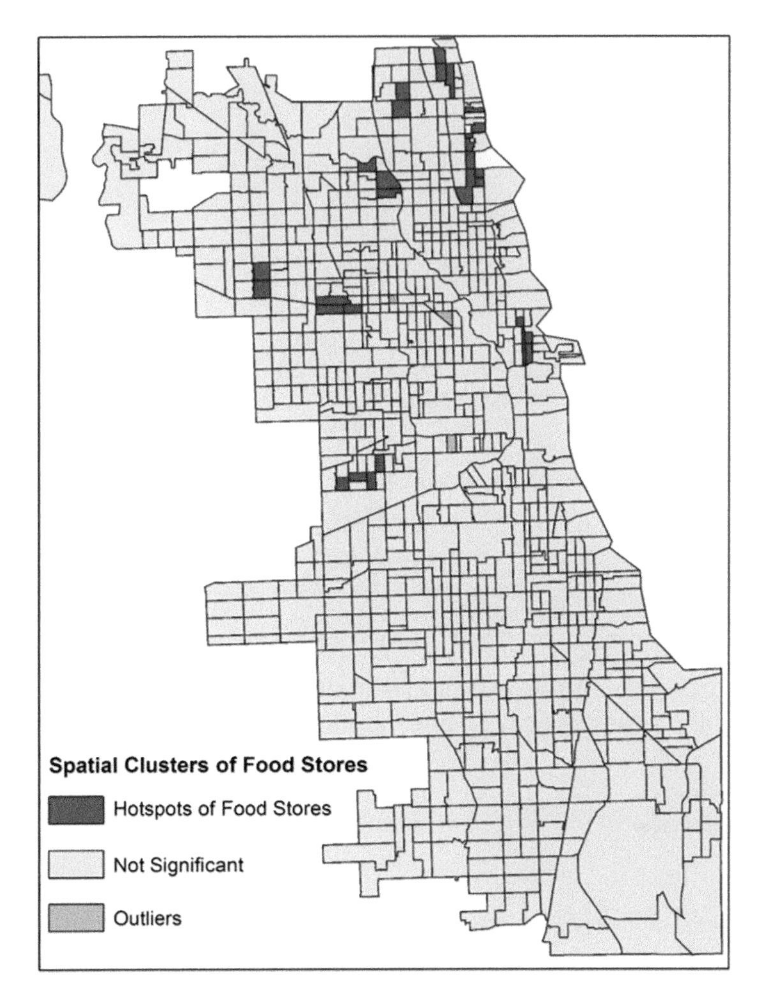

FIGURE 7.16
Hot spots of food access in the city of Chicago.

7. Do the methods listed in question 5 account for the spatial autocorrelation problem?
8. Describe the spatial patterns "hot spots" of food access in Figure 7.16. What insight do we get from the analysis of local clusters of food access in the city of Chicago?

Review and Study Questions

1. What are spatial weights? With the use of examples, explain how these are calibrated and integrated into the analysis of areal data.

2. The choice of analytical method for evaluating spatial autocorrelation in areal data is partly based on the measurement scale of the variable. With the use of examples, briefly explain what techniques are ideal for analyzing variables that are measured on each of the four scales.
3. With the use of examples, distinguish between global and local statistics in the analysis of aggregated spatial data. What would be the effect of MAUP on these two sets of statistics?
4. What are the similarities and differences between each of the following pairs of statistics?
   a. Join Count and Global Moran's *I*
   b. Global Moran's *I* and Getis-Ord *G*
   c. Global Moran's *I* and Global Geary's *C*
   d. Local Moran's *I* and Local Geary's *C*
5. With the use of examples from the statistics noted previously, explain the role of *Z*-scores in the spatial analysis of areal data.
6. What are LISA measures? Using one example from your research area, explain the benefits of these measures in exploratory analysis and visualization of spatial data.

Glossary of Key Terms

Getis-Ord *G*: This is another global measure that summarizes the pattern of spatial autocorrelation in the area. It is most applicable to ratio-scaled data and uses the distance between neighborhoods to assess the overall concentration (or lack thereof) of data values in study areas. The computed statistic, *G*, can be used to effectively delineate the location of hot spots and cold spots in a study area. It is compared to the expected value, and if *G* is larger, then there is a strong likelihood of hot spots with higher values clustering together in the region. If *G* is smaller than the expected value, then there is a strong likelihood of cold spots with low values clustering together in the distribution.

Global Geary's *C*: This is also a measure of spatial autocorrelation that produces a global statistic based on the sums of squared differences between pairs of actual data values in the distribution. The measure varies from 0 to 2, with 0 representing a clustered distribution with perfect positive autocorrelation, 1 representing a complete spatial randomness, and 2 indicating a perfect negative autocorrelation.

Global Moran's *I*: This is the most common measure of spatial autocorrelation that is derived from the sums of squared deviations from the means. It is applicable to interval- and ratio-scaled variables measured at either point locations or within areas. The statistic is a weighted correlation coefficient that ranges from −1 (representing a perfect negative correlation in which neighboring values are dissimilar and dispersed) through zero (complete spatial randomness) to +1 (perfect positive correlation that represents spatial patterns in which similar values [high or low] are clustered in space).

join count statistic: A measure that uses binary nominal data to assess the degree of clustering or dispersion among a set of spatially adjacent polygons.

LISA, Local Indicators of Spatial Autocorrelation: These belong to a suite of measures that disaggregate global measures of spatial autocorrelation into location-specific measures such as the Local Moran's *I*, Local *G*, and Local Geary's *C* coefficients. Unlike the global measures, these local measures enable a data scientist to hone in on individual spatial units and compare their data values relative to the neighboring units to assess the degree of similarity or dissimilarity. The end result can be a scatterplot or cluster map that can be used to effectively show spatial anomalies in the distribution. The aggregate value of LISA obtained by summarizing the measures for the individual units can be used as a global indicator of spatial autocorrelation.

spatial contiguity: This is a principle of adjacency or proximity between areal units that could lead to similarities in inherent properties within those units that are greater than units that are farther away.

Tobler's Law of Geography: This is often called the first law of geography where everything is similar to everything else; however, things that are closer are more similar than those that are farther away.

References

Anselin, L. 1995. Local indicators of spatial association: LISA. *Geographical Analysis* 27(2): 93–115.

Anselin, L. and D.A., Griffith. 1988. Do spatial effects really matter in regression analysis? *Papers in Regional Science* 65: 11–34.

Armhein, C. 1995. Searching for the elusive aggregation effect: Evidence from statistical simulations. *Environment and Planning A* 27(1): 105.

Boots, B. and A. Okabe. 2007. Local statistical spatial analysis: Inventory and prospect. *International Journal of Geographical Information Science* 21(4): 355–375.

Cliff, A.D. and J.K. Ord. 1969. The problem of spatial autocorrelation. In Scott, A.J. (ed.), *London Papers in Regional Science 1, Studies in Regional Science*. London, UK: Pion Limited, pp. 25–55.

Cliff, A.D. and J.K. Ord. 1973. *Spatial Autocorrelation, Monographs in Spatial Environmental Systems Analysis*. London, UK: Pion Limited.

Getis, A. 2010. Spatial autocorrelation. In Fischer, M.M. and A. Getis (eds), *Handbook of Applied Spatial Analysis: Software Tools, Methods and Applications*. Berlin, Germany: Springer, pp. 255–278.

Getis, A. and J. Aldstadt. 2004. Constructing the spatial weights matrix using a local statistic. *Geographical Analysis* 36(2): 90–104.

Getis, A. and J.K. Ord. 1992. The analysis of spatial association by use of distance statistics. *Geographical Analysis* 24(3): 189–206.

Goodchild, M.F. 1986. *Spatial Autocorrelation*, vol. 47. Norwich, UK: Geo Books.

Griffith, D.A. 1995. Some guidelines for specifying the geographic weights matrix contained in spatial statistical models. In Arlinghaus, S.L. (ed.), *Practical Handbook of Spatial Statistics*. Boca Raton, FL: CRC Press, pp. 65–82.

Kelejian, H.H. and I.R. Prucha. 2010. Specification and estimation of spatial autoregressive models with autoregressive and heteroskedastic disturbances. *Journal of Econometrics* 157: 53–67.

Moran, P.A. 1948. The interpretation of statistical maps. *Journal of the Royal Statistical Society. Series B (Methodological)* 10(2): 243–251.

Moran, P.A. 1950. Notes on continuous stochastic phenomena. *Biometrika* 37: 17–23.

Nakaya, T. 2000. An information statistical approach to the modifiable areal unit problem in incidence rate maps. *Environment and Planning A* 32: 1.

Odland, J. 1988. *Spatial Autocorrelation*, vol. 9. Newbury Park, CA: Sage.

Ord, J.K. and A. Getis. 1995. Local spatial autocorrelation statistics: Distributional issues and an application. *Geographical Analysis* 27(3): 286–306.

Oyana, T.J. and F.M. Margai. 2007. Geographic analysis of health risks of pediatric lead exposure: A golden opportunity to promote healthy neighborhoods. *Archives of Environmental and Occupational Health* 62(2): 93–104.

Oyana, T.J. and F.M. Margai. 2010. Spatial patterns and health disparities in pediatric lead exposure in Chicago: Characteristics and profiles of high-risk neighborhoods. *Professional Geographer* 62(1): 46–65.

8

Engaging in Geostatistical Analysis

Learning Objectives

1. Use exploratory tools to visualize and compute basic statistics.
2. Explore, describe, and characterize spatial structure using variograms.
3. Map, quantify, and incorporate spatial variability.
4. Perform and discover the best model for spatial prediction.
5. Account for secondary factors and make decisions on a spatial basis.

Introduction

Pioneering work in the field began in the 1950s with inspiration from the South African Danie Krige's work in geological mining. This work later expanded in the 1960s under the French Mathematician George Matheron's leadership and efforts to showcase the practical applications of the methods. Many disciplines, including engineering, hydrology, soil studies, medical geography, epidemiology, ecology, and environmental assessment now fully embrace geostatistical methodologies to solve spatial prediction and modeling problems (Goovaerts 1997, 1999, 2009; Haining et al. 2010; Barro and Oyana 2012; Birkin 2013; Noor et al. 2014). With the advent of geographic information system (GIS), spatial statistics and geostatistics have become virtually inseparable as computerized analytical and visualization approaches are developed to handle and display the large volume and variety of datasets representing both natural and anthropogenic phenomena in spatial modeling. These approaches are now fairly well established and integrated into leading software packages and are used in many scientific endeavors due to their analytical rigor and robustness. In a GIS context, the geostatistical approaches can be used to successfully analyze and integrate the different types of spatial data, measure spatial autocorrelation by incorporating the statistical distribution and spatial relationships between the sample data, perform spatial prediction, and assess uncertainty. Several scholars have also used these approaches (especially Poisson kriging and *p*-field simulation) to account for small number/population problems (Goovaerts and Jacquez 2004; Goovaerts 2005, 2006), account for uncertainty (Oyana 2004), and spatial prediction, as they are known to accurately predict better local estimates (Goovaerts 1997; Guo et al. 2006).

Rationale for Using Geostatistics to Study Complex Spatial Patterns

Modern geostatistics considers a variable of interest to be a random variable whose values are generated using a probability distribution structure. This branch of statistics was developed to overcome the challenges of applying traditional deterministic statistical approaches to address the inherent uncertainty of spatial data in a stochastic way (Cressie 1985; Robertson 1987; Isaaks and Srivastava 1989; Myers 1994a, b; Cromer 1996; Goovaerts 1997; Armstrong 1998; Mitas and Mitasova 1999; Naoum and Tsanis 2004; Yarus and Chambers 2006; Oliver 2010). The theory underlying geostatistical estimation is the regionalized variable theory, which is concerned with the variable distributions in space and their spatial support (such as the size and shape of the geographical units, or the physical size and dimensions in which the observations were recorded). For stochastic approaches such as kriging, the analysis is rooted in the fundamental assumption that both the actual and potential measurements of the variable are outcomes of the random process with an underlying element of uncertainty. Myers (1994b) describes various forms of uncertainty that are associated with spatial data. For example, one of the most common sources of uncertainty is linked to measurement errors that often introduce white noise in the modeling process. Uncertainty may be linked to the failure to operationally define and measure a latent or theoretical construct in a study. As Fisher (1999) explains, it could be caused by vagueness in the definition of objects or ambiguity or nonspecificity in the measurement. Another source of uncertainty in geostatistical analysis could arise from the random function itself that is unknown and has to be interpolated based on values that are measured at a finite set of sampled points. As Myers (1994b) rightly notes, this type of uncertainty can be effectively reduced or controlled if the function is known to have certain properties such as continuity or differentiability. Yet another form of uncertainty can be introduced during the model estimation or interpolation process, particularly when the sampled data points are irregularly spaced (which happens to be the case in most research studies). A core analytical goal in geostatistics, therefore, is to quantify the degree of element of uncertainty (using measures such as the variance of errors) and then choose the appropriate weights that will significantly minimize this uncertainty during the modeling process. Stochastic techniques such as kriging also acknowledge the underlying spatial structure and integrate the use of mathematical and statistical properties (or variogram parameters) of the measured sampling points to derive unbiased empirical estimates. In summary, there are two main reasons underlying the use of modern geostatistics to study complex spatial patterns: (1) a solid spatial statistical theory that is rooted in the need to minimize the variance of errors and (2) a flexible spatial weighting system that yields the best fitted variogram. Figure 8.1 outlines the chronological steps required to ensure a successful geostatistical estimation process. These include the following:

1. Start with the exploration of the spatial data by visualizing and describing the spatial patterns; use both traditional statistical descriptors and charts to present the results.
2. Identify spatial or temporal patterns through the use of variogram clouds.
3. Perform spatial modeling and prediction by selecting techniques that are most appropriate for the data.
4. Perform uncertainty analysis.
5. Review the model and incorporate secondary variables if necessary.

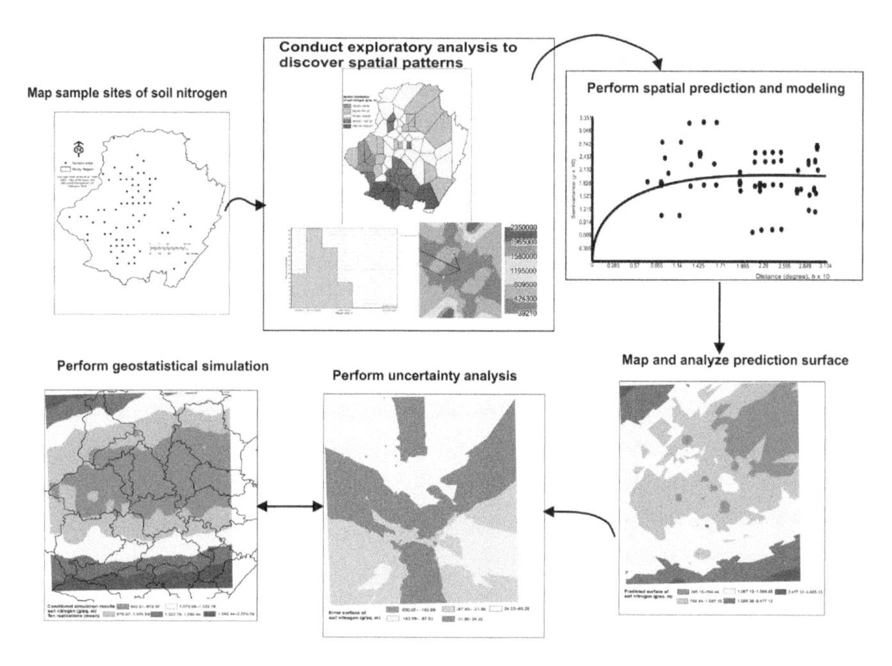

FIGURE 8.1
The workflow of a geostatistical estimation process.

6. Perform simulation, risk assessment, and management by predicting the most probable or possible spatial distribution of the phenomenon being studied.

In the subsequent kriging method and its theoretical framework section, we focus on stochastic techniques, specifically those belonging to the kriging family. We examine the core concepts and principles that underlie these approaches and, with the use of sample exercises, learn how to synthesize and interpret the results. Toward the end of the chapter, we also examine another commonly used spatial algorithm called inverse distance weighting (IDW), and discuss its strengths and limitations. This IDW will serve as an example of a deterministic interpolation method.

Basic Interpolation Equations

As noted previously, the field of geostatistics consists of deterministic and stochastic methods to interpolate spatial data based on information generated at known sampled points. The various forms of kriging, IDW, kernel estimators, splines, trend surfaces, and radial basis functions are all examples of these interpolation techniques. Myers (1994a) discusses these techniques at length and elaborates on how the dual nature of kriging and the positive definiteness property of the variogram connections can be shown between splines, kriging, and radial basis functions. When comparing these techniques, kriging is deemed the most logical choice in providing an unbiased optimal interpolator with optimality defined by the minimum expected error variance in the derived model. Kriging also offers many statistical advantages over the other techniques, including the ability to perform cross-validation of the model by using a fresh sample of observations. To illustrate a simple form of interpolation, Myers (1994a) presupposes that values of a function $f(x)$ are known at points $x_1, x_2, \ldots x_n$. In a one-dimensional case of points, the value of $f(x)$ for $x_{i-1} < x < x_i$ is of interest, and the continuity of $f(x)$ is sufficient to ensure that the linear interpolation is adequate when $x_i - x_{i-1} = a_i$ is small. Therefore,

$$f^*(x) = \left[\frac{(x - x_{i-1})}{a_i}\right] f(x_{i-1}) + \left[\frac{(x_i - x)}{a_i}\right] f(x_i)$$

is very close to $f(x)$. The estimation/interpolation error is $f^*(x) - f(x)$, and it is possible for other errors to exist. However, as rightly noted by Myers, this interpolation function has limited applications due to the lack of additional data locations required to further smooth out the data. Also, in higher-dimensional spaces with irregularly spaced data points, more complicated functions are required. Following is a description of the underlying theory and principles that guide the interpolation of more complex datasets in two- or three-dimensional space.

Spatial Structure Functions for Regionalized Variables

In geostatistics, knowledge of the regionalized random-variable theory and the fundamental concepts that guide the formulation of spatial structure functions is required prior to performing any kind of interpolation. The theory assumes that $Z(x)$ is a regionalized random variable that is associated with a true measurement, $z(x)$, that characterizes the quantity of a variable at point x. The two most important functions that are used to describe this regionalized variable are the spatial covariance and the variogram. For $Z(x)$, the spatial covariance describes how that variable is distributed across space focusing on the degree of similarity among pairs of data points. It also seeks to capture the underlying spatial structure by modeling the degree to which there is spatial autocorrelation with the belief that data values obtained at locations that are closer together are more likely to be similar, whereas values at locations farther apart are more likely to be independent (Tobler's law). This spatial autocorrelation structure informs the formulation of the random function. We can define the values of the random variable Z at two locations, $Z(x)$ and $Z(x + h)$, where h represents the distance (spatial lag) between a pair of sampling sites. There is also a set of assumptions that guide the mathematical formulation of this covariance. One is the basic assumption of stationarity (that certain attributes of the random process are the same everywhere). This effectively enables the inference of the stationary covariance. To derive the covariance and variogram functions, we also assume that each observation is independent under the weaker intrinsic hypothesis of geostatistics (Matheron 1963, 1965; Myers 1994a,b; Goovaerts 1997; Deutsch 2002; Oliver 2010). Using this principle, we can mathematically define the spatial covariance as follows:

$$C(h) = E[Z(x+h) \cdot Z(x)] - \mu^2$$

where μ is the stationary mean, normally estimated from the total number of data points (i.e., data points x_n in the area in which $z[x]$ is being estimated) approximately separated by the vector h. At $h = 0$, the stationary covariance $C(0)$ equals the stationary variance σ^2. We can rewrite this equation into a more standardized stationary correlation $\rho(h)$ as follows:

$$p(h) = C(h) / \sigma^2$$

Given that we are interested in a two-point measure of spatial correlation called the variogram, the equation (covariances) can be slightly modified by the expected squared differences as

$$2\gamma(h) = E\left\{\left[Z(x+h) - Z(x)\right]^2\right\}$$

In reality, however (if the process $Z[x]$ is second-order stationary, the variogram and covariance are equivalent), we would prefer more simplified forms of covariance and variogram as follows:

$$C(h) = \sigma^2 \cdot p(h) = \sigma^2 - \gamma(h)$$

$$\gamma(h) = \sigma^2 \{1 - p(h)\}$$

From the variogram function, we have the semivariance, $\gamma(h)$, which is defined as one-half of the mean squared difference between paired data points in the study area. To illustrate this further, suppose we have a sample of observations $Z(x_i)$, $i = 1,2,3 \ldots n$ where the mean is constant, and we can define the semivariogram as follows:

$$\gamma(h) = \frac{1}{2n(h)} \sum_{i=1}^{n(h)} \left[z(x_i + h) - Z(x_i)\right]^2$$

where n is the number of sample points, $Z(x_i)$ is the measured sample value at location x_i, $Z(x_i + h)$ is the sample value at location $x_i + h$, regionalized variable $Z(x)$, and $n(h)$ is the number of pairs of observations a distance h apart. The semivariogram is therefore a measure of one-half the mean square error produced by assigning the value of $Z(x_i + h)$ to the value $Z(x)$.

In sum, variogram analysis in geostatistics entails the derivation of three empirical measures as estimates of the true population parameters: the spatial covariance $C(h)$, the spatial correlation $\rho(h)$, and the semivariance $\gamma(h)$. The covariance and correlation both reflect the degree of similarity within the data, while the semivariance reflects the degree of dissimilarity with increasing distance among pairs of data points. Various plots can be produced using these three spatial structure functions. For example, a line plot of the spatial correlation against the lag h is called a spatial correlogram. Plotting the spatial covariance against the lag h produces the spatial covariance function. And the most commonly reported visualization is the semivariogram, a line plot that depicts the semivariance $\gamma(h)$ against the lag h (Figure 8.2). Although the technical term for this plot is a semivariogram (one-half of the mean square error), the terms *variogram* and *semivariogram* are used interchangeably in the literature to describe the plot, and we do likewise in this chapter.

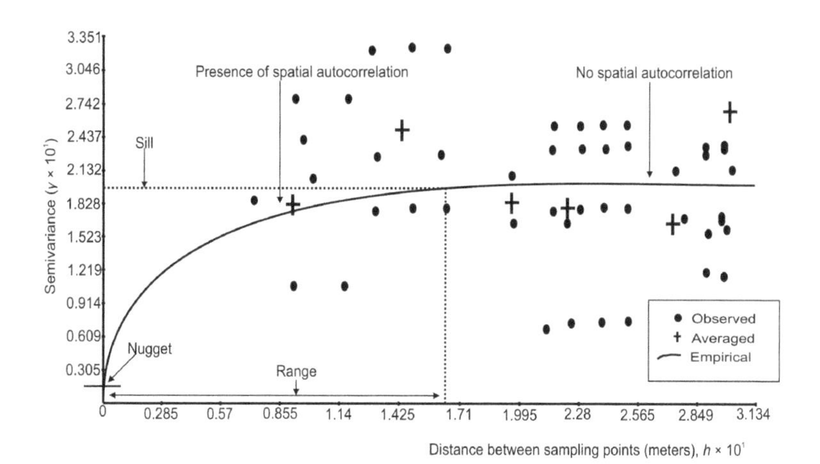

FIGURE 8.2
A semivariogram model and related concepts.

The shape of the variogram describes the degree of spatial autocorrelation that is present in the data. In Figure 8.2, you will find that as the lag distance h increases, the curve increases, and then it levels off at some point. There are three key properties illustrated in this diagram: sill, nugget, and range. The sill refers to the semivariance value at which the curve levels off. As shown in the figure, this is the point at which the $\gamma\ (h)$ value intersects with the range and becomes a constant value as the lag distance h increases. The nugget is a semivariance value $\gamma\ (h)$ that is significantly different from zero for lags that are very close to zero. This is not a measurement error; rather it is best characterized as white noise suggesting that even for data points that are close to one another, the measured values may not necessarily be identical. The range is the lag distance at which the variogram first reaches the sill and remains close to that level for subsequent distances. The variogram can also allow for anisotropy by incorporating both spatial dependence and how these variations change in different directions.

Overall, the variogram is a popular way to compute and visualize spatial autocorrelation, and it is highly recommended as a first step in geostatistical modeling. The relations captured in the equation and the visual depiction provide the foundation for modeling spatial autocorrelation. The procedures for synthesizing and interpreting the results may be summarized in three key points: (1) the sill of the variogram corresponds to the point where there is zero autocorrelation, (2) the autocorrelation between $Z(x)$ and $Z(x + h)$ is positive when the variogram is less than the sill, and (3) the autocorrelation between $Z(x)$ and $Z(x + h)$ is negative when the variogram exceeds the sill (not depicted in Figure 8.2). Once the variogram is developed, it is incumbent on the researcher to choose the statistical model with weights that best represent the data and to share those values, including the variogram, in the statistical results. The selected model will significantly impact the next stage of the analysis that entails the prediction of the unknown values across the study area.

Kriging Method and Its Theoretical Framework

As stated earlier, kriging belongs to a subset of geostatistical methods that rely on the stochastic process in developing predictive surfaces. Named after Danie Krige's pioneering work, the approach is an optimal, unbiased, and generalized least-squares spatial interpolation method that minimizes the estimation from a fitted variogram model. It offers a far better understanding of the spatial structure of the variable in a set of observations, and then provides unbiased estimates for unmeasured locations using the semivariogram model (Goovaerts 1997; Deutsch 2002; Oyana and Margai 2010; Asa et al. 2012). Several kriging methods exist in the geostatistical literature, and they are broadly classified into either linear or nonlinear approaches. The former includes simple kriging (SK), ordinary kriging (OK), universal kriging (UK), Bayesian kriging, and factorial kriging, while the latter includes lognormal kriging, multi-Gaussian kriging, disjunctive kriging, indicator kriging (IK), probability kriging, and rank kriging (Asa et al. 2012).

Asa et al. (2012) outlined four basic assumptions of kriging estimators: (1) the unknown sample data $z(x)$, and the n sample values belong to the regionalized variables, $Z(x)$, and $Z(x_1),\ldots,Z(x_n)$, and no measurement or positional errors exist; (2) for any two points x_1 and x_2 in the area over which $z(x)$ is being estimated, the covariance $\text{Cov}(Z[x_1], Z[x_2])$ of the associated regionalized variables $Z(x_1)$ and $Z(x_2)$ are known; (3) K, the nonnegative matrix of covariances between measured variables (data) at the sample point, is positive definite; and (4) the trend in the area of interest is homogenous. As a result, the mean of the regionalized variables will be the same for the data points x_n in the area in which $z(x)$ is being estimated. If a trend exists in the area of interest, the stationarity of the local mean is relaxed, and a nonstationary random function is employed to represent the mean (kriging with a trend or universal kriging). The random functions adopting $Z(x)$, in the kriging equations, will define the kriging method.

The basic equation of kriging estimators is given in Goovaerts (1997) as follows:

$$Z^*(x) - m(x) = \sum_{i=1}^{n(x)} w_i \left[Z(x_i) - m(x_i) \right]$$

where x and x_i are location vectors for the estimation point and one of the neighboring data points indexed by i; $n(x)$ is the number of data points in a local neighborhood used for estimation of $Z^*(x)$; $m(x)$, $m(x_i)$ are expected mean values of $Z(x)$ and $Z(x_i)$; and w_i is the kriging weight assigned to $Z(x_i)$ for estimation location x. As noted earlier, the goal of kriging is to minimize the variance of the estimator $\sigma_E^2(x) = Var\left\{Z^*(x) - Z(x)\right\}$ under the circumstance $E\left\{Z^*(x) - Z(x)\right\} = 0$ by determining weights w_i.

The random variable $Z(x)$ consists of two components: the residual component ($R(x)$) and the trend component ($m(x)$): $Z(x) = R(x) + m(x)$.

Following the basic assumptions, there are two crucial steps in fitting a semivariogram model and kriging:

1. Measuring the degree of spatial autocorrelation among the measured data points, i.e., description and modeling of spatial patterns (described in the preceding section).
2. Interpolating values between measured points based on the degree of spatial autocorrelation encountered, i.e., prediction of local estimates.

After this, we can also account for secondary factors using co-kriging, a method that integrates multiple variables associated with the primary variable into the analysis. Following are the conceptual descriptions and some illustrated examples of simple, ordinary, universal, and indicator kriging.

Simple Kriging

Simple kriging (SK), identified by the subscript SK, is an estimate that is derived from the modification of the mean. The mean value $m(x)$ of the stationary random variable in an SK equation is assumed to be constant and known throughout the study area. The global mean assumption is contingent upon the SK estimator being unbiased and having a minimal variance of the error of estimation. The SK estimator is derived using this equation:

$$Z^*_{SK}(x) = \sum_{i=1}^{n} w_i \cdot Z(x_i) + \left[1 - \sum_{i=1}^{n} w_i(x)\right] m$$

where $Z(x)$ is the random variable at the location x, all x_i values are equal to n data locations, $m(x) = E\{Z(x)\}$ is equal to location-dependent expected values of random variable $Z(x)$, Z^*_{SK} is the linear regression estimator, $w_i(x)$ is the weight, and $m(x)$ is the mean.

Ordinary Kriging

Ordinary kriging (OK), identified by the subscript OK, is a very powerful and widely used geostatistical method for modeling spatial data (Cressie

1985; Isaaks and Srivastava 1989; Goovaerts 1997; Armstrong 1998; Deutsch 2002). It assumes that the local means are not necessarily closely related to the population mean and will use only the samples in the local neighborhood for an estimate. Simply stated, the local mean $m(x)$ is unknown, but it is assumed to be constant within the search area. OK relies on the spatial correlation structure of the data to determine the weighting values, and the correlation between data points determines the estimated value at an unsampled point. It makes the assumption of normality among the data points. The method is based on three basic ideas:

1. A search is only conducted within a local neighborhood and only samples drawn from this neighborhood are used for estimation. As a result of this process, OK is able to account for the local variation.
2. Weight assignment relies on spatial variability within each local neighborhood.
3. Computation of the average weight is based on each local neighborhood, which is then used to derive the local neighborhood estimate.

OK is derived using this equation:

$$Z^{*}_{OK}(x) = \sum_{i=1}^{n} w_i(x) \cdot Z(x_i) + \left[1 - \sum_{i=1}^{n} w_i(x)\right] m(x)$$

where Z^{*}_{OK} is the linear regression estimator, and the others are as defined in the SK equation presented earlier.

In this equation, we assume the mean $m(x_i) = m(x)$ for each nearby data value $Z(x_i)$, so that $Z^{*}(x) = m(x) + \sum_{i=1}^{n(x)} w_i(x)\left[Z(x_i) - m(x)\right] = \sum_{i=1}^{n(x)} w_i(x) \cdot$ $\leq Z(x_i) + \left[\left(1 - \sum_{i=1}^{n(x)} w_i(x)\right] m(x)\right]$

As $\sum_{i=1}^{n(x)} w_i(x) = 1$, an OK kriging estimator can be calculated from $Z^{*}_{OK}(x) = \sum_{i=1}^{n(x)} w_i^{OK}(x) Z(x_i)$ with $\sum_{i=1}^{n(x)} w_i^{OK}(x) = 1$

The estimator based on a set of variables $Z = \{Z_1, Z_2, \ldots Z_k\}$ can be rewritten as

$$\hat{z}_0(s_0) = \sum_{i=1}^{n} \sum_{j=1}^{k} w_{ij} z_j(s_i)$$

TASK 8.1 CALCULATING THE ORDINARY KRIGING ESTIMATOR

Let us illustrate the calculation of the OK estimator using Burrough and McDonnell's example dataset presented in Figure 8.3 (Burrough and McDonnell 1998).

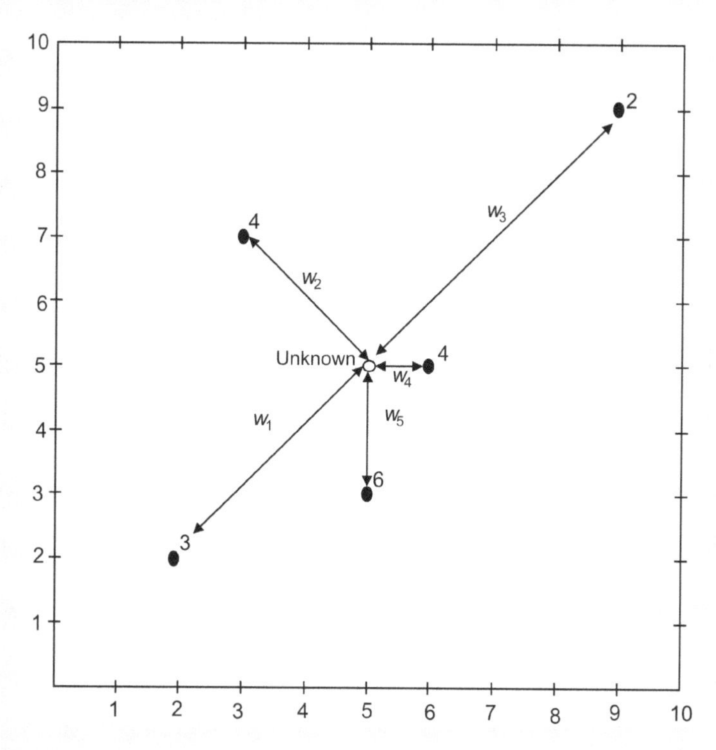

FIGURE 8.3
An example of a dataset to illustrate the kriging estimator.

The sample data in Figure 8.3 has five sampled sites with coordinates (x, y) and values (z), and we will predict the value for the coordinate (5,5). The OK model is based on a constant mean ($m[x]$) and no trend for the data as follows:

$$Z^*_{OK}(x) = m(x) + \varepsilon(x_i)$$

where $x_i = (x, y)$ for each sampled location, and $Z(x_i)$ represents the value of each sampled location and random errors $\varepsilon(x_i)$ with spatial dependence. We will predict the value for unknown point z ($x_i = o$) at coordinates ($x = 5$, $y = 5$). We will apply a spherical variogram model (the equation is given later) to compute the spatial variation of the data sampled at the five locations based on the following parameters: nugget $(C_0) = 2.5$, sill $(C_1) = 7.5$, range a=10.

$$\begin{cases} 2.5+(7.5-2.5)\left(\frac{3h}{20}-\frac{h^3}{2000}\right) & 0 \le h \le 10 \\ 0 & h=1 \\ 7.5 & \text{otherwise} \end{cases}$$

$$\gamma(h) = \sigma^2 - C(h)$$

$$C(h) = 7.5 - 2.5 + (7.5 - 2.5)\left(\frac{3h}{20} - \frac{h^3}{2000}\right)$$

The OK predictor is formed as a weighted sum of the data as follows:

$$Z^*_{OK}(x) = \sum_{i=1}^{n(x)} w_i^{OK}(x) Z(x_i) \text{ with } \sum_{i=1}^{n(x)} w_i^{OK}(x) = 1$$

$$\hat{\sigma}_e^2 = \sum_{i=1}^{n} \lambda_i \gamma(x_i, x_0) + \phi$$

where $\gamma(x_i, x_0)$ is the semivariance between sampled point x_i, and unsampled point $x_0 \cdot \phi$ is a Lagrange multiplier required for the minimalization.

We have to solve the following equation:

$$A^{-1} \cdot b = \begin{bmatrix} w \\ \phi \end{bmatrix}$$

where A is the matrix of semivariances between pairs of data points, b is the vector of semivariances between the predicted point and each sampled data point and w is the vector of weights, and ϕ is a Lagrangian. A distance matrix for the data points is given by

i	1	2	3	4	5
1	0	5.099	9.899	5	3.162
2	5.099	0	6.325	3.606	4.472
3	9.899	6.325	0	5	7.211
4	5	3.606	5	0	2.236
5	3.162	4.472	7.211	2.236	0

The distance vector of covariances for the data points and the unknown point is given by

i	o
1	4.243
2	2.828
3	5.657
4	1.0
5	2.0

We can substitute these numbers to the variogram to obtain the corresponding semivariances:

$A = i$	1	2	3	4	5	6
1	2.5	7.739	9.999	7.656	5.939	1
2	7.739	2.5	8.667	6.381	7.196	1
3	9.999	8.667	2.5	7.656	9.206	1
4	7.656	6.381	7.656	2.5	4.936	1
5	5.939	7.196	9.206	4.936	2.5	1
6	1	1	1	1	1	0

$b = i$	o
1	7.151
2	5.597
3	8.815
4	3.621
5	4.720
6	1

Obtain the inverse matrix:

$A^{-1} = i$	1	2	3	4	5	6
1	−0.172	0.05	0.022	−0.026	0.126	0.273
2	0.05	−0.167	0.032	0.077	0.007	0.207
3	0.022	0.032	−0.111	0.066	−0.01	0.357
4	−0.026	0.077	0.066	−0.307	0.19	0.03
5	0.126	0.007	−0.01	0.19	−0.313	0.134
6	0.273	0.207	0.357	0.003	0.134	−6.873

w as weights:

1	0.0175
2	0.2281
3	−0.0891
4	0.6437
5	0.1998
6	0.1182----ϕ

Minimization of the error variance:

The predicted value at $z(x_i = o) = 0.0175 \times 3 + 0.2281 \times 4 - 0.0891 \times 2 + 0.6437 \times 4 + 0.1998 \times 6 = 4.560$

Estimation variance

$$\sigma_e^2 = 0.0175 \times 7.151 + 0.2281 \times 5.597 - 0.0891 \times 8.815$$
$$+ 0.6437 \times 3.621 + 0.1998 \times 4.720 + \phi = 3.890 + 0.1182 = 4.008$$

Since our standard error is 2.002, we derive the predicted interval at a 95% confidence interval, which ranges from 0.636 to 8.484 (4.56 ± 1.96*2.002).

Universal Kriging

Universal kriging (UK) is kriging with a trend and is similar to ordinary kriging. However, a UK deals with situations where the local mean is variable over the study area. Although local mean $m(x)$ is unknown just like in OK, UK models this as a linear combination of functions of coordinates. Simply stated, it accommodates a nonstationary mean where the expected value of $Z(x)$ is a linear or high-order (deterministic) function of the (x, y) coordinates of the data points. Caution is required when fitting complex models.

The random function, $Z(x)$, is a combination of a trend component with a deterministic variation, $m(x)$, and a residual component, $R(x)$. UK is derived as follows:

$$Z(x) = m(x) + R(x) \text{ and } m(x) = E\{Z(x)\} = \sum_{k=0}^{n} \mu_k \lambda_k(x)$$

where $\lambda_k(x)$ is the known basic function, and μ_k represents the fixed, unknown coefficients.

Indicator Kriging

IK is a method used with categorical data or data converted from continuous data to categorical data. IK is a least-squares estimator of the cumulative distribution function at a threshold, z_k. IK employs the samples in a neighborhood to estimate the probability that data points in a given area exceed a defined threshold. Transformed indicator values (0, 1) are coded 1 if they exceed a defined threshold, and are coded 0 if below the threshold. The semivariogram of indicator data is computed as follows:

$$\gamma(\boldsymbol{h}; z_k) = \frac{1}{2n(h)} \sum_{i=1}^{n(h)} \left[i(x_i; z_k) - i(x_i + \boldsymbol{h}; z_k) \right]^2$$

The local probability at x by kriging of indicator values is given by this equation:

$$\left[i(x; z_k) \right] = E\{I(x; z_k \,|\, (n))\}^* = Prob^*\{Z(x) \le z_k \,|\, (n)\}$$

where n is the conditional information available in the neighborhood of location x. A declustering algorithm is used to decluster the sample data if the z data values are clustered.

Key Points to Note about the Geostatistical Estimation Using Kriging

1. Modeling decisions are driven by insights acquired during the exploratory phase of the geostatistical analysis using the histogram, variogram cloud/*h*-scatterplot, or covariance cloud. The histogram provides a useful tool for confirming normality, a key assumption in geostatistical analysis.
2. The selection of the most appropriate kriging equation/semivariogram model that fits your data is typically based on the preceding steps and results from the exploratory analysis/description of the spatial patterns, and the prediction error analysis.
3. Each kriging equation is designed to meet certain requirements/ assumptions, but all the kriging equations honor the data characteristics, preserve the mean, and preserve the spatial correlation structure.
4. Final decisions are based on the uncertainty analysis using the cross-validation approach. You should select the best kriging option after carefully reviewing the five types of prediction errors: (1) mean prediction error, (2) standardized mean prediction error, (3) root mean squared prediction error, (4) predicted standardized root mean square error (RMSE), and (5) average standard error. An optimal kriging model is one in which both the mean prediction error and the standardized mean prediction error are close to zero. For an optimal model, the predicted RMSE should be as small as possible, of approximately the same magnitude as the average standard error. The predicted standardized root mean square error (RMSES) is dimensionless and must be close to 1 for an optimal model. Use these prediction errors to construct an error decision matrix and pick the semivariogram model with the least error.

5. Kriging technique is not applicable to datasets that have spikes, abrupt changes, or datasets that do not have a spatial autocorrelation structure. Data that lack adequate spatial coverage and are not fully representative of the study region may also result in poorly fitted models.

We apply a set of five geostatistical methods and tools to achieve these two objectives.

Exploratory Data Analysis

Figures 8.4 and 8.5 show the spatial distribution of soil carbon and soil nitrogen in the study region. Soil carbon has a mean of 12.5 kg/m^2 and standard deviation of 13.6 kg/m^2 (Figure 8.4c), while soil nitrogen has a mean of 907.8 g/m^2 and a standard deviation of 631.6 g/m^2 (Figure 8.5c). About 100 of the sampled locations have soil carbon values of less than 20 $kg/m^{2\%}$, and 67% of the entire sample has values that are less than the mean value.

The frequency distribution of soil carbon is positively skewed with a sharp peak. Three of the sampled locations have a very high content of soil carbon (54.8, 76.3, and 113 kg/m^2). The frequency distribution of soil nitrogen is moderately skewed with a sharp peak. About 50% of the sampled locations of soil nitrogen have values that are less than the mean value. Most of the values of sampled locations of soil nitrogen range from 180 to 2362 $g/m^{2\%}$, and 15% of the sampled locations have zero values. There is one sampling location of soil nitrogen with more than 4000 g/m^2 that is located in the south (Figure 8.5b). Sampled locations with the highest content of soil nitrogen are generally located in the south, the medium values are generally located in the central area, and the lowest values are generally located in the east and north east. Spatial patterns of soil carbon (Figure 8.4b) are similar to soil nitrogen except

TASK 8.2 KRIGING SOIL SAMPLE DATA, SYNTHESIS, AND INTERPRETATION OF RESULTS

The sample data presented in this chapter are based on 115 sampling sites for organic soil carbon and nitrogen in southern Africa. The soil database was compiled by SAFARI 2000 but was originally sourced from soil surveys by Zinke et al. (1986) and soil survey literature. There are two objectives that we wish to achieve using this data:

1. Characterize soil organic carbon and soil nitrogen with the intention to understand the interaction of these two soil properties and environmental factors.
2. Use the spatial dependence structure to predict soil organic carbon and soil nitrogen estimates at unsampled locations.

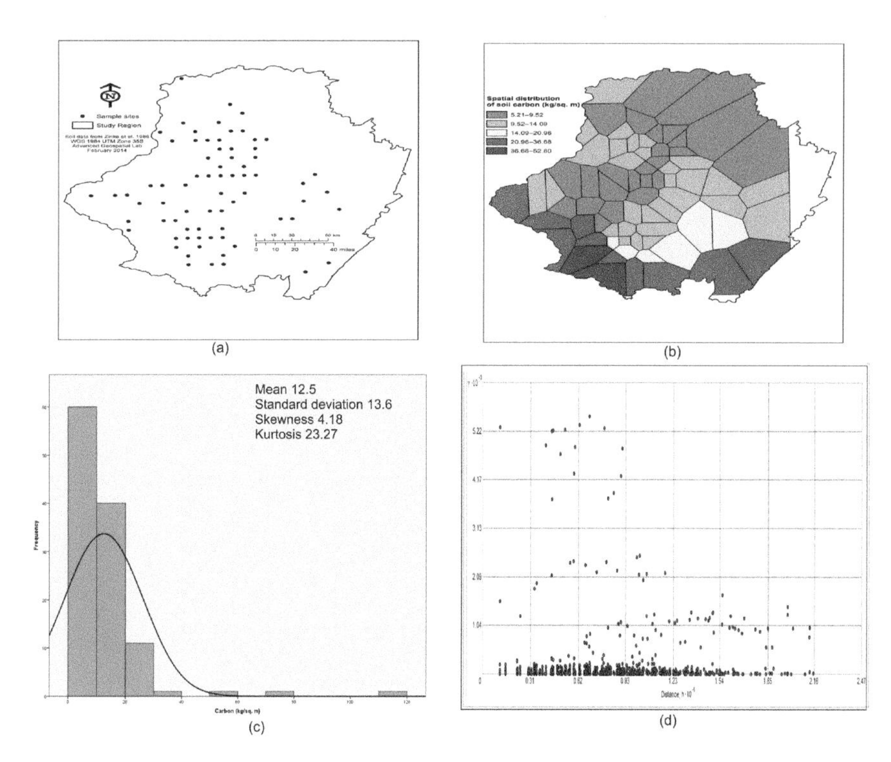

FIGURE 8.4
Distribution and visualization of the soil carbon datasets: (a) sampling locations, (b) Voronoi map, (c) histogram for raw data, and (d) semivariogram cloud. Paired distances of soil carbon in the lower part of the plot in (d) indicate spatial clustering of similar values in the neighboring areas and small semivariances.

that there is a clear pattern in the distribution of soil carbon, especially where there are high, medium, and low values (Figure 8.5b).

The semivariogram cloud (Figures 8.4d and 8.5d) is a form of h-scatterplot that gives the semivariance estimates of paired distances (the distance between sampling points) for soil carbon and soil nitrogen. The paired distances that are closer together suggest spatial dependence, and vice versa. In both sampled locations of soil carbon and soil nitrogen, the paired distances in the lower part of the semivariogram clouds show closeness suggesting the presence of a spatial dependence structure and a small semivariance among values. However, there is evidence of outliers in the upper-left and bottom-right corners of the two semivariogram clouds suggesting large semivariance estimates or wide distances between some paired sampled points.

Spatial Prediction and Modeling

The fitted soil carbon semivariogram models are provided in Figure 8.6a through 8.6d, while the models for soil nitrogen are given in Figure 8.7a

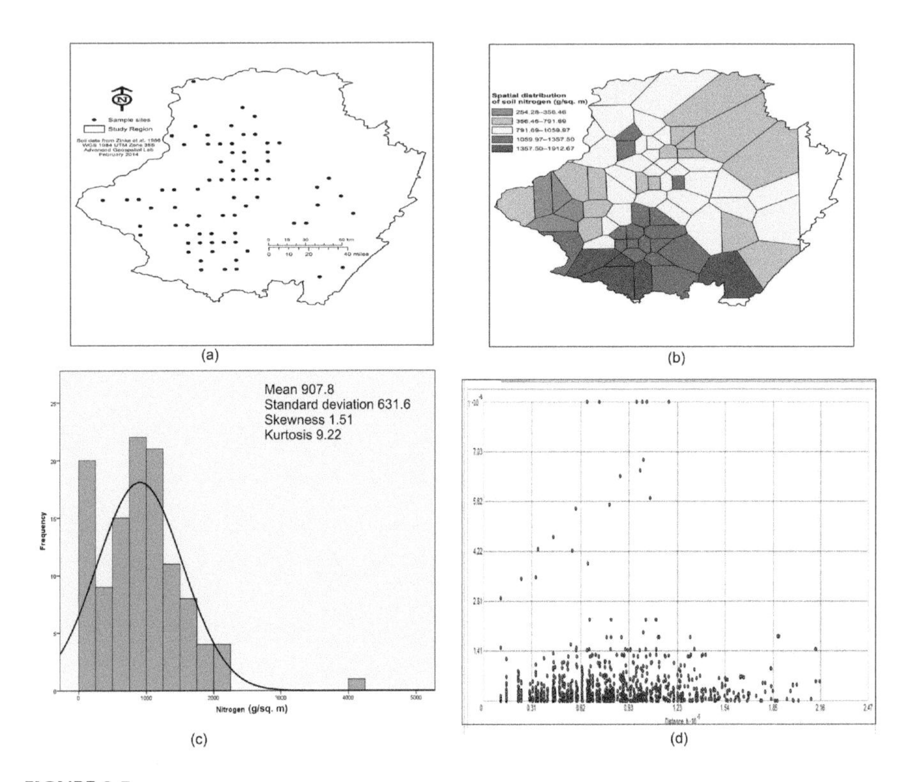

FIGURE 8.5
Distribution and visualization of the soil nitrogen datasets: (a) sampling locations, (b) Voronoi map, (c) histogram for raw data, and (d) semivariogram cloud. Paired distances of soil nitrogen in the lower part of the plot in (d) indicate spatial clustering of similar values in the neighboring areas and small semivariances.

through 8.7d. Table 8.1 summarizes the semivariogram models for the parameters/coefficients of soil nutrients. Note that the UK model for soil nitrogen has a very large nugget suggesting that sampled locations are randomly distributed in the study region. The predicted areas from these models are presented in Figures 8.8 and 8.9. The spatial patterns of UK and IK estimators are more reflective of the raw (soil carbon) data than those from OK and SK estimators (Figure 8.8a through 8.8d).

In Figure 8.8a, the low values are located in the central area of the study region. In the immediate surroundings, there are medium values, and the high values are located in the lower-left corner of the study region. Spatial patterns in Figure 8.8c and d are similar with slight differences evident in their spatial extents. The indicator semivariogram model gives a probability surface map for soil carbon where there is not an exceedance of an optimal threshold of 10.15 kg/m^2 (Figure 8.8d).

The spatial patterns for the prediction surface of OK, UK, and IK estimators are more reflective of the raw data of soil nitrogen than the SK estimator. In Figure 8.9a and b, the low and medium values of soil nitrogen are mainly located in the central area and toward the northeast. However, spatial

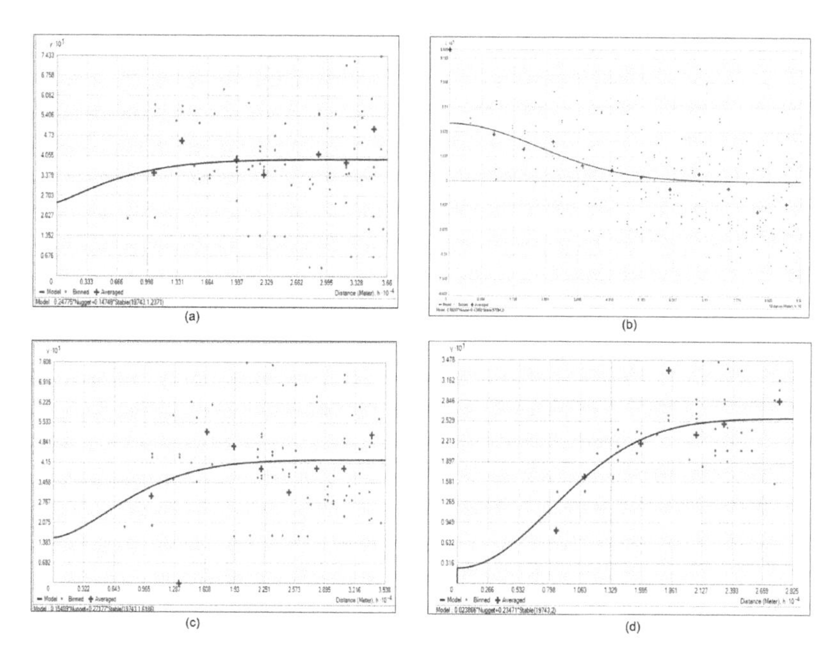

FIGURE 8.6
Semivariogram models for estimating soil carbon: (a) OK estimator, (b) SK estimator, (c) UK estimator, and (d) IK estimator.

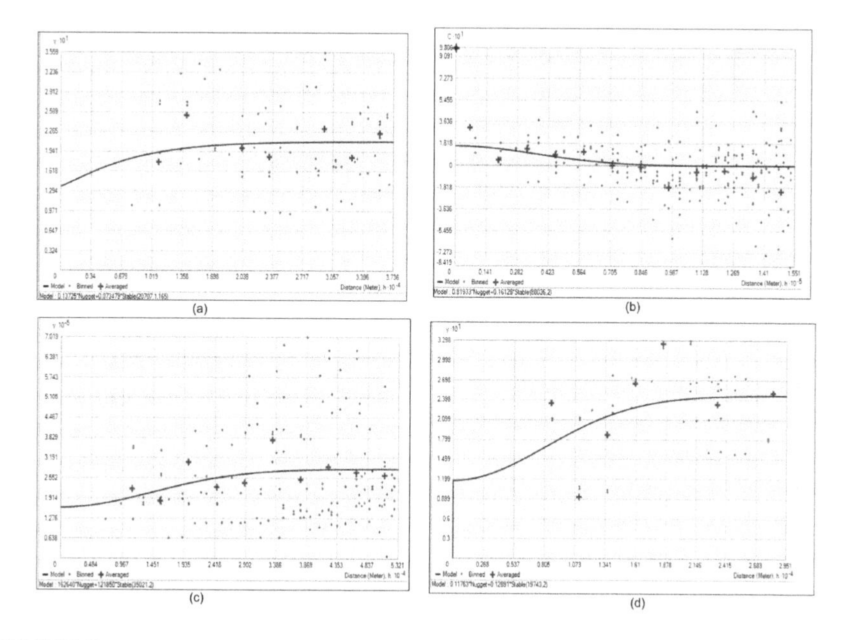

FIGURE 8.7
Semivariogram models for estimating soil nitrogen: (a) OK estimator, (b) SK estimator, (c) UK estimator, and (d) IK estimator.

TABLE 8.1

Fitted Semivariogram Models for Soil Carbon and Soil Nitrogen

Kriging Estimator	Soil Carbon	Soil Nitrogen
OK	$\gamma = 0.2478*C_0+0.1475*$Stable (19743, 1.371)	$\gamma = 0.1373*C_0+0.0735*$Stable (20707, 1.165)
SK	$\gamma = 0.5529*C_0+0.4296*$Stable (57294)	$\gamma = 0.8193*C_0+0.1613*$Stable (88036, 2)
UK	$\gamma = 0.1549*C_0+0.2718*$Stable (19743, 1.6186)	$\gamma = 162640*C_0+121850*$Stable (35021, 2)
IK	$\gamma = 0.0234*C_0+0.2347*$Stable (19743, 2)	$\gamma = 0.1176*C_0+0.1289*$Stable (19743, 2)

patterns in Figure 8.9d are more pronounced than those from other models. The indicator semivariogram model shows a probability surface map for soil nitrogen where there is not an exceedance of an optimal threshold of 1070.5 g/m^2 (Figure 8.9d).

Uncertainty Analysis

Having reviewed the preliminary models, we now need to judge the most appropriate semivariogram model for soil carbon and soil nitrogen. The

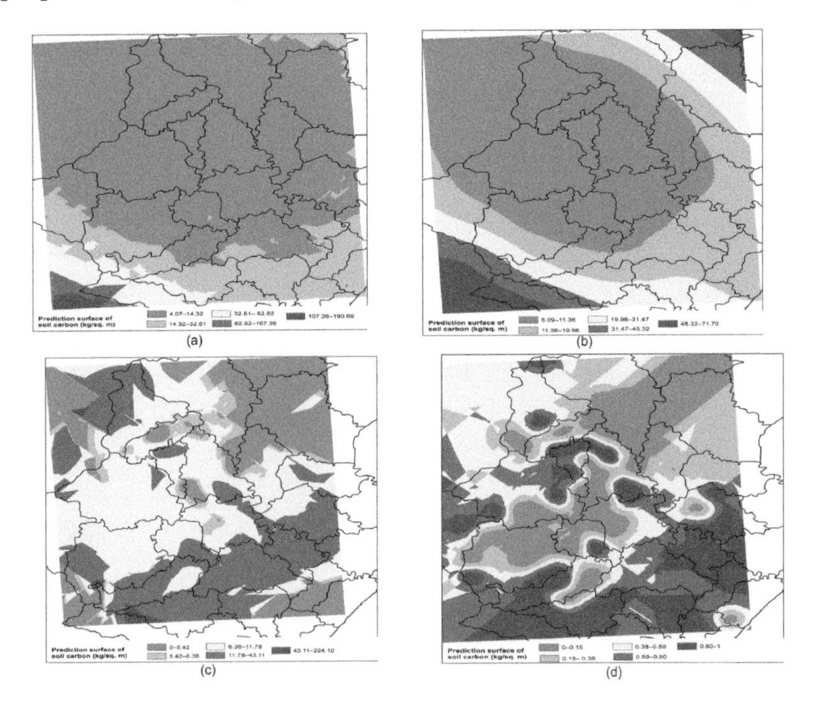

FIGURE 8.8

Prediction surface maps for soil carbon: (a) OK estimator, (b) SK estimator, (c) UK estimator, and (d) IK estimator. The spatial patterns for UK and IK estimators are more reflective of the raw data of soil carbon than those from the OK and SK estimators.

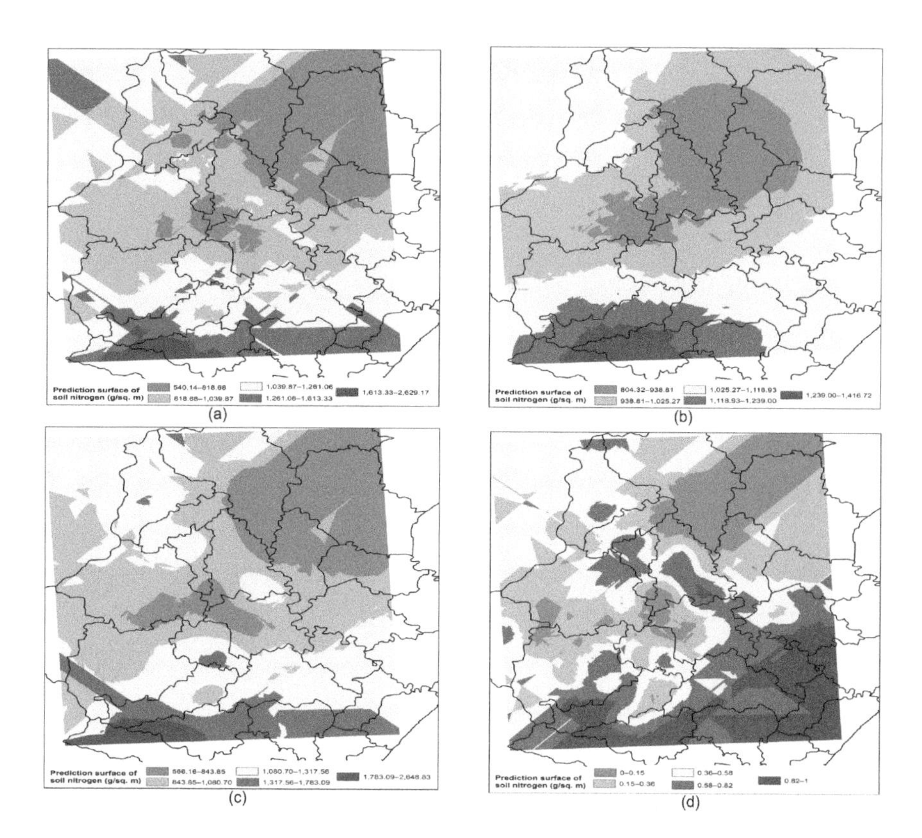

FIGURE 8.9
Prediction surface maps for soil nitrogen: (a) OK estimator, (b) SK estimator, (c) UK estimator, and (d) IK estimator. The spatial patterns for OK, UK, and IK estimators are more reflective of the raw data of soil nitrogen than the SK estimator.

cross-validation plots are presented in Figures 8.10 and 8.11 for this purpose. Decision matrices for the performance of the five sets of geostatistical methods are presented in Tables 8.2 and 8.3. A critical examination of the cross-validation plots reveals the following observations. In Figure 8.10a through d, if we use an error cutoff point of ±2 in both directions, we can identify sampling sites of soil carbon with under- or overprediction. In the OK model, 29% and 48% show under- and overprediction, respectively (SK [25%, 52%]; UK [29%, 39%]; and IK [21%, 28%]). An in-depth scrutiny of the standardized errors (if we use a cutoff of ±2 in both directions) in the validated sites shows there is a major reduction in underprediction with OK (9.7%), SK (4.2%), and IK (7%), with the exception of UK (32%). Overprediction also reduced in a number of validated sites with OK having none, SK and IK having 3% and 4.2%, respectively, with the exception of UK (4.2%). Most of the sites have a 50% probability of having less than the threshold of soil carbon 10.15 kg/m^2.

In Figures 8.11a through d, the OK model shows 40% and 57% under- and overprediction, respectively (SK [42%, 57%]; UK [40%, 49%]; and IK [33%, 40%]).

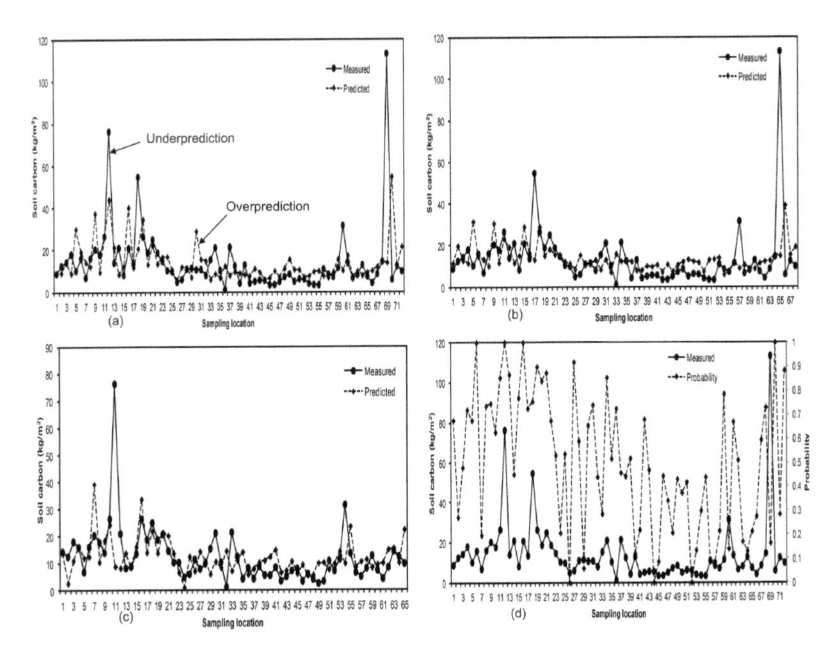

FIGURE 8.10
Cross-validation plots for estimating soil carbon: (a) OK estimator, (b) SK estimator, (c) UK estimator, and (d) IK estimator.

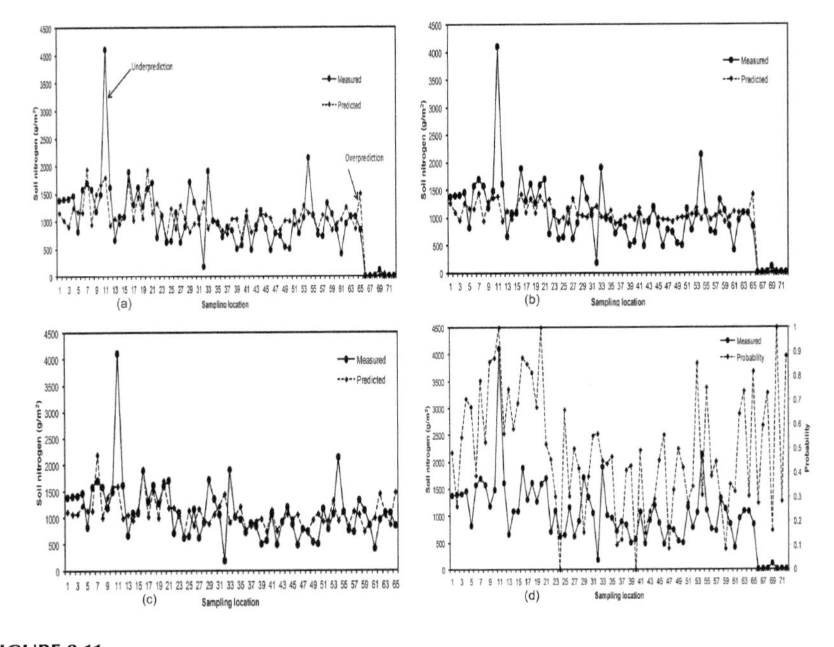

FIGURE 8.11
Cross-validation plots for estimating soil nitrogen: (a) OK estimator, (b) SK estimator, (c) UK estimator, and (d) IK estimator.

TABLE 8.2

Cross-Validation Statistics for Five Interpolation Methods for Soil Carbon

Statistics	Decision Criteria	OK	SK	UK	IK*	IDW
Mean	Near zero	0.021	0.3233	0.289	−0.0007	−0.197
Standardized mean	Near zero	−0.114	−0.0004	0.324	−0.008	
Root mean square error (RMSE)	Very small	15.959	15.11	5.684	0.491	9.294
Standardized root mean square error (RMSES)	Near 1	1.303	1.422	8.613	1.027	
Average standard error (ASE)	Very small	13.512	12.41	0.676	0.481	
Ranking (the best*)		5	4	3	1	2

An in-depth analysis of the standardized errors in the validated sites of soil nitrogen shows a major reduction in underprediction as well as overprediction: OK (5.6%, none); SK (5.6%, 1.4%); UK (5.6%, 1.4%); and IK (none, none). Most of the sites have a 67% probability of having less than the threshold of soil nitrogen (1070.5 g/m^2).

Based on these cross-validation statistics, the indicator semivariogram models for soil carbon and soil nitrogen provide the best and most appropriate fit for data. The two models have overcome the presence of extreme values (outliers) by coding each of the sampling location values into two groups using targeted thresholds of soil carbon 10.15 kg/m^2 and soil nitrogen 1070.5 g/m^2. The probability maps and cross-validation plots represent spatial variability of soil carbon and soil nitrogen in the study region. We are able to discern sampling locations where the probability values were not greater than the targeted thresholds.

Following the uncertainty analysis and verification of the measured and predicted estimates, we can make the following conclusions about the spatial prediction and modeling of soil carbon and soil nitrogen:

1. It is evident that the indicator semivariogram models (IK) of soil carbon and soil nitrogen are superior to the other models.

TABLE 8.3

Cross-Validation Statistics for Five Interpolation Methods for Soil Nitrogen

Statistics	Decision Criteria	OK	SK	UK	IK*	IDW
Mean	Near zero	−9.979	−12.526	−20.626	−0.0063	−9.373
Standardized mean	Near zero	−0.038	−0.011	−0.035	−0.0114	
Root mean square error (RMSE)	Very small	503.25	520.73	513.40	0.494	512.70
Standardized root mean square error (RMSES)	Near 1	0.808	0.938	1.024	0.984	
Average standard error (ASE)	Very small	597.87	523.48	502.56	0.504	
Ranking (the best*)		5	4	3	1	2

2. The spatial distributions of prediction surfaces of soil carbon and soil nitrogen are evidently similar; for example, high values are located in the south. Most of the sampled carbon sites had less than 20 kg/m^2, while most of the soil nitrogen values ranged from 180 to 2400 g/m^2.
3. Sampled data are lacking in the far north and other areas, which is problematic for the predictive ability of the models at these locations. Not surprisingly, the highest prediction errors were observed in these areas. Putting aside this limitation, overall, the indicator semivariogram provided the best clues about the spatial variability of soil carbon and soil nitrogen. Efforts to revise and improve the models will require the establishment of more sampling sites for use in collecting additional soil measurements in carbon and nitrogen.

Conditional Geostatistical Simulation

Geostatistical simulation provides us with a practical mechanism for drawing multiple equally probable realizations from the random function model. A single realization at each location is derived from a random variable function (Myers 1994a). Realizations are an embodiment of the spatial variability in a sample because they honor data characteristics, preserve the spatial correlation structure, and preserve the mean and marginal distributions (Myers 1994a). The simulation process, which is normally encoded in a computer algorithm, represents "equally likely" values at each of the sample locations based on preserving the core spatial structure (invariant properties) of the data values without smoothening it. A conditional geostatistical simulation provides us with the capability to produce practical realizations that reflect the spatial structure and relationships among a variety of informative factors. We can also express the simulated results in probabilistic terms, thus enabling the quantification of uncertainty and providing important input for risk analysis and management. Although several interpolations can be used with conditional simulation, we have only used SK to illustrate the significance of these algorithms in geostatistics.

Figure 8.12 presents mean and standard deviation of ten realizations for carbon and soil nitrogen. The simulated results give us further insights about the spatial structure and relationships of soil and areas where there is uncertainty. In Figure 8.12a, the low to medium simulated mean values of soil carbon are located in the central and the surrounding areas. However, in Figure 8.12b, there are several pockets of areas (eight clusters) with a varied standard deviation. In Figure 8.12c, the low to medium simulated mean values of soil nitrogen are also located in the central and the immediate surroundings. However, in Figure 8.12d, there are many scattered pockets showing varied standard deviations throughout the study region.

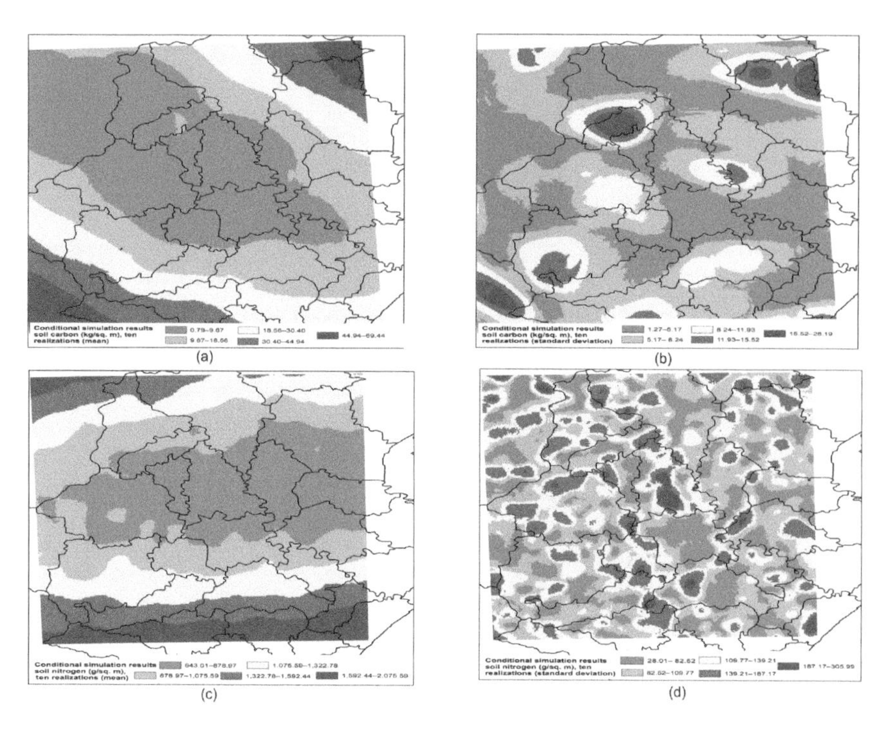

FIGURE 8.12
Conditional geosimulation maps from ten realizations: (a) soil carbon realizations based on mean, (b) soil carbon realizations based on standard deviation, (c) soil nitrogen realizations based on mean, and (d) soil nitrogen realizations based on standard deviation.

Inverse Distance Weighting

IDW is a deterministic interpolation technique that estimates the values of unsampled points according to the values at nearby locations weighted only by distance. IDW is based on an assumption that the relationship between nearby location and interpolation location is closer. We illustrate the IDW interpolation technique using the sample datasets presented in Figure 8.3. We predict the value for coordinate (5, 5) through a linearly weighted combination of the equation:

$$z(x) = \frac{\sum_{i=1}^{n} w_i z_i}{\sum_{i=1}^{n} w_i}$$

$$w_i = \frac{1}{d_i^2}$$

where $z(x)$ is the value for unknown point x, w_i is weight for sampled point x_i, z_i is the value for sampled point x_i, and d_i is the distance from point x_i to point x.

The distance between sampled points x_i and unknown point x_0 is

x_i	x_0
1	4.243
2	2.828
3	5.657
4	1.0
5	2.0

Derive the weights:

x_i	w_i
1	0.056
2	0.125
3	0.031
4	1.0
5	1.25

$$\sum_{i=1}^{n} w_i = 1.462$$

Derive the $z(x)$ value for the unknown point:

$$\sum_{i=1}^{n} w_i z_i = 0.056 \times 3 + 0.125 \times 4 + 0.031 \times 2 + 1 \times 4 + 0.25 \times 6 = 6.229$$

$$z(x) = \frac{\sum_{i=1}^{n} w_i z_i}{\sum_{i=1}^{n} w_i} = \frac{6.229}{1.462} = 4.261$$

IDW provides an easy way to predict values of continuous variables at locations where measurement is unavailable. However, it is not sensitive to areas of peaks or pits and would lead to undesirable results.

Conclusions

In this chapter, we learned about geostatistical analysis as a growing field of advanced statistical techniques that characterize spatial dependence among naturally occurring phenomena and use the results to model spatial continuity

TASK 8.3 APPLYING INVERSE DISTANCE WEIGHTING TO SOIL SAMPLE DATA, SYNTHESIS AND INTERPRETATION OF RESULTS

The same soil used in fitting the previous kriging methods will be applied to the IDW interpolation technique.

Figure 8.13 presents IDW prediction surfaces and the cross-validation plots for the soil carbon and soil nitrogen. In Figure 8.13a, most of the areas have low to medium values of soil carbon with the exception of the south. Figure 8.13b has a few pockets of areas with medium to high scattered values of soil nitrogen throughout the study region. The largest spatial extent with high values is presently located in the south. The examination of cross-validation plots reveals under- and overprediction to be 25% and 44%, respectively. However, the errors for nitrogen are quite large.

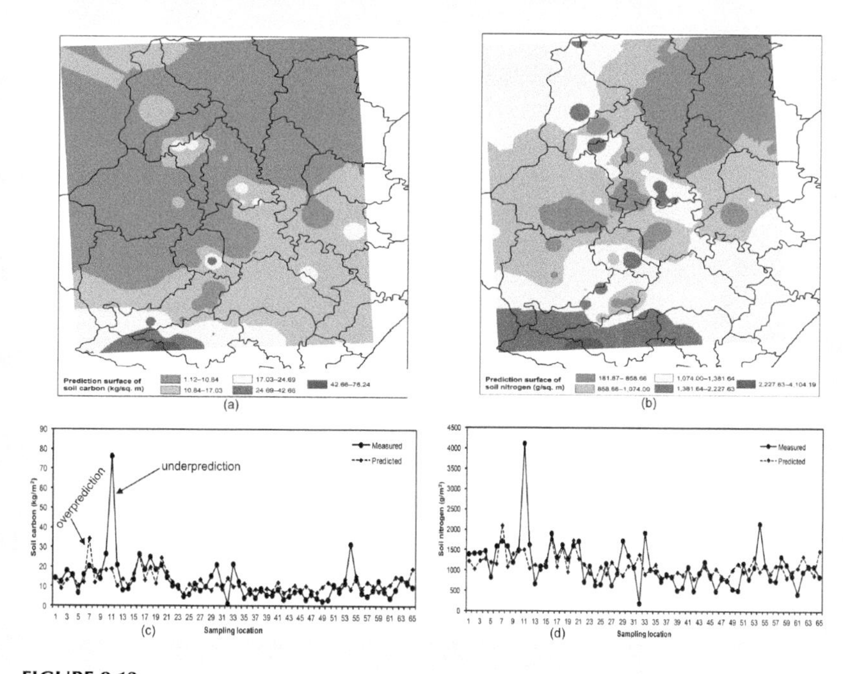

FIGURE 8.13
Prediction surface maps and cross-validation plots using inverse distance weighting interpolation technique: (a) surface for soil carbon, (b) surface for soil nitrogen, (c) measured versus predicted soil carbon estimates, and (d) measured versus predicted soil nitrogen estimates.

in a study area. The analytical approaches consist of both deterministic and stochastic approaches such as IDW and kriging. Our focus in this chapter was primarily on the stochastic approaches. These are governed by the regionalized random variable theory that underlies the formulation of spatial structure functions such as covariance and variogram. We explored the use of the variogram in capturing spatial autocorrelation and the corresponding weights that are derived for spatial interpolation. Using a series of tasks, the chapter also demonstrated the computation and interpretation of estimators for different types of kriging methods and the measures that are used to derive optimal models. Working through the following challenge exercises will help solidify the concepts that were introduced in this chapter and will set you well on your way to becoming proficient in these approaches.

Worked Examples in R and Stay One Step Ahead with Challenge Assignments

The overarching objective of this problem set is to analyze the impact of ambient pollution/environmental exposure on the communities living in two study regions. One of the study regions contains O'Hare International Airport (i.e., noise exposure), and the other is located within areas surrounding air pollution monitoring sites in California (i.e., nitrogen dioxide and ozone exposure). Datasets and materials to be used to complete the problem set include (1) Average Day/Night Sound (DNL), measured in decibels (dB), summarized from 34 Permanent Noise Monitor Locations near the airport; the dataset covers a 7-year study period (2004–2010), O'Hare Noise Compatibility Commission (ONCC); (2) wind data, National Renewable Energy Laboratory (NREL); (3) elevation, U.S. Geological Survey (USGS); and (4) nitrogen dioxide and ozone, California Air Resources Board, and the U.S. Environmental Protection Agency.

In solving these geostatistical problems, we use the R code, ArcGIS, and MS Excel. Tasks 8.4 through 8.9 are solved using ArcGIS and MS Excel. However, the R code for solving these tasks is presented in the Chapter8-R-Code document. As part of your Challenge Assignment, try them out. This will really enhance your geostatistical knowledge and skills.

A sample R-code for exploring spatial datasets using a geostatistical package is presented as follows.

```
# set your workspace directory:
setwd("../K24901_Data_Folders/Chapter8_Data_Folder")

# list and view the files in the workspace directory (It
should list 22 files):
list.files(path=".")

library(sf)
## Linking to GEOS 3.6.1, GDAL 2.2.3, PROJ 4.9.3
```

```
# select data (…/K24901_Data_Folders/Chapter8_Data_Folder)
# Now read in the datasets of interest; ca_ozone_pts.shp
(Ozone), ca_outline.shp and # ca_NO2_pts.shp (Nitrogen dioxide)

# Explore the datasets:
ca_NO2_pts <- st_read('ca_NO2_pts.shp')

ca_ozone_pts <- st_read('ca_ozone_pts.shp')

ca_outline <- st_read('ca_outline.shp')

# Let's plot the Nitrogen dioxide and Ozone air pollution
monitoring sites: Now visualize the Nitrogen dioxide pollution
point measurements over California.

library(ggplot2)
```

Now visualize the ozone pollution point measurements over California.

```
ggplot() + geom_sf(data=ca_outline) + geom_sf(data=ca_ozone_
pts, mapping=aes(col=OZONE)) + ggtitle("Ozone Exposure")

ggplot() + geom_sf(data=ca_outline) + geom_sf(data=ca_NO2_pts,
mapping=aes(col=NO2AAM)) + ggtitle("Nitrogen dioxide Exposure")

Refer to Chapter8-R-Code document for the full code.
```

TASK 8.4 USING EXPLORATORY TOOLS TO VISUALIZE AND COMPUTE BASIC STATISTICS

1. Add the datasets using the "Add Data" button on the Standard toolbar.
2. Navigate to the data folder, hold down the Ctrl key, then click and add the ca_*ozone_pts* (Ozone), ca_*outline*, and ca_*NO2_pts* (nitrogen dioxide, NO_2) datasets. They are measured in ppm, parts per million (by volume).
3. Open the ca_*No2_pts* attribute table and explore elevation and NO2AAM fields using the Statistics tool. How many observations are there? In a table, summarize the minimum, maximum, sum, mean, and standard deviation values for elevation and NO2AAM fields. How many elevation records have zero values? Examine the distribution based on these statistical summaries.
4. Open the ca_*ozone_pts* attribute table and explore elevation and ozone fields using the Statistics tool. How many observations are there? In a table, summarize the minimum, maximum, sum, mean, and standard deviation values for elevation and ozone fields. How many elevation records have zero values?
5. Examine the distribution based on these statistical summaries.

TASK 8.5 FINDING AND UNDERSTANDING SPATIAL AND TEMPORAL PATTERNS

1. Ensure that the Geostatistical Analyst extension is enabled before starting this task. Click on the Geostatistical Analyst toolbar > Explore Data > Histogram. Click and explore the following attributes elevation and NO2AAM from the ca_ *NO2_ pts* layer; elevation and ozone from ca_*ozone_pts*. Are the data normally distributed? Capture and present histogram screenshots for NO_2 and Ozone. Describe the shape and distribution of the data as depicted in these histograms.
2. Select the ca_*NO2_pts* and ca_*ozone_pts* layers and fill in the missing values in Table 8.4. The values in the histogram have been rescaled by a factor of 10, so you need to look up (brush) the selected records in the attribute table and the map. Identify and examine the locations of sample measurements of NO_2 and Ozone in California (include two screenshots highlighting the exploration/brushing of the data). Identify the elevation outliers in both ambient sources.
3. For pollution monitoring purposes, the critical thresholds should be greater than 0.09 and 0.025 ppm (EPA standard is 0.053 ppm) for ozone and nitrogen dioxide, respectively. These ambient levels have adverse health effects. Identify using the Histogram tool and examine locations with the critical thresholds.
4. Identify any global trends in ambient exposure data. Click on the Geostatistical Analyst toolbar > Explore Data > Trend Analysis. Explore both NO2AAM and Ozone to determine if

TABLE 8.4

Distribution of Sampled Measurements of NO_2 and Ozone

NO_2		Ozone	
Elevation (m)	**Observations (n)**	**Elevation (m)**	**Observations (n)**
<189		<1	
194–369		1–232	
381–870		244–1052	
1006–1440		1244–1905	
ppm		**ppm**	
0.0012–0.02329		0.0465–0.1463	
0.0251–0.04809		0.1506–0.1736	

there are nonrandom components of the surface that can be represented by a mathematical formula. Explore the location angle at which NO2AAM and Ozone express a mathematical trend. (This will be apparent when you rotate the trend surface to an angle at which the green trend line represents a U-shaped parabola.) What type of mathematical trend does this parabola represent (think back to high school algebra what type of equation creates a parabolic line….)? Record the angle at which the green trend line becomes a U-shape. Do the same for the x-axis trend line (blue line); record that location angle. Sometimes, these types of trends do not occur, but they do here. Later in this lab, when you run your geostatistical model, you will need to know whether or not a particular type of trend needs to be accounted for within your geostatistical model. Accounting for this trend will help to stabilize your final model.

5. Explore the spatial autocorrelation influence. Click on the Geostatistical Analyst toolbar > Explore Data > Semivariogram/Covariance Cloud. Each of the points in the semivariogram represents a pair of points. The position of a "paired point" describes both the level of spatial autocorrelation between the pair of points and the distance between the pair of points. Increases in the y-axis illustrate a decrease in spatial autocorrelation, and increases in the x-axis illustrate increased distance between the paired points. For spatial modeling purposes, we expect nearby points to display higher spatial autocorrelation; large deviations from this modeling perspective represent inaccuracy within the semivariogram. On the semivariogram, identify where the cloud flattens out by using Select Features by Rectangle tool. Also, identify values that have a higher semivariogram and determine whether these pairs of locations are inaccurate. Put a few screenshots of your semivariogram analysis in your lab, showing the flattened section of the semivariogram and the higher section of the semivariogram.
6. Explore directional influences. Determine whether NO_2 and Ozone are isotropic (without directional influence) or anisotropic (with directional influence). If it is anisotropic, what is the direction of better continuity for ca_*ozone_pts* and ca_*NO2_pts* datasets?

TASK 8.6 MAPPING, QUANTIFYING, AND INCORPORATING SPATIAL AND TEMPORAL VARIABILITY IN A MODEL

1. Use Geostatistical wizard to create a prediction map of ca_*NO2* using OK and the default settings. Compile screenshots of the semivariogram and parameters (from layer properties) for submission.
2. Click on the Geostatistical Analyst toolbar > Geostatistical Wizard.
3. Click Kriging/Cokriging in the Methods list box.
4. Click the Input data drop-down arrow and click ca_*ozone_pts*.
5. Click the Attribute drop-down and click the *OZONE* attribute.
6. Click next. By default, OK type and Prediction output type will be selected.
7. From the exploratory analysis, we discovered a global trend, and during investigation the second-order polynomial seemed reasonable. Click the order of trend removal drop-down arrow and click Second. Click Next.
8. There is a directional influence in the Ozone distribution with a northwest-southeast influence. It is possible this is due to the buildup of ozone between the mountains and the coast. Other contributing factors could be elevation, prevailing wind direction, and high concentration of human population and activities, including industries, greenhouses, automobile, residential emissions, and so on. We call these secondary factors. Under Model #1 box, click on the drop-down list for anisotropic and set this to True. Capture a Screenshot of this model and place in your word document. An illustrative diagram of a fitted semivariogram model is given in Figure 8.14.
9. Click next. The next window allows the fitting and searching of specific neighborhoods. Explore this.
10. Click next. The next window allows the saving of cross-validation tables for further analysis, and it provides different prediction errors that must be compiled. Export and save this table. Click Finish and remember to capture a screenshot or copy of the Model Report . Evaluate these results.

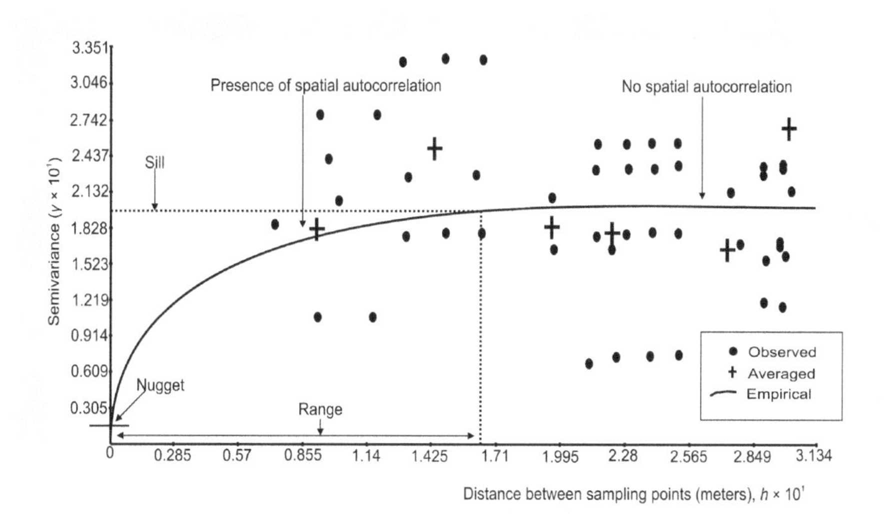

FIGURE 8.14
Fitting a semivariogram model. An illustrated semivariogram fitted with an exponential model (three derived quantities: nugget, sills, and range are highlighted). The summary/average of the semivariance of all points within a particular spatial lag is also given.

TASK 8.7 PERFORMING AND DISCOVERING THE BEST MODEL FOR SPATIAL PREDICTION

1. Now we run an SK and IDW on ca_*ozone_pts* with ozone as your attribute as we did in Task 8.6.
2. Compile in a table the results of three models showing the following prediction errors: Mean, RMS, Mean Standardized, RMS Standardized, and Average Standard Error.
3. Create three prediction maps of ca_*NO2_pts* using the three models (IDW, OK, and SK).
4. Create two tables (from Table 8.5) and use this information to select the best-performing geostatistical methods.
5. Complete the following short essay: In your own words, describe the results, maps, model, and spatial patterns of ambient nitrogen dioxide and ozone levels in the state of California.
6. Load the ambient noise-levels dataset in ArcMap > Explore the ambient noise levels in areas surrounding O'Hare International Airport. Repeat the steps outlined for Tasks 8.4 through 8.7.

7. Compile the ambient noise-level information in the same way you did for nitrogen dioxide and ozone. Use this information to complete the following short essay: In your own words, describe the results, maps, model, and spatial patterns of ambient noise levels.

TABLE 8.5

A Decision Matrix for the Performance of Three Sets of Geostatistical Methods

Statistics	Decision Criteria	IDW	OK	SK
Mean prediction error (Mean)	Near zero			
Standardized mean predictor error (SM)	Near zero	N/A		
Root mean square error (RMSE)	Very small			
Standardized root mean square predictor error (RMSES)	Near 1	N/A		
Average standard error (ASE)	Very small	N/A		
Total				
Ranking				

Abbreviations: IDW, inverse distance weighting; OK, ordinary kriging; SK, simple kriging.

TASK 8.8 FITTING A NOISE DISTANCE DECAY MODEL USING MS EXCEL

Let us begin this task by a practical example to illustrate the distance decay model. Consider exposure/disturbance caused by the noise of the train in the city of Carbondale, Illinois. People who live closer to the train station/railroad experience more disturbance than those who live farther. As we move away from the train station, the disturbance intensity continuously decreases to reach such a level that we do not experience the pollution effect. The effect of the noise pollution can then be modeled as a function of distance. This is known as distance decay, which is a mathematical representation of the effect of distance on a variable of interest. The model expresses a negative relationship between the variable of interest and the increase of the distance using a power function or an exponential function.

1. Create a scatterplot of the ambient noise-level data (the *Noise_Project* file). Analyze the structure of the data, and give an interpretation of the data trend. The distance (*Dist_Feet*) should

be in the x-axis, and the average sound levels data (dB) of the 7-year period should be in the y-axis.

2. What is the relationship between ambient noise levels and distance? How do we model this relationship mathematically?

TASK 8.9 ACCOUNTING FOR SECONDARY FACTORS AND MAKING DECISIONS ON A SPATIAL BASIS

In Task 8.7, you created a continuous map of nitrogen dioxide (NO_2) and other attributes, which represents a prediction of the concentration levels of NO_2 at unsampled locations. Recall that in the kriging process, you have used only one variable (concentrations of NO_2 at sample locations). In reality, the dispersion of NO_2 concentration levels may depend on other contributing factors (natural or anthropogenic). For example, wind direction may have an effect on the dispersion of NO_2, in which case, the dispersion is said to be anisotropic. If we had more information about wind direction when NO_2 data were being collected, we could predict the dispersion of NO_2 along that direction. Also, there might be some situations where temperature or rainfall spatial variability is influenced by topography. Many examples of spatial relationships exist between two or more natural or anthropogenic continuous phenomena. Therefore, it is necessary to account for secondary factors, when its use is justified, to predict other continuous variable values at unsampled locations. In this task, create a prediction map of the average day/night sound levels taking into account a secondary variable—the elevation data.

1. Use ArcGIS Geostatistical wizard to create a prediction map of *Noise_ Project* using OK and the default settings. Compile screenshots of the semivariogram and parameters (from layer properties) for submission.
2. On the Geospatial Analyst toolbar > Click on the Geostatistical Analyst toolbar > Launch the Geostatistical Wizard.
3. Click Kriging/Cokriging in the Methods list box.
4. Click the Input data drop-down arrow for Dataset and click *Noise_Project* > Click the Attribute drop-down and click the Ildem attribute.
5. Click the Input data drop-down arrow for Dataset and click *Noise_Project* > Click the Attribute drop-down and click the All_averag attribute.

6. For Dataset #1, Model Global Trend as a third one.
7. For Dataset #2, Model Global Trend as a second one.
8. Under Model #1 box, click on the drop-down list for anisotropic and set this to True.
9. Run your model as you did before, and generate the results.
10. Now, repeat the process and exclude the Trend analysis. What is the difference in the two results?

Use this information to complete the following short essay: In your own words, describe the results, maps, model, and spatial patterns of ambient noise levels relative to elevation as a secondary factor. How can we improve these results?

TASK 8.10 STAY ONE STEP AHEAD WITH CHALLENGE ASSIGNMENTS: REPORT WRITING AND RELATING FINDINGS TO EXISTING EMPIRICAL EVIDENCE

1. Write a short abstract for the spatial patterns of ambient ozone and NO_2 in California. Your abstract should not be more than 350 words. It should have the following components: a background/problem statement, objectives and hypothesis, data and methods, results, and conclusions and implications.
2. Write a short abstract for the spatial patterns of ambient noise levels in Chicago's O'Hare International Airport. Your abstract should not be more than 350 words. It should have the following components: a background/problem statement, objectives and hypothesis, data and methods, results, and conclusions and implications. You may also revisit previous analysis on the dataset and use it in this abstract.

Review and Study Questions

1. One of the core goals of geostatistics is to quantity spatial uncertainty. Provide examples of spatial uncertainty and describe how this can be quantified and integrated into subsequent steps in the spatial interpolation process.

2. Geostatistical methods are effective only when certain key assumptions and statistical properties are met. Describe at least two of these properties and explain how you would go about checking and validating these assumptions during the spatial modeling process.
3. What are the benefits of deriving and plotting a variogram function in geostatistics? Draw a sample variogram and explain the key properties to a layperson.
4. The accuracy and optimality of a geostatistical model can be assessed through cross-validation and a series of statistical measures derived during the analysis. Explain these measures and how they can help with the interpretation of your results.
5. Choose any three of the following methods and explain the similarities and differences in their use to interpolate spatial data generated in your research area:
   a. IDW versus OK
   b. OK versus UK
   c. OK versus Cokriging
   d. OK versus IK

Glossary of Key Terms

anisotropic (semi)variogram: This is when the spatial pattern is strongly biased toward a specific direction. This phenomenon is also at times referred to as directional variograms because the weighting scheme depends on distance and direction.

isotropic (semi)variogram: This is when the spatial pattern is identical in all directions. In this case, the fitting of the semivariogram model will heavily depend on the (Euclidean) distance between locations.

kriging: The process of fitting the best linear unbiased estimate of a value at a point or of an average over a volume. Kriging provides a powerful tool to model spatial autocorrelation in the data and a means to use this resulting knowledge to predict precise, unbiased estimates of data pairs within the sampling unit. It could be simply stated that kriging facilitates the quantification of spatial variability.

nugget effect: The vertical height of the discontinuity at the origin. It is the combination of (1) short-scale variations that occur at a scale smaller than the closest sample spacing and (2) sampling error due to the way the samples were collected, prepared, and analyzed.

range: the distance at which the variogram reaches the sill.

sill: The plateau that the variogram reaches; in the variogram context, it is the average squared difference between paired data values, and it is approximately equal to twice the variance of the data.

variogram: An *h*-scatterplot for characterizing the spatial continuity of the variable.

References

Armstrong, M. (ed.) 1998. *Basic Linear Geostatistics.* Berlin, Germany: Springer-Verlag.

Asa, E., M. Saafi, J. Membah, and A. Billa. 2012. Comparison of linear and nonlinear kriging methods for characterization and interpolation of soil data. *Journal of Computing in Civil Engineering* 26(1): 11–18.

Barro, A.S. and T.J. Oyana. 2012. Predictive and epidemiological modeling of the spatial risk of human onchocerciasis using biophysical factors. *Spatial and SpatioTemporal Epidemiology* 3: 273–285.

Birkin, M. 2013. Big data challenges for geoinformatics. *Geoinformatics and Geostatistics: An Overview* 1: 1.

Burrough, P.A. and R. McDonnell. 1998. *Principles of Geographical Information Systems,* vol. 333. Oxford, UK: Oxford University Press.

Cressie, N. 1985. Fitting variogram models by weighted least squares. *Mathematical Geology* 17: 563–586.

Cromer, M.V. 1996. Geostatistics for environmental and geotechnical applications: A technology transferred. In Srivastava, R.M., S. Rouhani, M.V. Cromer, A.J. Desbarats, and A.I. Johnson (eds.), *Conference Proceedings American Society for Testing and Materials 1283.*

Deutsch, C.V. 2002. Geostatistics. In Meyers R. (ed.),. *Encyclopedia of Physical Science and Technology,* 3rd ed. Cambridge, MA: Academic Press/Elsevier, pp. 697–707.

Fisher, P.F. 1999. Models of uncertainty in spatial data. In Longley, P.A., M.F. Goodchild, D.J. McGuire, and D.D. Rhind (eds.), *New Developments in Geographic Information Systems: Principles, Techniques, Management and Applications.* New York, NY: John Wiley & Sons.

Goovaerts, P. 1997. *Geostatistics for Natural Resources Evaluation.* New York, NY: Oxford University Press.

Goovaerts, P. 1999. Geostatistics in soil science: State-of-the-art and perspectives. *Geoderma* 89(1): 1–45.

Goovaerts, P. 2005. Geostatistical analysis of disease data: Estimation of cancer mortality risk from empirical frequencies using Poisson kriging. *International Journal of Health Geographics* 4: 31.

Goovaerts, P. 2006. Geostatistical analysis of disease data: Visualization and propagation of spatial uncertainty in cancer mortality risk using Poisson kriging and p-field simulation. *International Journal of Health Geographics* 5: 7.

Goovaerts, P. 2009. Medical geography: A promising field of application for geostatistics. *Mathematical Geosciences* 41(3): 243–264.

Goovaerts, P. and G.M. Jacquez. 2004. Accounting for regional background and population size in the detection of spatial clusters and outliers using geostatistical filtering and spatial neutral models: The case of lung cancer in Long Island, New York. *International Journal of Health Geographics* 3: 14.
Guo, D., R. Guo, C. Thiart, T.J. Oyana, D. Dai, and S.L. Hession. 2006. GM (1,1)-Kriging prediction of soil dioxin pattern. *Proceedings of the GIS Research UK 14th Annual Conference, GISRUK 2006, School of Geography, University of Nottingham,* April 5–7.
Haining, R., R. Kerry, and M.A. Oliver. 2010. Geography, spatial data analysis and geostatistics: An overview. *Geographical Analysis* 42(1): 7–31.
Isaaks, E.H. and R.M. Srivastava. 1989. *An Introduction to Applied Geostatistics.* New York, NY: Oxford University Press.
Krige, D.G. 1951. A statistical approach to some basic mine problems on the Witwatersrand. *Journal of the Chemical, Metallurgical and Mining Society of South Africa* 52: 119–139.
Myers, D.E. 1994a. Spatial interpolation: An overview. *GeoDerma* 62: 17–28.
Myers, D.E. 1994b. Statistical methods for interpolating spatial data. *Journal of Applied Science and Computations* 1(2): 283–318.
Matheron, G. 1963. Principles of geostatistics. *Economic Geology* 58: 1246–1266.
Matheron, G., 1965. *Les variables régionalisées et leur estimation.* Paris: Masson.
Mitas, L. and H. Mitasova. 1999. Spatial interpolation. In *Geographical Information Systems: Principles, Techniques, Management and Applications,* p. 481. New York, NY: John Wiley & Sons.
Naoum, S. and I. K. Tsanis. 2004. Ranking spatial interpolation techniques using a GIS-based DSS. *Global Nest: The International Journal* 6(1): 1–20.
Noor, A.M., D.K. Kinyoki, and C.W. Mundia. 2014. The changing risk of *Plasmodium falciparum* malaria infection in Africa: 2000–10: A spatial and temporal analysis of transmission intensity. *The Lancet* 383(9930): 1739–1747.
Oyana, T.J. 2004. Statistical comparisons of positional accuracies of geocoded databases for use in medical research. In Egenhofer, M.J., C. Freksa, and H.J. Miller (eds.), *Proceedings of the Third International Geographic Information Science,* October 20–23, pp. 309–313. Regents of the University of California.
Oyana, T.J. and F.M. Margai. 2010. Spatial patterns and health disparities in pediatric lead exposure in Chicago: Characteristics and profiles of high-risk neighborhoods. *Professional Geographer* 62(1): 46–65.
Oliver, M.A. (ed.) 2010. *Geostatistical Applications in Precision Agriculture.* Dordrecht, Netherlands: Springer.
Robertson, G.P. 1987. Geostatistics in ecology: Interpolating with known variance. *Ecology* 68(3): 744–748.
Yarus, J.M. and R.L. Chambers. 2006. Practical geostatistics: An armchair overview for petroleum reservoir engineers. Distinguished Author Series, JPT, Society of Petroleum Engineers. *Journal of Petroleum Technology* 58(11): 78–86.
Zinke, P.J., A.G. Stangenberger, W.M. Post, W.R. Emanuel, J.S. Olson, R.E. Millemann, and T.A. Boden. 1986. Worldwide Organic Soil Carbon and Nitrogen Data (1986) (NDP-018). ORNL/CDIC-018, NDP-018. Oak Ridge National Laboratory, Oak Ridge, Tennessee, United States. doi:10.3334/CDIAC/LUE.NDP018.

9

Data Science: Understanding Computing Systems and Analytics for Big Data

Learning Objectives

1. Define and describe data science concepts.
2. Manage and process big geospatial data.
3. Effectively use, explore, analyze, and synthesize big geospatial data.
4. Develop actionable knowledge and information from big geospatial data.
5. Effectively implement emerging methods, programming languages and algorithms, and tools for big geospatial data.

Introduction to Data Science

Data science is both a new concept and a recent field that has evolved with the concurrent growth of large-scale datasets and emerging technologies to handle the volume and variety of information from multiple sources and formats. The field draws heavily from several existing disciplines that we discussed in this book: mathematics, statistics, computer science, GIS, visualization, and more, including engineering, physics, psychology, cognitive science, operations research, business, and artificial intelligence (AI). The primary aims entail the development and application of scientific approaches for the systematic exploitation, organization, management, analysis, and use of large amounts of data for decision-making. Data science utilizes traditional or novel tools, methods, and strategies, which are tailored toward the discovery of complex patterns in high-dimensional data through visualizations, simulations, and various types of model building (Kelling et al. 2009). It is being fueled by the critical need to design efficient, scalable, and reliable systems, tools, and programs that can easily handle "big data."

Given the nascent stages of the field of data science, the notion of what constitutes big data is still up for debate. The term *big data* has been generally used to describe very large amounts of datasets that are complex, heterogeneous, and hard to process using traditional statistical and computational tools (Jacob 2009; Loukides 2010; Helbing and Balietti 2011; Allen et al. 2012). Increasingly, many scholars use five attributes to characterize big data, notably the "5Vs": volume, variety, velocity, veracity, and value. The large volume, variety, and increasing speeds at which data are being generated are driving creativity and the development of new analytic methods (Kelling et al. 2009; Schadt et al. 2010), ranging from statistical packages/tools to sophisticated data-mining algorithms. At the same time, there is the ongoing need, as with traditional datasets, to ensure that these data are reliable and valid, following which meaningful techniques can be applied and the results used to generate new knowledge and value-added information for decision-making.

The development of these new analytic methods and strategies enables the processing, management, visualization, and presentation of big datasets in usable and actionable knowledge formats. Intensive search of patterns in extremely large datasets provides many exciting opportunities for designing and testing hypotheses and the creation of data products (Loukides 2010). Large datasets also provide facts and clear evidence that have a potential to significantly advance science. Many observers are truly optimistic and confident that the analysis of big data will yield new objective knowledge that will advance our understanding of phenomena.

Within the context of spatial analysis, recent improvements in sensor technology, reduction in data storage costs, and improvement in data collection methods have led to an explosion in the amount of geospatial data collected and available to organizations (Loukides 2010; Barnes et al. 2011; Longley 2012; Wang et al. 2013). The datasets exist in three main formats: structured, semistructured, and unstructured. These are usually stored in large-scale server farms at a data center, where they can be mined or analyzed to support the decision-making process. The greatest challenge of our time, however, is how to effectively make sense of these big datasets or turn them into meaningful and informative products in a timely manner. The holistic approach to big data analysis is what differentiates data science from traditional statistics. Data science integrates methods from several disciplines to gain fundamental insights from the data. Some of the core goals of data science are to simplify the data and make them accessible to those who need them in a timely manner. To accomplish this, organizations need to address three principal areas: data management, analytics and strategies, and communication of the results/reporting applications.

Rationale for a Big Geospatial Data Framework

Data science is a data-driven discovery and prediction process with the principal aim of making sense of big data and using the results to increase our understanding. Data-driven discovery provides the basis of producing actionable knowledge. The big data framework consists of three essential

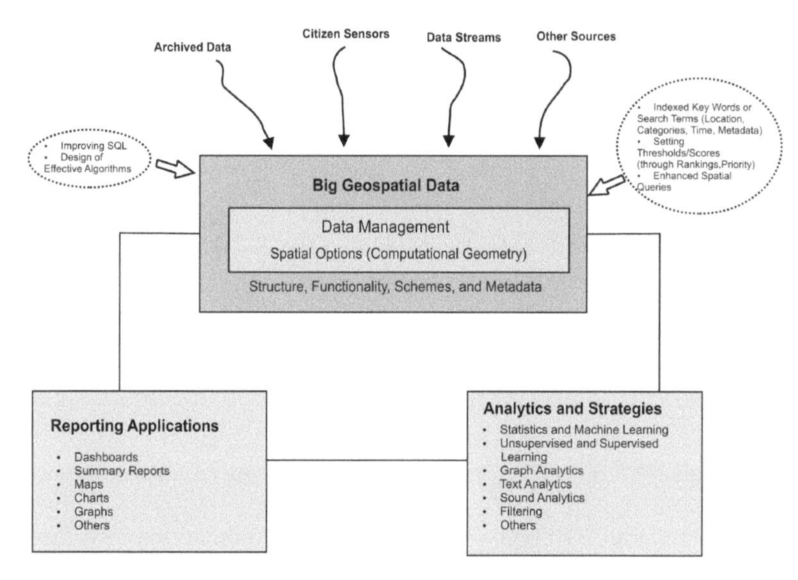

FIGURE 9.1
A visual representation of components of data science.

components: data management, analytics and strategies, and reporting applications (Figure 9.1).

First, the data management component entails data processing of large quantities of data in a database. Massive geospatial datasets are currently generated through internet activities; portable, wearable, and mobile devices; citizen sensors; instrumentation; simulations; satellite and global positioning system (GPS)–equipped vehicles; government agencies; and other research and development institutions. For big geospatial data, it simply means the managing of data through the use of spatial databases and computational geometry. Managing data requires deep knowledge and skills in the design and use of spatial databases, especially in Structured Query Language (SQL) manipulation using relational algebra/spatial query processing. Additional areas include algorithms and in-database analytics.

Second, the analytics and strategies component provides the analytical/statistical basis for the development of interactive tools and systems with a core capability that will allow the exploration, visualization, summarization, classification, identification, and extraction of existing patterns or trends in large-scale datasets. Most of the spatial techniques and methods discussed in earlier chapters still apply and can be used; however, to be able to harness the new sources, types, and large geospatial data, we must think beyond them. For example, algorithms and models, new knowledge, facts, and reasoning can be used to develop rules to support analytical reasoning or can be used for making predictions about future events.

Third, the communication of a results component entails the production of reports with interpretable summaries, synthesized outcomes, facts,

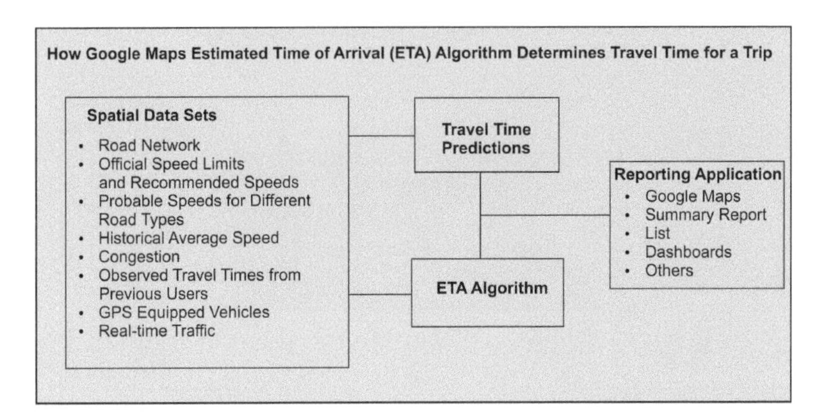

FIGURE 9.2
Google Maps Estimated Time of Arrival algorithm and a list of big geospatial datasets for travel time.

rules, and knowledge. Graphs, plots, and other visualizations introduced in Chapter 4 are especially useful, because they not only provide a foundation for discovering the basic characteristics of raw data but also help tell a story about the data.

Effective presentation of results as a data product or in a report to the target audience is a crucial element of big data exploitation. The communication formats can range from the use of text to audio or images. For the reports to be effective, they should be kept simple and accessible to a wide audience. Let us take a look at an example of a simple reporting system for estimating travel time using big data. Although the estimation of dynamic travel time is based on point-based or trip-based approaches, the well-known Google Maps Estimated Time of Arrival (ETA) algorithm uses a variety of massive data sources for traffic data to make their travel time predictions. These predictions differ from one area to another, because they depend on the data available in a particular area. Figure 9.2 presents a list of spatial datasets that Google's ETA algorithm uses and the travel time prediction process. The ETA algorithm is constantly sharpened through the comparison of current estimates with actual historical travel time in various traffic conditions. Google then comes up with the best prediction that can be made from these massive datasets, which is presented to the users in a very simple but accessible format.

Data Management

The ever-increasing amounts of geospatial data are problematic to many organizations; without the right tools, they are not able to get value out of these data. Recent advances in data capture and computation methods have transformed the way organizations handle and process data. The rate at which geospatial data is generated exceeds the ability to organize and analyze them to extract patterns critical for understanding the constantly changing world. For example, Google generates about 25 PB of data per day, with a significant portion of it being geospatial data. Although the computational and analytical methods are not moving as fast as the rate of increase in geospatial data, there has been a lot of progress in this area. To analyze these data efficiently, the management and retrieval processes must be organized and centralized into accessible storage. Recent innovations have led to an increase of new data management solutions, for example, Globus Online (GO), rsync algorithm, YouSendIt, Dropbox, BitTorrent, Content Distribution Networks, and PhEDEx data service (Allen et al. 2012). Figure 9.3 illustrates the elements of data management, from the first stage of combining data from multiple sources through its presentation. The centralization of data management and retrieval is referred to as data warehousing, whereas the actual analysis of the data is referred to as data mining. In this chapter, the details of these terms are discussed.

Data Warehousing

Kimball and Ross (2013) describe data warehouses as a complete ecosystem for extracting, cleaning, integrating, and delivering data to decision makers, and it therefore includes the extract-transform-load(ETL) and business intelligence (BI) or analysis functions.

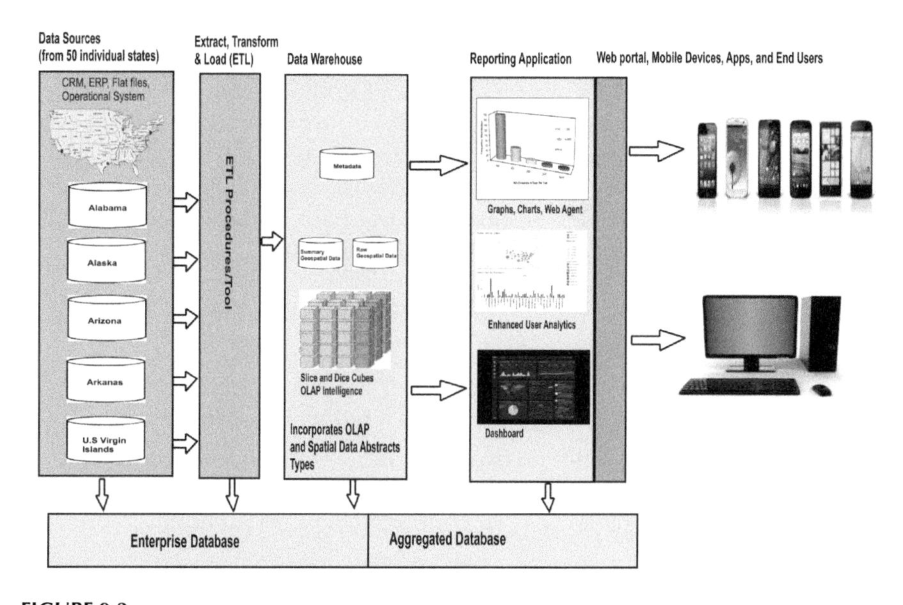

FIGURE 9.3
Elements of data management workflow showing different platforms, software infrastructure, tools, and methods.

Data Sources, Processing Tools, and the Extract-Transform-Load Process

The first component is the extraction of data from each of the individual sources (these can include historical data in the form of flat files or operational databases) into a temporary staging area where data integration takes place. Data extraction methods can be divided into two categories:

- *Logical extraction*: Could be a full extraction of the complete dataset from the source or an incremental extraction (change data capture) of the data changes in a specified time period
- *Physical extraction*: Can be done online, directly from the source system, or offline from a system staged explicitly outside the original source system

Data transformations are usually the most complex and time-consuming part of the ETL process. They range from simple data conversions to extremely complex data scrubbing techniques. Data can be transformed in two ways:

- *Multistage transformation*: Data are transformed and validated in multiple stages outside the database before being inserted into the warehouse tables.
- *Pipelined data transformation*: The database capabilities are utilized, and data are transformed while being loaded into the database.

Using data quality tools, one can ensure that the correct data and format are loaded into the warehouse. This process can be done manually using code created by programmers or automated by the use of ETL tools available in the market. Some of the popular tools include Oracle Warehouse Builder, Data Integrator and Services by SAP, and IBM Information Server. The result of this process is metadata and standardized data, which are then loaded into a data warehouse. *Metadata* is "data about the data," which may include mapping rules, ETL rules, description of source data, and precalculated field rules. Some of the benefits of this ETL process include

- *One source of truth*: All the data are stored in the same format, ensuring their consistency and accuracy.
- *Reduction of resources*: The reduction of interface programs used to access the consolidated data results in a reduction of resources.
- *Improved planning and decision-making*: Strategic planning and organization-wide decision-making are greatly improved.
- *More timely data*: Having data in one location speeds up the access and processing time and reduces problems related to timing discrepancies.

Data integration is the process of combining data from multiple sources into one common representation, with the goal of providing the users with one version of truth. This is a very important process of data warehousing since the quality of the data fed into the system determines the accuracy and reliability of the resulting business decisions.

Data Integration and Storage

In a data warehouse, data are subject oriented, integrated, nonvolatile, time variant, and process oriented. Spatial data warehouses host data for analysis, separating them from transaction workload and thus enabling organizations to consolidate data from multiple sources. The primary purpose of a spatial data warehouse is to organize these data according to the organization's business model to support management decision-making. Many decisions consider a broader view of the business and require foresight beyond the details of day-to-day operations. Spatial data warehouses are built to view businesses over time and spot trends, which is why they require large amounts of data from multiple sources. The analysis capability of a data warehouse enables users to view data across multiple dimensions. The use of a single repository for an organization's data promotes interdepartmental coordination and greatly improves data quality. The spatial data warehouse may contain metadata, summary data, and raw data of a traditional transactional system. Summaries are very valuable because they precompute long operations in advance, which improve query performance. In cases where organizations need to separate their data by business function, data marts can be included for this purpose.

Spatial data warehouses read trillions of bytes of data and therefore require specialized databases that can support this processing. Most data warehouses are bimodal and have a batch of windows (usually in the evenings) when new data are loaded, indexed, and summarized. To accommodate these shifts in processing, the server must be able to support parallel, large-table, full-table scans for data aggregation and have on-demand central processing unit (CPU) and random-access memory (RAM) resources, and the database management system must be able to dynamically reconfigure its resources. Overall, data warehouses provide many advantages to the end user including, but not limited to, improved data access and analysis, increased data consistency, and reduction in costs for accessing historical data.

Data-Mining Algorithms for Big Geospatial Data

Data mining, also referred to as knowledge discovery, is the process of analyzing centralized integrated data to find correlations or patterns to aid decision-making. Centralization of these data is needed to maximize user access and analysis. Data mining is supported by analytical software tools for analyzing data. Some of the open-source tools available include KNIME (http://www.knime.org/); GeoDa (https://geodacenter.asu.edu/projects/opengeoda); and CLAVIN, a package for document geotagging and geoparsing that uses context-based geographic entity resolution (http://clavin.bericotechnologies.

Computational Resources for Handling Big Geospatial Data

Main Types of Computing Platforms

Cluster Computing: Computers are linked through a fast local area network and function as a single unit.

Cloud Computing: Computers are linked together through the Internet to provide a shared pool of computing resources for accessing and storing data and programs.

Grid Computing: A loosely coupled network of computers from multiple locations that work together on common computing tasks.

Heterogenous Computing: Specialized computing system that uses more than one kind of processor; for example, central processing units and graphics processing units.

A List of Currently Available Software Kits

Spatial Analytical Tools and Methods	GISolve	GeoDa/ PySAL	Open-Topography	PGIST	pd-GRASS	R
Agent-Based Modeling	X					X
Choice Modeling				X		
Domain-Specific Modeling	X	X				X
Geostatistical Modeling	X					X
Local Clustering Detection	X	X				X
Spatial Interpolation	X	X				X
Spatial Econometrics		X				X
Visualization and Map Operations	X	X	X	X	X	X
Spatial Middleware	X					
Generic Cyberinfrastructure Capabilities	X	X			X	X
Online Problem-solving	X	X	X			X

Compiled from Schadt et al. (2010) and Wang et al. (2013)

FIGURE 9.4
Computing resources for conducting complex spatial analytical work. (Compiled from Schadt, E.E. et al. 2010. *Nature Reviews Genetics* 11: 647–657 and Wang, S. et al., 2013. *International Journal of Geographical Information Science* 27[11]: 2122–2145.)

com/). Wang et al. (2013) have documented recent CyberGIS spatial analysis and visualization software toolkits including GISolve, GeoDa/PySAL, OpenTopography, PGIST, pd-GRASS, and R (Figure 9.4).

Data mining consists of two major elements:

1. Data analysis using BI application software
2. Presentation and visualization

Tools, Algorithms, and Methods for Data Mining and Actionable Knowledge

Geospatial data analysis includes manipulation and transformation of data into useful information to support decision-making and reveal patterns and anomalies that are not immediately obvious. It focuses on measuring properties and relationships, taking into account the spatial localization of the study attributes, as discussed in earlier chapters. The idea is to incorporate space or location into the analysis. The opportunity to mine big geospatial data, for example, from major social media networks (Facebook, Instagram, LinkedIn, Twitter, Pinterest, and Google+), has provided substantial advantages in three areas: it (1) reduced gaps of knowledge and understanding of human activities, (2) enabled a greater understanding of human activities because we are able to predict situations, and (3) fueled knowledge discovery and improvement in decision-making.

The core tools, algorithms, and methods for data mining have two major components: software to (1) store the data over thousands of machines in a data center and (2) retrieve and perform computation with data spanned over thousands of machines in a data center (Helbing and Balietti 2011). Some of the tools available to perform these data-mining tasks are MapReduce and Hadoop. There are four main categories of large-scale computing platforms (Figure 9.4) for processing, managing, and analyzing big geospatial data; they include cluster computing, cloud computing, grid computing, and heterogeneous computing (Schadt et al. 2010).

To succeed in the use of these computational resources for mining large-scale geospatial data, it is important to keep the following checklist in mind:

1. Know the nature, magnitude, and complexity of geospatial data.
2. Determine memory requirements.
3. Determine network bandwidth requirements.
4. Know about data management services (data movement tools, access, storage, security, performance, and scalability).
5. Understand processing, analysis, or simulation methods and tools.
6. Know about reporting applications.

Business Intelligence, Spatial Online Analytical Processing, and Analytics

Computer applications have advanced greatly and do a very good job of processing data, but they still cannot effectively tell stories about the data. This is where they fall short in communicating with the consumers. Data should be organized in a manner that engages the way human brains actually work; then, we can process a larger amount of data (Ideas Economy: Information Forum 2013).

BI is the application of knowledge and experience to data to produce valuable business information. BI applications enable users to get an insight of the knowledge in the data. The combination of geospatial data analysis and BI application is known as location intelligence (LI). The ability to visualize geospatial data and understand relationships between specific locations helps organizations make more strategic business decisions. LI is more than just mapping; it includes advanced analysis related to spatial relationships. GIS is at the heart of LI, and it is clear that business data need to be location enabled. Spatial analysis allows you to ask "where and why?" questions, and when combined with Spatial Online Analytical Processing (SOLAP) in your BI systems, the location component can be the dimension in the analysis that leads to more focused decision-making. Figures 9.5 and 9.6 illustrate the application of SOLAP where users are able to drill into details within the Tableau mapping application.

Figure 9.7 shows a sample script for running and loading all data definition language (DDL) tables and procedures into Oracle or Tableau acceptable format (a step-by-step instruction manual is presented in the ChapterData/Chapter9_Data_Folder). The visual power of maps reveals trends, patterns,

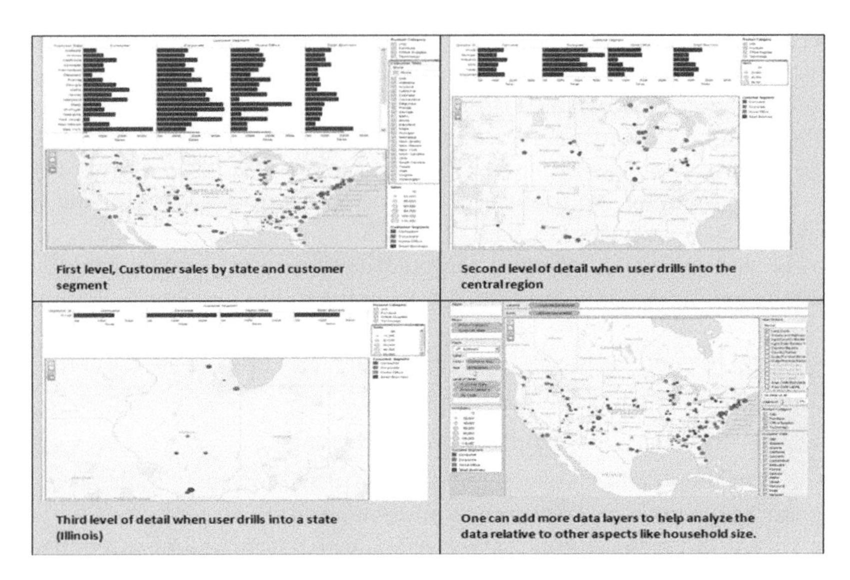

FIGURE 9.5
Dashboard screenshots showing actionable knowledge derived by leveraging a visual analytics of an integrated business intelligence and locational intelligence platform.

and insights that are not as easily detected in other data presentation formats such as tabular views, or bar and pie charts. Because of customer demand, BI application vendors have incorporated location-based intelligence technology in their core BI platforms, for example,

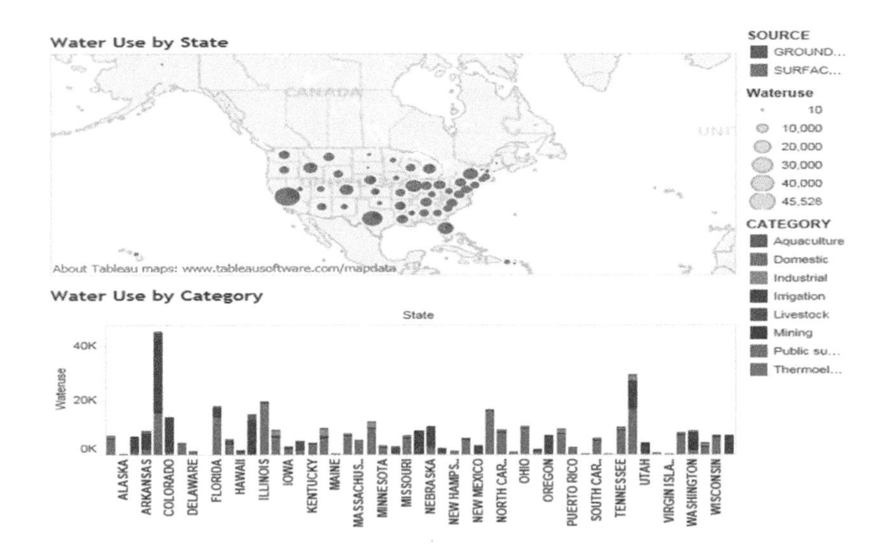

FIGURE 9.6
Visual analytics from Tableau showing water use by state.

A sample SQL script for running and loading all DDL tables and procedures into Oracle or Tableau acceptable format

```
DROP TABLE ENROLLMENT;
--CREATE TABLE TO INCLUDE THE PIVOTED DATASET

CREATE TABLE ENROLLMENT (
ColumnName   VARCHAR2(300),
ALABAMA  NUMBER,
ALASKA   NUMBER,
ARIZONA   NUMBER,
ARKANSAS   NUMBER,
CALIFORNIA   NUMBER,
COLORADO   NUMBER,
CONNECTICUT   NUMBER,
DELAWARE   NUMBER,
DISTRICTOFCOLUMBIA   NUMBER,
FLORIDA   NUMBER,
GEORGIA   NUMBER,
HAWAII   NUMBER,
IDAHO   NUMBER,
ILLINOIS   NUMBER,
INDIANA   NUMBER,
IOWA   NUMBER,
KANSAS   NUMBER,
KENTUCKY   NUMBER,
LOUISIANA   NUMBER,
MAINE   NUMBER,
MARYLAND   NUMBER,
MASSACHUSETTS   NUMBER,
MICHIGAN   NUMBER,
MINNESOTA   NUMBER,
MISSISSIPPI   NUMBER,
MISSOURI   NUMBER,
MONTANA   NUMBER,
NEBRASKA   NUMBER,
NEVADA   NUMBER,
NEWHAMPSHIRE   NUMBER,
NEWJERSEY   NUMBER,
NEWMEXICO   NUMBER,
NEWYORK   NUMBER,
NORTHCAROLINA   NUMBER,
NORTHDAKOTA   NUMBER,
OHIO   NUMBER,
OKLAHOMA   NUMBER,
OREGON   NUMBER,
PENNSYLVANIA   NUMBER,
RHODEISLAND   NUMBER,
SOUTHCAROLINA   NUMBER,
SOUTHDAKOTA   NUMBER,
TENNESSEE   NUMBER,
TEXAS   NUMBER,
UTAH   NUMBER,
VERMONT   NUMBER,
VIRGINIA   NUMBER,
WASHINGTON   NUMBER,
WESTVIRGINIA   NUMBER,
WISCONSIN   NUMBER,
WYOMING   NUMBER)
;

--IMPORT THE PIVOTED SPREADSHEET

--CREATE TABLE FOR THE FINAL DETAILS
CREATE TABLE ENROLLMENTFINAL
(State         VARCHAR2(100),
Enrollment  NUMBER,
Grade         NUMBER,
Gender        VARCHAR2(10),
Race           VARCHAR2(200),
Year NUMBER);

--CREATE TABLE FOR THE STATES
CREATE TABLE ERSTATES (STATE VARCHAR2(100));

--IMPORT THE STATES FROM THE DISTINCT STATES INICLUDED IN THE DATA TABLE

--Run the procedure below after creating the objects and importing the data.

CREATE OR REPLACE PROCEDURE ENROLL_LOAD IS


CURSOR C_CNAMES IS
   select state ste, REPLACE(state,' ','') st
   from ERSTATES;

v_cnames         c_cnames%ROWTYPE;
v_cname   VARCHAR2(80);
v_cnamer   VARCHAR2(80);
v_SQLString    VARCHAR2(3000);

BEGIN

-- Truncate table
DELETE FROM ENROLLMENTFINAL;

-- Insert data for each state

FOR v_cnames IN c_cnames LOOP
      v_cname := v_cnames.st;
      v_cnamer := v_cnames.ste;


 v_SQLString :=
   'INSERT INTO ENROLLMENTFINAL ('||
  ' STATE,'||
  ' ENROLLMENT,'||
  ' GRADE,'||
  ' GENDER,'||
  ' RACE,' ||
  ' YEAR) ' ||
' SELECT ' ||''''||v_cnamer||'''' ||' , '||--'ALASKA' ST,
  'ER.'|| LTRIM(RTRIM(v_cnames.st))||','||
  ' TO_NUMBER(substr(columnname, 7,1)),'||
  ' substr(columnname, instr(columnname,''-'',1,2)+2,6 ),'||
  ' substr(columnname,20, instr(substr(columnname,20),''-'',1,1)-2 ),'||
  ' substr(columnname, instr(columnname,'']'',1,1)+2,4 )'||
  '  FROM ENROLLMENT ER'||
  ' WHERE ER.'|| LTRIM(RTRIM(v_cnames.st)) ||' = '||v_cnames.ST ||'';

  IF v_cname = v_cnames.st  THEN

        EXECUTE IMMEDIATE v_SQLString;
        --ELSE EXECUTE IMMEDIATE v_SQLString2;
   END IF;

    COMMIT;
 END LOOP; --FOR v_cnames IN c_cnames LOOP
   COMMIT;

/*--UDPATE THE GRADE
  UPDATE ENROLLMENTFINAL
  SET GRADE = 6
  WHERE GRADE IS NULL;*/

--REMOVE THE EXTRA SIGN ON THE GENDER
UPDATE ENROLLMENTFINAL
SET GENDER = REPLACE(GENDER, '[','')
WHERE GENDER LIKE '%[%';

END;
```

FIGURE 9.7
A sample script for running and loading all data definition language tables and procedures into Oracle or Tableau acceptable format.

1. Pitney Bowles Enterprise Location Intelligence includes Geocoding Modules, Routing Modules, Location Intelligence, and Spectrum Spatial Modules.
2. Tableau Mapping Software.
3. SAS Business Analytics patterned with Environmental Systems Research Institute Inc. (ESRI).
4. SAP embedded Google's mapping APIs within BusinessObjects BI/ EIM 4.1.
5. MapInfo's LI component is offered as a plug-in tool Integration with ESRI GIS by APOS Systems Inc.

Following is a sample SQL script for running and loading all DDL tables and procedures into Oracle or Tableau acceptable format:

```
DROP TABLE ENROLLMENT;
--CREATE TABLE TO INCLUDE THE PIVOTED DATASET
CREATE TABLE ENROLLMENT (
ColumnName VARCHAR2(300),
ALABAMA NUMBER,
ALASKA NUMBER,
ARIZONA NUMBER,
ARKANSAS NUMBER,
CALIFORNIA NUMBER,
COLORADO NUMBER,
CONNECTICUT NUMBER,
DELAWARE NUMBER,
DISTRICTOFCOLUMBIA NUMBER,
FLORIDA NUMBER,
GEORGIA NUMBER,
HAWAII NUMBER,
IDAHO NUMBER,
ILLINOIS NUMBER,
INDIANA NUMBER,
IOWA NUMBER,
KANSAS NUMBER,
KENTUCKY NUMBER,
LOUISIANA NUMBER,
MAINE NUMBER,
MARYLAND NUMBER,
MASSACHUSETTS NUMBER,
MICHIGAN NUMBER,
MINNESOTA NUMBER,
MISSISSIPPI NUMBER,
MISSOURI NUMBER,
MONTANA NUMBER,
NEBRASKA NUMBER,
NEVADA NUMBER,
```

```
NEWHAMPSHIRE NUMBER,
NEWJERSEY NUMBER,
NEWMEXICO NUMBER,
NEWYORK NUMBER,
NORTHCAROLINA NUMBER,
NORTHDAKOTA NUMBER,
OHIO NUMBER,
OKLAHOMA NUMBER,
OREGON NUMBER,
PENNSYLVANIA NUMBER,
RHODEISLAND NUMBER,
SOUTHCAROLINA NUMBER,
SOUTHDAKOTA NUMBER,
TENNESSEE NUMBER,
TEXAS NUMBER,
UTAH NUMBER,
VERMONT NUMBER,
VIRGINIA NUMBER,
WASHINGTON NUMBER,
WESTVIRGINIA NUMBER,
WISCONSIN NUMBER,
WYOMING NUMBER)
;
--IMPORT THE PIVOTED SPREADSHEET
--CREATE TABLE FOR THE FINAL DETAILS
CREATE TABLE ENROLLMENT FINAL
(State     VARCHAR2(100),
Enrollment       NUMBER,
Grade NUMBER,
Gender VARCHAR2(10),
Race       VARCHAR2(200), Year NUMBER);
--CREATE TABLE FOR THE STATES
CREATE TABLE ERSTATES (STATE VARCHAR2(100));
--IMPORT THE STATES FROM THE DISTINCT STATES INICLUDED IN THE
DATA TABLE
--Run the procedure below after creating the objects and
importing the data.
CREATE OR REPLACE PROCEDURE ENROLL_LOAD IS
CURSOR CCNAMES IS
select state ste, REPLACE(state,″ ″,″) st from ERSTATES;
v_cnames c_cnames%ROWTYPE; v_cname VARCHAR2(80); v_cnamer
VARCHAR2(80);
v_SQL String VARCHAR2(3000);
BEGIN
-- Truncate table
DELETE FROM ENROLLMENTFINAL;
-- Insert data for each state
FOR v_cnames IN c_cnames LOOP v_cname := v_cnames.st; v_
cnamer := v_cnames.ste;
v_SQL String :=
```

```
'INSERT INTO ENROLLMENTFINAL ('||
' STATE,'||
' ENROLLMENT,'||
' GRADE,'||
' GENDER,'||
' RACE; ||
' YEAR)" ||
' SELECT " ||""||v_cnamer||"" ||", "||--'ALASKA" ST,
'ER.'|| LTRIM(RTRIM(v_cnames.st))||",'||
' TO_NUMBER(substr(columnname, 7,1)),'||
' substr(columnname, instr(columnname,"-",1,2)+2,6),'||
' substr(columnname,20,
instr(substr(columnname,20),"-",1,1)-2),'||
' substr(columnname, instr(columnname,"]",1,1)+2,4)'||
' FROM ENROLLMENT ER'||
' WHERE ER.'|| LTRIM(RTRIM(v_cnames.st)) ||"="||v_cnames.ST
||";
IF v_cname=v_cnames.st THEN
EXECUTE IMMEDIATE v_SQLString;
--ELSE EXECUTE IMMEDIATE v_SQLString2; END IF;
COMMIT;
END LOOP; --FOR v_cnames IN c_cnames LOOP COMMIT;
/*--UDPATE THE GRADE
UPDATE ENROLLMENTFINAL
SET GRADE=6
WHERE GRADE IS NULL;*/
--REMOVE THE EXTRA SIGN ON THE GENDER
UPDATE ENROLLMENTFINAL
SET GENDER=REPLACE(GENDER,'[',")
WHERE GENDER LIKE "%[%"; END;
```

Analytics and Strategies for Big Geospatial Data

Although there are a number of spatial analytical methods, tools, and strategies for handling big geospatial data, scientists within academia and industry are investigating the most efficient ways of doing so. New strategies require a complete rethinking of the existing computational framework. The challenge is even greater, for example, in spatial science, where we deal with spatial data that are of a multidimensional nature with sophisticated data structures. To successfully analyze such datasets, we have to systematically search, assemble, process, and manage large-scale spatial databases. In spite of technological transfer challenges, the spatial science community has been proactive in finding new solutions for distinct data centers with different standards, dashboards, and new web tools. For example, recent innovations in Web Tools 2.0 enable users to work together on the same collaborations, providing them with the right privilege to annotate, comment, and generally enrich the data repository by adding tags and metadata (Barnes et al. 2011). The community is deepening their computing knowledge and gradually adopting new computational environments, such as cloud computing and heterogeneous computational environments, which are relatively recent inventions that address many of the limitations of data transfer, access control, data management, standardization of data formats, and advanced model building (Schadt et al. 2010; Wang et al. 2013). Compared to general-purpose processors (GPPs), heterogeneous systems can deliver a tenfold increase or greater in peak arithmetic throughput for a few hundred U.S. dollars. It also optimizes peak performances. Cloud computing can make large-scale computational clusters available on a pay-as-you-need basis. It is low cost and flexible (Schadt et al. 2010; Wang et al. 2013).

Current research work is aimed at tailoring advanced transformation methods toward large-scale computations, data processing, and analysis using available computational resources. For example, Oyana (2010) has focused on a number of useful algorithms for the representation and transformation of large-scale geospatial data. This work has entailed the investigation of cognitive and visual interpretation capabilities that enable the exploration of invariant topographic and geometric properties of a spatial dataset. Also relevant has been the development of several algorithms that focus on the mathematically improved learning-self-organizing map (MIL-SOM); Improved Genetic Algorithm; and Fast, Efficient, and Scalable k-means (FES k-means) Algorithm (Oyana et al. 2004, 2006; Dai and Oyana 2006; Oyana 2006; Oyana and Scott 2008; Oyana et al. 2012; Zhu et al. 2012).

The increased urgency and demand for new methods, algorithms, and analytical strategies is further fueled by the availability of big geospatial data and powerful computing platforms. Several algorithms for big geospatial datasets with linear or nonlinear features already exist in the literature. Examples of such algorithms that deal with the interpretation of massive data include multidimensional scaling (MDS), self-organizing maps (SOMs), k-means, genetic algorithm, graph representations, locally linear embedding (LLE), Isomap, and others. However, some problems relating to their ability to transform data remain. These include whether they (1) can successfully perform a strong recovery of original topological structures, (2) have fast convergence, (3) can take advantage of the most appropriate metric space, (4) can quickly and systematically search through a massive dataset, and (5) can maintain topological stability and preserve geometric properties. Although numerous solutions for these problems have been proposed, such as the techniques based on wavelets and manifold diffusion (Coifman and Lafon 2006), for better geometric preserving properties, little information is available for modeling dynamic features or transforming spatiotemporal datasets. This is further compounded by increased size, nature, and complexity of spatial databases or data streams that require clustering methods to detect variously oriented clusters more reliably, accurately, and efficiently.

Although current algorithms are able to discover compact representations and expose hidden patterns and complex relationships within multivariate datasets, there is still significant demand for more powerful methods and analytical strategies that can easily transform data from high dimensionality to low dimensionality without destroying the original topological structures. The framework for designing such efficient algorithms should consist of three core phases: algorithm design, structure, and functionality; code development and implementation; and performance evaluations. The algorithm design workflow for transforming and modeling dynamic features of large-scale spatiotemporal datasets should entail the following aspects: (1) formulation of mathematical algorithms that effectively transform complex dynamic systems and enable visual exploration of large-scale spatiotemporal datasets; (2) formulation of analytical reasoning and efficient rules with a capability to transform, visualize, and analyze disparate spatial datasets within the subfields of GIS, remote sensing, health care, and medical image processing; (3) scaling methods and tools for existing and future computing platforms; (4) wide dissemination of new methods and tools to increase the exploitation of large-scale spatiotemporal datasets; and (5) continued research efforts and support to improve or develop better analytical methods, tools, and strategies.

Spatiotemporal Data Analytics

Let us now review some of the existing spatiotemporal data analytics and knowledge gaps. Dynamic aspects of spatial data are critical to our understanding of spatial structures and processes. Available dynamic models (e.g., time series and time series combined with variogram-based models) are not versatile enough to deal with complex patterns of large-scale spatiotemporal dynamics, yet the current demand for such models has increased. This increased interest is due to the existence of powerful computational platforms and availability of digital repositories of diseases, demographics, and remotely sensed images. Most recent work is inspired by previous work in complex patterns of spatiotemporal dynamics in ecology (Durrent and Levin 1994; Hastings and Harrison 1994; Bascompte and Sole 1995; Stroud et al. 2001) and transportation geography (Kwan 2000a, b; Wang and Cheng 2001; Peuquet 2002; Yu 2006, 2007; Yu and Shaw 2008). From these reports, there are two central ways of conceptualizing and modeling the complex patterns of spatiotemporal dynamics: (1) continuous space and time models and (2) discrete space-time models. Bascompte and Sole (1995) noted the use of reaction-diffusion mathematical models/partial differential equations in the representation of continuous space and time models. Coupled map lattices are used to represent discrete space-time models. Cellular automata (CA) is the most popular discrete dynamic system to date; but serious shortcomings exist in this system in terms of type of grid and its state, neighborhood definition, distance function (metrics), and quality/complexity of rules. Drawing from these basic concepts and principles, we can write sophisticated rules to represent and model the complexity of spatiotemporal dynamics. Activity pattern algorithms to quantify or simulate activity levels over space and time can be derived using CA and agent-based modeling.

Classification Algorithms for Detecting Clusters in Big Geospatial Data

A common problem in exploring very large-scale spatiotemporal datasets is how to extract relevant, interesting patterns and, more importantly, how to derive a lower-dimensional representation of the original data without significant loss of information. Most clustering algorithms are based on the "frequentist framework" in which the data are used repeatedly to converge on acceptable clusters. A number of new-generation clustering algorithms use the Bayesian approach in which the prior probability distribution (e.g., the probability that a data object belongs to a given cluster) is systematically improved by evaluating the posterior probability (i.e., probability that a data object belongs to a given cluster provided another data object is known to belong to that cluster) (Ben-Hur et al. 2001). Wu et al. (2008) published a report about the top ten data-mining algorithms that were identified by the IEEE International Conference on Data Mining. They were C4.5, k-means, SVM, Apriori, EM, PageRank, AdaBoost, kNN, Naïve Bayes, and CART.

Current clustering algorithms are based on geometric concepts and the notion of a distance metric between two data objects. The distance metric may not have physical meaning for some data entries. For example, a distance metric describing the difference between a neighborhood with and without bike paths does not make physical sense. For very large datasets, the computation of distance metrics is very intensive or prohibitive. This is more so with clustering, since the latter involves iterative evaluation of distance functions. Moreover, geometric clustering bases cluster membership on how close a data object is to a reference data object; "how close" depends on a distance parameter, epsilon. The clustering results are often very sensitive to the error term, epsilon. To address the issue of data size (i.e., number of data objects), data dimension (i.e., number of entries in a data object), and epsilon, we need to incorporate topological aspects of the data. We cluster using data entries for which a distance metric makes physical sense. This leads to dimensionality reduction. Data entries for which the distance metric does not make sense are treated using topological concepts.

Table 9.1 presents a set of plausible topological rules that could be built into the database. Let us illustrate this thinking with the following scenario. Suppose we have used five variables (namely, age, gender, height, weight, and location information at both census-tract level and county level) to do the clustering using the two states of Florida and Mississippi. Suppose also that one of the clusters is made up of two neighborhoods, namely, neighborhood A from Florida and neighborhood B from Mississippi. Plausible questions are (1) is this a viable cluster? and (2) what characteristics do the two noncontiguous neighborhoods share? To address these questions, we can look at a topological rule that says "if a neighborhood is well lit and has pedestrian pathways, then it is likely that the residents will exercise after dinner." If the two neighborhoods share this characteristic, then our level of confidence in the

TABLE 9.1

Topological and Geometric Rules for Geographic Information System Database-Derived Behaviors/Methods

Topological Rules with No Low-Level Noise	Geometric Rules with Low-Level Noise
• Neighborhood has body mass index (BMI) either healthy or not • High socioeconomic status (SES) neighborhoods are surrounded by high SES neighborhoods • BMI-healthy neighborhoods are surrounded by BMI-healthy neighborhoods • Non-BMI-healthy neighborhoods are surrounded by non-BMI-healthy neighborhoods • High density of walking spaces in a neighborhood is adjacent to another high density of walking spaces in a neighborhood • High density of recreational facilities in a neighborhood is adjacent to another high density of walking spaces in a neighborhood • A well-lit neighborhood is adjacent to another well-lit neighborhood • A non-well-lit neighborhood (light information can be derived from remotely sensed data) does not promote physical activities at dusk • A low-lying altitude neighborhood is adjacent to another low-lying altitude neighborhood • A short commuting/travel time neighborhood is adjacent to another short commuting neighborhood • A grid-like land use mix with short block lengths is adjacent to another grid-like land use mix with short block lengths • A grid-like street pattern with short block lengths is adjacent to other grid-like street patterns with short block lengths • A bike path neighborhood is adjacent to another bike path neighborhood	• Neighborhood in close proximity to healthy food outlets correlates with low risk • Neighborhood with a higher density of healthy food correlates with low risk • Neighborhood in close proximity to restaurants correlates with BMI—intermediate • High density of recreational facilities within a neighborhood correlates with BMI-healthy • High density of walking spaces within a neighborhood correlates with BMI-healthy • A well-lit neighborhood (light information can be derived from remotely sensed data) promotes physical activities at dusk • A low-lying altitude neighborhood correlates with low energy consumption • A short commuting/travel time correlates with low risk • A grid-like land use mix with short block lengths correlates with low risk • Grid-like street patterns with short block lengths correlates with low risk • Neighborhood with bike paths correlates with low risk • Neighborhood in close proximity to recreational facility correlates with low risk • Neighborhood with long walking spaces correlates with low risk • Neighborhood with grid-like street patterns and short block lengths correlates with low risk • Neighborhood with long bike paths correlates with low risk • Neighborhood with nonpedestrian walkways correlates with high risk

cluster will improve. We can capture this increased confidence by estimating the posterior probability of the cluster being viable—using Bayesian analysis.

We can initially assign a low prior probability, indicating that we are very cautious. If the answer is no, then we will look at other topological rules. We can consecutively apply each of our topological rules to the cluster in question. At the end, we can have a posterior probability for a viable cluster. We formalize this approach using the following algorithm.

Embedding Solutions/Algorithm with Topological Considerations

Step 1: Obtain a number of clusters.

Step 2: Assign a low prior probability to each cluster's viability.

Step 3: For each cluster, use topological rules sequentially to calculate the posterior probability that the cluster is viable.

Step 4: Analyze clusters with low probabilities. For each nonviable cluster, reassign members to viable clusters.

Step 5: Place unassigned cluster members into a temporary cluster—designate as "unassigned."

Repeat steps 3–5 until the number of viable clusters is stable. If the cluster unassigned is nonempty, investigate the most violated topological rules to see if they can be softened.

The concept of the sense-making loop model (Card et al. 1999; Pirolli and Card 2005) is essential in the creation of sound rules. Available empirical knowledge about obesity and type 2 diabetes is instrumental in the development of sound rules. The rules can be built in as behaviors/methods. Potential critical rules that represent topological and geometric properties of obesity and type 2 diabetes data are presented in Table 9.1. These rules are a result of recent efforts to extract interesting spatial patterns of obesity and type 2 diabetes.

Graph and Text Analytics

A number of methods, tools, and strategies have been developed to facilitate the visualization and analysis of massive social media content (Fink et al. 2009; Beltran et al. 2013; Ghosh and Guha 2013; Lee et al. 2013; Liu et al. 2013; Yin et al. 2013). Content retrieval, sharing, and analysis are common internet activities; but the most exciting feature that has generated a lot of interest among data scientists is their capacity to explore, mine, and acquire fundamental spatial and temporal insights or any practical insights.

Some of the known strategies that are used to search and understand unstructured text information include topic modeling (Ghosh and Guha 2013); spatial and spatiotemporal modeling taking advantage of automated geolocation services, geotargeting markers, place names, or any other explicit and implicit markers (Fink et al. 2009; Lee et al. 2013); and identifying semantics, trending themes, sentiments, events, or influences (Beltran et al. 2013; Liu et al. 2013; Yin et al. 2013). Examples of commonly used text-mining tools include ATLAS.ti, Textalyser.net, QDA Miner, SAS Text Miner, and SPSS Text Analysis for Surveys (Figure 9.8).

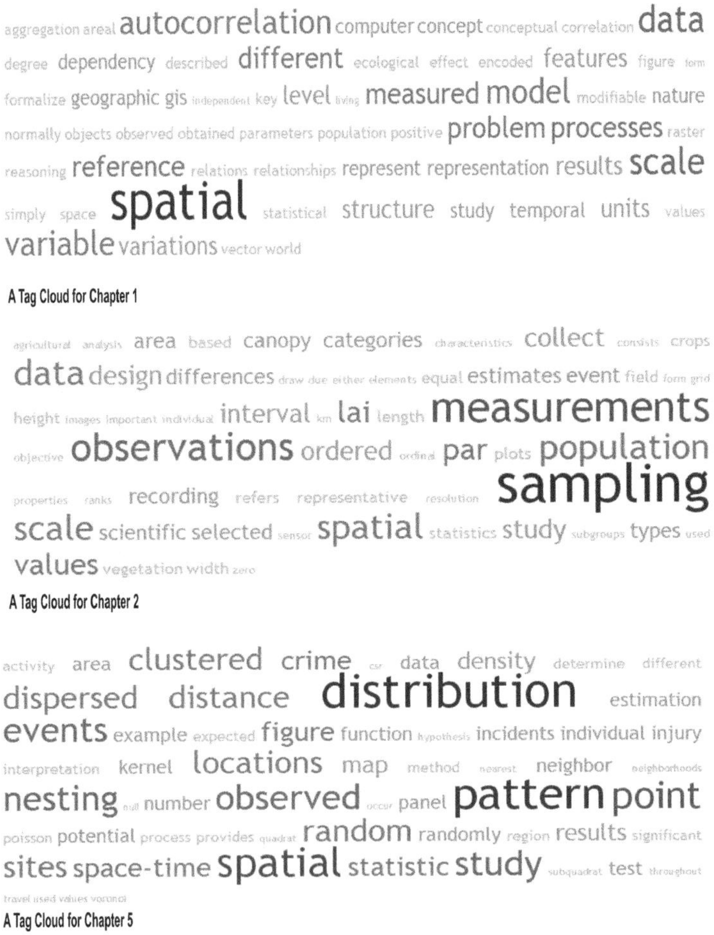

aggregation areal autocorrelation computer concept conceptual correlation data degree dependency described different ecological effect encoded features figure form formalize geographic gis independent key level living measured model modifiable nature normally objects observed obtained parameters population positive problem processes raster reasoning reference relations relationships represent representation results scale simply space spatial statistical structure study temporal units values variable variations vector world

A Tag Cloud for Chapter 1

agricultural analysis area based canopy categories characteristics collect consists crops data design differences draw due either elements equal estimates event field form grid height images important individual interval km lai length measurements objective observations ordered ordinal par plots population properties ranks recording refers representative resolution sampling scale scientific selected sensor spatial statistics study subgroups types used values vegetation width zero

A Tag Cloud for Chapter 2

activity area clustered crime csr data density determine different dispersed distance distribution estimation events example expected figure function hypothesis incidents individual injury interpretation kernel locations map method nearest neighbor neighborhoods nesting null number observed occur panel pattern point poisson potential process provides quadrat random randomly region results significant sites space-time spatial statistic study subquadrat test throughout travel used values voronoi

A Tag Cloud for Chapter 5

FIGURE 9.8
An example of tag clouds of Chapters 1, 2, and 5 of this textbook mined using TagCrown.

TASK 9.1 EXPLORE GOOGLE TRENDS

Google Trends provides data analytics reports on what is trending as people around the world perform searches. The analytical reports display relative search volume across geographies, time trends, and queries that people wish to know about. For this task, we are going to explore Google Trends.

Conclusions

As we prepare to wrap up the chapters in this book, it is only appropriate that we explore the emerging trends and future directions for spatial analysis and related disciplines. As shown in this chapter, one of the bright spots with significant opportunities that lie ahead is the emerging field of data science. The ability to effectively and efficiently process the volume and variety of big geospatial data and to report the findings in clear and simplified terms for the consumer is one of the most important skill sets that a geospatial data scientist must have in the twenty-first century. Completing the chapters in this book to gain knowledge of the traditional analytical approaches and the underlying challenges associated with these spatial data goes a long way toward gaining these professional skills.

Worked Examples in R and Stay One Step Ahead with Challenge Assignments

Chapter 9 R code provides a very practical way of developing your analytical skills and knowledge on the topic of data science, sometimes referred to as "big data analytics." The full code is available in the Chapter9-R-Code document. The primary objective of the exercises presented here is to provide hands-on experience for those working with big geospatial data. To accomplish this goal, we explore Google's predictive analytics for text mining, data reduction and classification algorithms, building footprints, and building footprints and street network data presented in Bing and Google's map services, and R-statistical tools.

1. Open your internet browser and paste this link: http://www.google.com/trends/. We will explore some of the searches and their origins in depth. List six top topics that are currently trending and where they are coming from.
2. Select any two topics/themes of your choice. Describe briefly the spatial and temporal patterns (from 2004 to date) of these topics. List

the top ten countries associated with this search. Describe the spatial and temporal changes over this period. Turn on the forecast button and conduct predictive analytics of these topics.

3. Submit three representative maps and time series charts (capture screen shots).
4. Open your Internet browser and paste this link: http://www.google.org/flutrends/. We explore in depth the two sets of activity data.
   a. Using the flu trends map, briefly describe the spatial and temporal patterns of this epidemic around the world.
   b. Using the dengue trends map, briefly describe the spatial and temporal patterns of this epidemic around the world.

For this task, we consider results from two key algorithms: MIL-SOM and FES k-means.

1. *MIL-SOM*: The MIL-SOM consists of a regular, usually two-dimensional (2D), grid of map units, or it can be defined as a spatial organization of map units. The MIL-SOM learning procedure closely follows a biological understanding of how neurons in the human brain function as they process, organize, and store incoming and outgoing information.
2. *FES k-means algorithm*: This uses a hybrid approach that comprises the k-d tree data structure, nearest neighbor query, the original k-means algorithm, and a better adaptation rate. The primary function of the FES k-means algorithm is to partition data into *k* disjoint subgroups, and then the quality of these clusters is measured via different validation methods.

As part of this task, we are going to perform visual comparisons and analyses between two datasets: a trained and an untrained synthetic dataset. The datasets have already been classified for you using the newly improved versions of SOM (MIL-SOM) and k-means (FES k-means) algorithms. One dataset is classified as it is using the FES k-means algorithm without

TASK 9.2 VISUALIZE SPATIAL DATA

In geography, the use of clustering algorithms, for example, SOMs, principal component analysis (PCA), k-means, and MDS, to solve geographical problems is now widespread. Dimensionality reduction techniques provide generalized methods for data simplification. The ability to transform large, high-dimensional, and structured datasets (untrained) into lower-dimensional representations (trained) is important for the generation of visual representations.

dimensionality reduction, and the second dataset has its dimensionality reduced based on the MIL-SOM algorithm, and then it is classified using the FES k-means algorithm.

The following datasets are provided in the Chapter9_Data_Folder\Data_algorithm folder:

1. *Original synthetic dataset*: A large, high-dimensional structured dataset is directly classified using the improved version of k-means (FES k-means) named as "untrained_FES_kmeans_data."
2. *Trained synthetic dataset*: A dataset reduced from the original synthetic dataset using improved versions of k-means (FES k-means) and SOM (MIL-SOM) algorithms is named as "trained_MILSOM_ FES_kmeans_data." Using the two datasets, conduct the following tasks. To complete these tasks, you may use MS Excel, ArcGIS, SPSS, or any open-source statistical software:
   a. Open/import untrained_FES_kmeans_data and trained_MILSOM_FES_kmeans_data dataset text files and convert them in a format that you can use.
   b. Rename variables as follows: VAR1: Y_axis; VAR2: X_axis; and VAR3: Cluster_Class.
   c. Create a three-dimensional (3D) scatterplot for each of the two datasets.
   d. Create boxplots for each of the Cluster_Class variables for two datasets.

Tips: Identify and describe the spatial distribution of the clusters in the two datasets. Also, compare and contrast the untrained and trained datasets. On completion, prepare a short report and description of the results. Select the most appropriate visual artwork and captions for your report.

Based on these results, answer the following questions:

Question 1: Compare plots of original and synthetic datasets.

Question 2: Analyze the clusters formed in both datasets.

Question 3: Discuss the differences and similarities of the clusters.

Question 4: Discuss the consistency of observed clusters (untrained versus trained data).

TASK 9.3 EXPLORE BING AND GOOGLE MAP SERVICES

1. For this task, you will use both Bing and Google Map services to search for Points of Interests.

2. Find directions between two locations of your choice. Compile this information in a short report. Suppose you wanted to modify your directions to other Points of Interests. Which of the tools would you use and why?
3. Find restaurants near Lakeshore Drive in Chicago, Illinois. Select by top reviewers or type. On average, estimate how far the restaurants are from the main access roads (use the scale bar to estimate the distance). Describe the locations of the restaurants relative to the main access roads.
4. Find gas stations near Lakeshore Drive in Chicago.
5. Compare both map services in terms of ease of use, efficiency, and quality of information.

TASK 9.4 STAY ONE STEP AHEAD WITH CHALLENGE ASSIGNMENTS: EXPLORE AND PERFORM PREDICTIVE ANALYTICS ON FLU AND DENGUE ACTIVITY DATA

1. Download flu and dengue activity data from https://www.google.org/flutrends/ and https://www.google.org/denguetrends/, respectively. The link for the datasets is located below the displayed maps. These need preprocessing before any exploration and analysis can be done. You may use MS Excel, SPSS, ArcGIS, or R-statistical tools.
2. Explore the two activity datasets. Are there any insights?
3. Use any insights to develop some hypotheses for confirmation or further investigations.
4. Create some maps and charts to show the distribution and trends of two activity datasets.
5. Create a trend/predictive analytics model using the two activity datasets.
6. Prepare a short report covering the background, materials and methods, results and discussion, and implications of your findings.

Review and Study Questions

1. The term *big data* has been described within the context of the "5Vs." What are these five properties? In your opinion, do these fully capture the core elements of big data? Are there some missing elements that you would wish to add to the core properties of big data?
2. What is data warehousing? With the use of examples, explain the sources, processes involved, as well as benefits of data warehousing.
3. Distinguish between BI and LI. Use concrete examples from your research area to illustrate the application of one of these analytical strategies.
4. Explain the role of data mining in the analysis of big geospatial data. Choose a software toolkit such as GISolve, GeoDa/PySAL, OpenTopography, PGIST, pd-GRASS, or R. Research the basic functionalities and report your findings.
5. Data scientists are increasingly required to conduct large-scale analyses of graphics and textual data embedded in social media and other databases. Choose one of the following text-mining tools. Research their basic functionalities and report on your findings:
   a. ATLAS.ti
   b. Textalyser.net
   c. QDA Miner
   d. SAS Text Miner
   e. SPSS Text Analysis for Surveys

Glossary of Key Terms

big data: Massive and varied amounts of information that are produced at such speed, variety, and volume that traditional analytical approaches are no longer adequate for processing and visualizing them. These data call for new analytics.

data integration: The process of combining data from varied and sometimes incompatible sources into a unified format within a data warehouse for use in analysis, visualization, reporting, and decision-making.

extract-transform load: This is a standardized/computerized process of extracting relevant data, transforming the data, and cleaning and integrating the data before uploading into a data warehouse. The process ensures consistency and accuracy in the information that is supplied to all users.

metadata: This describes all of the primary features of a dataset. It is a valuable piece of information for technicians engaged in data warehousing, as well as end users of the data. The metadata captures the data lineage and sources, table and column names, entity/attribute definitions, currency of the information and updating schedules, reports/query tools that are available, report distribution information, and help desk/contact information.

self-organizing map: This is a pattern recognition process that relies on unsupervised learning algorithms to produce visual representations of high-dimensional data. The analytical process typically entails two phases, a training phase followed by a prediction phase.

standardized query language: This is a Structured Query Language (SQL) that is used for searching and manipulating data within a relational database management system (RDBMS). The SQL environment contains several features including a catalog, DDL, data manipulation language, and data control language that includes commands that guide the control of the data and administrative privileges.

References

Allen, B., J. Bresnahan, L. Childers, I. Foster, G. Kandaswamy, R. Kettimuthu, J. Kordas, M. Link, S. Martin, K. Pickett, and S. Tuecke. 2012. Software as a service for data scientists. *Communications of the ACM* 55(2): 81–88.

Barnes, C.R., M.M.R Best, L. Pautet, and Pirenne. 2011. Understanding earth–ocean processes using real-time data from NEPTUNE, Canada's Widely Distributed Sensor Networks, Northeast Pacific. *Geoscience Canada* 38(1): 21–30.

Bascompte, J. and R.V. Sole. 1995. Rethinking complexity: Modeling spatiotemporal dynamics in ecology. *Tree* 10(9): 361–366.

Beltran, A., C. Abargues, C. Grabell, M. Nunez, L. Diaz, and J. Huerta. 2013. A virtual globe tool for searching and visualizing geo-referenced media resources in social networks. *Multimedia Tools Application* 64: 171–195.

Ben-Hur, A., D. Horn, H.T. Siegelmann, and V. Vapnik. 2001. Support vector clustering. *Journal of Machine Learning Research* 2: 125–137.

Card, S.K., J.D. Mackinlay, and B. Shneiderman 1999. *Readings in Information Visualization: Using Vision to Think.* San Francisco, CA: Morgan Kaufmann.

Coifman, R.R. and S. Lafon. 2006. Diffusion map. *Applied Computational Harmonic Analysis* 21: 5–30.

Dai, D. and T.J. Oyana. 2006. An improved genetic algorithm for spatial clustering. In *Proceedings of the 18th IEEE International Conference on Tools with Artificial Intelligence (ICTAI-2006)*, pp. 371–380. Arlington, VA; Washington, DC. November 13–15.

Durrent, R. and S.A. Levin. 1994. The importance of being discrete (and spatial). *Theoretical Population Biology* 46: 363–394.

Fink, C., C. Piatko, J. Mayfield, and D. Chou. 2009. The geolocation of web logs from textual clues. In *Proceedings of the IEEE International Conference on Computational Science and Engineering.* Volume 4, pp. 1088–1092. Vancouver, BC: IEEE. doi: 10.1109/CSE.2009.584

Ghosh, D. and R. Guha. 2013. What are we "tweeting" about obesity? Mapping tweets with topic modeling and geographic information system. *Cartography and Geographic Information Science* 40(2): 90–102.

Hastings, A. and S. Harrison. 1994. Metapopulation dynamics and genetics. *Annual Review of Ecology and Systematics* 25: 167–188.

Helbing, D. and S. Balietti 2011. From social data mining to forecasting socio-economic crisis. *The European Physical Journal–Special Topics,* 195(1):3–68.

Ideas Economy: Information Forum. 2013. *Finding Value in Big Data.* June 4. San Francisco, CA: Yerba Buena Center for the Arts.

Jacob, A. 2009. The pathologies of big data. *Communications of the ACM* 52(8): 36–44.

Kelling, S., M.W. Hochacka, D. Fink, M. Riedewald, R. Caurana, G. Ballard et al. 2009. Data-intensive science: A new paradigm for biodiversity studies. *BioScience* 59(7): 613–620.

Kimball, R. and M. Ross. 2013. *The Data Warehouse Toolkit: The Definitive Guide to Dimensional Modeling,* 3rd ed. Indianapolis, IN: John Wiley and Sons.

Kwan, M.-P. 2000a. Human extensibility and individual hybrid-accessibility in spacetime: A multi-scale representation using GIS. In Janelle, D. and D. Hodge (eds.), *Information, Place, and Cyberspace: Issues in Accessibility,* pp. 241–256. Berlin, Germany: Springer-Verlag.

Kwan, M.-P. 2000b. Interactive geovisualization of activity-travel patterns using three-dimensional geographical information systems: A methodological exploration with a large data set. *Transportation Research Part C* 8: 185–203.

Lee, K., R. Ganti, M. Srivatsa, and P. Mohapatra. 2013. Spatio-temporal provenance: Identifying location from unstructured text. In *Proceedings of the 2013 IEEE International Conference on Pervasive Computing and Communications Workshops,* pp. 499–504. San Diego, CA. March 18–22.

Liu, X., Y. Hu, S. Northm, and H. Shen. 2013. CompactMap: A mental map preserving visual interface for streaming text data. In *Proceedings of the 2013 IEEE International Conference on Big Data,* pp. 48–55. Silicon Valley, CA. October 6–9.

Longley, P.A. 2012. Geodemographics and the practices of geographic information science. *International Journal of Geographical Information Science* 26(12): 2227–2237.

Loukides, M. 2010. *What is Data Science?* Sebastopol, CA: O'Reilly Media.

Oyana, T.J. 2006. Introducing an improved fast and computationally-efficient SOM algorithm—MIL-SOM*: Issues and benefits for large-scale geospatial data. In Raubal, M., H. Miller, A. Frank, and M. Goodchild (eds.). *Geographic Information Science. Proceedings of the Fourth International Conference, GIScience,* pp. 141–145. Münster, Germany. September 20–23.

Oyana, T.J. 2010. A new-fangled FES-k-means clustering algorithm for disease discovery and visual analytics. *EURASIP Journal on Bioinformatics and Systems Biology* 2010(1): 746021.

Oyana, T.J., L.E.K. Achenie, and J. Heo. 2012. The new and computationally efficient MIL-SOM algorithm: Potential benefits for visualization and analysis of a large-scale high-dimensional clinically acquired geographic data. *Computational and Mathematical Methods in Medicine* 2012, Article ID 683265: 14 pages doi:10.1155/2012/683265

Oyana, T.J., L.E.K. Achenie, E. Cuadros-Vargas, P.A. Rivers, and K.E. Scott. 2006. A mathematical improvement of the self-organizing map algorithm. Chapter 8: ICT and mathematical modeling, pp. 522–531. In Mwakali, J.A. and G. Taban-Wani (eds.), *Advances in Engineering and Technology*, p. 847. London, UK: Elsevier.

Oyana, T.J., P. Rogerson, and J.S. Lwebuga-Mukasa. 2004. Geographic clustering of adult asthma hospitalization and residential exposure to pollution sites in Buffalo neighborhoods at a U.S.-Canada border crossing point. *American Journal of Public Health* 94(7): 1250–1257.

Oyana, T.J. and K.E Scott. 2008. A geospatial implementation of a novel delineation clustering algorithm employing the k-means. In Lars, B., F. Anders, and P. Hardy (eds.). *The European Information Society, Taking Geoinformation Science One Step Further Series, Lecture Notes in Geoinformation and Cartography*, p. 447. Heidelberg, Germany: Springer.

Peuquet, D. 2002. *Representations of Space and Time*. New York, NY: Guilford Press.

Pirolli, P. and S. Card. 2005. Sensemaking processes of intelligence analysts and possible leverage points as identified through cognitive task analysis. In *Proceedings of the 2005 International Conference in Intelligence Analysis*. McLean, VA.

Schadt, E.E., D.M. Linderman, J. Sorenson, L. Lee, and P.G. Nolan. 2010. Computational solutions to large scale data management and analysis. *Nature Reviews Genetics* 11: 647–657.

Stroud, J.R., P. Muller, and B. Sanso. 2001. Dynamic models for spatiotemporal data. *Journal of the Royal Statistical Society, Series B (Statistical Methodology)* 63(4): 673–689.

Wang, D. and T. Cheng. 2001. A spatio-temporal data model for activity-based transport demand modeling. *International Journal of Geographical Information Science* 15(6): 561–585.

Wang, S., L. Anselin, B. Bhaduri, C. Crosby, M.F. Goodchild, Y. Liu, and T.L. Nyerges. 2013. CyberGIS Software: A synthetic review and integration roadmap. *International Journal of Geographical Information Science*, 27(11): 2122–2145.

Wu, X., V. Kumar, J.R. Quinlan, J. Ghosh, Q. Yang, H. Motoda et al. 2008. Top 10 algorithms in data mining, *Knowledge and Information Systems* 14: 1–37.

Yin, H., B. Cui, H. Lu, Y. Huang, and J. Yao. 2013. A unified model for stable and temporal topic detection from social media data. In *IEEE 29th International Conference on Data Engineering (ICDE)*, pp. 661–672. Brisbane, Queensland, Australia. doi: 10.1109/ICDE.2013.6544864

Yu, H. 2006. Spatio-temporal GIS design for exploring interactions of human activities. *Cartography and Geographic Information Science* 33(1): 3–19.

Yu, H. 2007. Visualizing and analyzing activities in an integrated space-time environment: Temporal GIS design and implementation. *Transportation Research Record: Journal of the Transportation Research Board* 2024: 54–62.

Yu, H. and S.L. Shaw. 2008. Exploring potential human activities in physical and virtual spaces: A spatio-temporal GIS approach. *International Journal of Geographic Information Science* 22(4): 409–430.

Zhu, M., G. Wang, and T.J. Oyana. 2012. Parallel spatiotemporal autocorrelation and visualization system for large-scale remotely sensed images. *The Journal of Supercomputing* 59(1): 83–103.

Index